Lecture Notes in Mathematics

Edited by A. Dold and B. Eckmann

844

Groupe de Brauer

Séminaire, Les Plans-sur-Bex,
Suisse 1980

Edité par M. Kervaire et M. Ojanguren

Springer-Verlag
Berlin Heidelberg New York 1981

Editeurs

Michel Kervaire
Institut de Mathématiques
Université de Genève
2 Rue du Lièvre
Genève 1211/Suisse

Manuel Ojanguren
Institut de Mathématiques
Université de Lausanne
1015 Lausanne/Suisse

AMS Classifications (1980): 12A62, 12E15, 13A20, 14C35, 14G05, 16A27

ISBN 3-540-10562-X Springer-Verlag Berlin Heidelberg New York
ISBN 0-387-10562-X Springer-Verlag New York Heidelberg Berlin

CIP-Kurztitelaufnahme der Deutschen Bibliothek. Groupe de Brauer: séminaire, Les Plans-sur-Bex, Suisse, 1980 / éd. par M. Kervaire et M. Ojanguren. – Berlin; Heidelberg; New York: Springer, 1981. (Lecture notes in mathematics; Vol. 844)
ISBN 3-540-10562-X (Berlin, Heidelberg, New York)
ISBN 0-387-10562-X (New York, Heidelberg, Berlin)
NE: Kervaire, Michel [Hrsg.]; GT

Printing and binding: Beltz Offsetdruck, Hemsbach/Bergstr.
2141/3140-543210

P R E F A C E

Les articles de ce volume reproduisent une partie des exposés qui ont été donnés lors d'un séminaire sur le groupe de Brauer qui a eu lieu aux Plans-sur-Bex (Vaud) du 16 au 22 mars 1980.

On trouvera en outre dans ce volume le texte de la thèse de O. Gabber qui est publiée ici avec la permission de l'auteur.

Le séminaire faisait partie du programme d'enseignement du Troisième Cycle Romand de Mathématiques qui en a assuré le financement et l'organisation scientifique.

Au nom de tous les participants, nous exprimons nos plus vifs remerciements à Monsieur et Madame André Amiguet dont le sens de l'organisation, la bonne humeur et l'énergie inépuisable permettent la réalisation annuelle et le succès des Rencontres Internationales des Plans-sur-Bex.

Nous remercions également Madame Cl. Brugger pour son excellent travail de dactylographie qui a permis de publier ces comptes rendus dans les délais les meilleurs.

M. Kervaire

M. Ojanguren

TABLE DES MATIERES

PARTICIPANTS

AYOUB G.	(Lausanne)	KERVAIRE M.	(Genève)
FLUCKIGER-BAYER E.	(Genève)	KNUS M.	(Zürich)
BLOCH S.	(Chicago)	KRATZER Ch.	(Lausanne)
BOECHAT J.	(Lausanne)	LANG Fr.	(Lausanne)
BOILLAT J.	(Berne)	MATHEY J.B.	(Neuchâtel)
BOVET H.	(Lausanne)	OJANGUREN M.	(Lausanne)
CIBILS C.	(Genève)	PINO-ORTIZ O.	(Genève)
COLLIOT-THELENE J.L.	(Orsay)	PROCESI Cl.	(Roma)
CONTI B.	(Fribourg)	REICHENTHAL S.	(Lausanne)
CORAY D.	(Genève)	ROSSET S.	(Tel Aviv)
DEFFERRARD J.M.	(Neuchâtel)	SANSUC J.J.	(Paris)
DRAXL P.	(Bielefeld)	SCHACHER M.	(UCLA)
ELIAHOU S.	(Genève)	SIGRIST F.	(Neuchâtel)
FEIN B.	(Oregon State University)	SPALTENSTEIN N.	(Lausanne)
GERBER A.	(Berne)	STÄMPFLI-ROLLIER N.	(Lausanne)
GONIN J.B.	(Lausanne)	STEINER Ph.	(Genève)
GRIVEL P.P.	(Genève)	TADDEI G.	(Genève)
HABEGGER N.	(Genève)	TANNENBAUM A.	(Zürich & Tel Aviv)
DE LA HARPE P.	(Genève)	THEVENAZ J.	(Lausanne)
HATT-ARNOLD D.	(Genève)	TIGNOL J.P.	(Louvain)
HURLIMANN W.	(Zürich)	VUST Th.	(Genève)
JONES V.	(Genève)		

CORPS A INVOLUTION
NEUTRALISES PAR UNE EXTENSION ABELIENNE ELEMENTAIRE

J.-P. Tignol
Université Catholique de Louvain
B-1348 Louvain-la-Neuve, Belgique.

Les corps considérés dans ce travail sont de caractéristique différente de 2. Une *involution* d'un anneau est un anti-automorphisme dont le carré est l'identité. Si les éléments du centre sont invariants par l'involution, on dit que celle-ci est *de première espèce*; dans le cas contraire, on dit que l'involution est *de seconde espèce*.

A.A.Albert [2,Th.10.19] a démontré qu'une algèbre simple A de rang fini sur son centre F possède une involution de première espèce si et seulement si son exposant est 1 ou 2. (Par définition, l'*exposant* de A est l'ordre de sa classe de similitude dans le groupe de Brauer Br(F); une algèbre d'exposant 1 est donc une algèbre de matrices.) Il en résulte que le degré d'un corps à involution de première espèce, c'est-à-dire la racine carrée de son rang sur le centre, est une puissance de 2.

Les algèbres simples de degré 2, appelées *algèbres de quaternions*, et plus généralement les produits tensoriels de telles algèbres, possèdent une involution de première espèce; un problème classique, énoncé ici sous forme de conjecture, est de déterminer si tout élément d'ordre 2 dans le groupe de Brauer est la classe d'un tel produit:

Conjecture principale: tout corps à involution de première espèce, de rang fini sur son centre, est semblable à un produit tensoriel de corps de quaternions.

Peu de résultats étayent cette conjecture; cependant, on ne connaissait naguère aucun contre-exemple à la conjecture suivante, dont l'énoncé est pourtant sensiblement plus fort:

Conjecture forte: tout corps à involution de première espèce, de rang fini sur son centre, est un produit tensoriel de corps de quaternions.

La conjecture forte est évidemment vraie pour les corps de degré 2; elle l'est également pour les corps de degré 4, d'après un théorème de A.A.Albert [1]. En 1977, L.H.Rowen [12,Th.6.2] démontre que tout corps à involution de première espèce, de degré 8, (de caractéristique différente de 2), contient un sous-corps commutatif maximal qui est une extension abélienne élémentaire du centre; ce résultat permet de prouver [17] que la conjecture principale est vraie pour les corps à involution de degré 8; cependant, S.A.Amitsur, L.H.Rowen et l'auteur [3] ont montré que la conjecture forte n'est pas vraie: il existe un corps à involution de première espèce, de degré 8, ayant pour centre le corps des fractions rationnelles en quatre indéterminées sur le corps des nombres rationnels, qui ne se décompose pas en produit tensoriel de trois corps de quaternions.

La démonstration de la conjecture principale pour les corps à involution de degré 8 et la construction de contre-exemples à la conjecture forte (ainsi qu'une démonstration, due à M.Racine [10] , du théorème d'Albert sur les corps à involution de degré 4) reposent sur certaines propriétés d'un complexe de groupes abéliens associé à une extension abélienne d'exposant 2 de corps commutatifs. Le but du présent travail est d'introduire ce complexe et d'en donner quelques propriétés. Diverses applications à la théorie des formes quadratiques, dues à R.Elman, T.Y.Lam, A.R.Wadsworth, D.B.Shapiro et l'auteur, ne sont pas abordées ici; elles font l'objet d'articles en préparation.

Pour préciser le contenu de ce travail, il convient d'introduire quelques notations: si F est un corps commutatif, on note $F^{\times}$ le groupe multiplicatif de F et $\hat{F}$ le quotient $F^{\times}/F^{\times 2}$; on désigne par $Br_2(F)$ le groupe des éléments de Br(F) dont l'ordre divise 2. Soit σ l'homomorphisme "symbole quaternionien" de $\hat{F} \otimes \hat{F}$ dans $Br_2(F)$, défini par:

$$\sigma(\Sigma\; x \otimes y) = \otimes(x,y/F),$$

où (x,y/F) est la classe de similitude de l'algèbre de quaternions sur F dont les éléments i, j de la base usuelle satisfont les relations: $i^2 = x$; $j^2 = y$. (Dans la suite, l'élément (x,y/F) de Br(F) est souvent noté (x,y), lorsqu'aucune confusion n'est à craindre.) Soit $(\ker \sigma)'$ le sous-groupe de $\hat{F} \otimes \hat{F}$ engendré par les éléments décomposables (c'est-à-dire de la forme $x \otimes y$) contenus dans le noyau de σ. La conjecture principale peut s'énoncer comme suit: *pour tout corps (commutatif)* F, *la suite:*

$$(\ker \sigma)' \to \hat{F} \otimes \hat{F} \xrightarrow{\sigma} Br_2(F) \to 1 \qquad (0.1)$$

est exacte en $Br_2(F)$; par ailleurs, l'exactitude de la suite ci-dessus en $\hat{F} \otimes \hat{F}$ est équivalente à l'injectivité du "symbole quaternionien" du groupe k_2F de Milnor dans $Br_2(F)$, car, d'après [16,p.266], le quotient $\hat{F} \otimes \hat{F}/(\ker \sigma)'$ s'identifie à k_2F.

Le complexe C(M/F) associé à une extension abélienne d'exposant 2 s'obtient en remplaçant, dans la suite (0.1), l'un des facteurs $\hat{F}$ du produit tensoriel par le noyau de l'homomorphisme canonique de $\hat{F}$ dans $\hat{M}$, et en modifiant les autres termes en conséquence. (La définition précise est donnée au §1.) Les groupes d'homologie du complexe sont notés $N_i(M/F)$.

Lorsque le rang de M sur F ne dépasse pas 4, le complexe C(M/F) est acyclique; on montre dans les §§2 et 3 que ce résultat permet de démontrer la conjecture forte pour les corps de degré 4 et la conjecture principale pour les corps de degré 8.

Lorsque l'extension M/F est telle que $N_2(M/F)$ ne soit pas trivial, alors certains produits croisés abéliens génériques construits à partir de l'extension M/F sont des contre-exemples à la conjecture forte (voir

[18]). Dans le §4, on montre que, pour une extension M/F de rang 8, le groupe $N_2(M/F)$ est isomorphe à un quotient de sous-groupes du groupe multiplicatif $F^{\times}$; cet isomorphisme permet d'interpréter géométriquement la trivialité du groupe $N_2(M/F)$. On donne également, à la fin de ce §, un exemple d'extension de rang 8 telle que $N_2(M/F) \neq 1$; cet exemple est dû à S.A.Amitsur, L.H.Rowen et l'auteur.

L'objet du §5 est de montrer que, si K est une extension de F de rang impair, alors les groupes d'homologie du complexe C(M/F) sont appliqués de manière injective dans ceux du complexe C(M.K/K), par extension des scalaires.

Enfin, dans le §6, on démontre que le complexe associé à une extension abélienne d'exposant 2 d'un corps complet pour une valuation discrète non dyadique a même homologie que celui qui est associé à l'extension résiduelle. Ce résultat permet de construire, pour tout entier $n > 3$, une extension abélienne élémentaire de rang 2^n telle que $N_2(M/F)$ soit non trivial.

De nombreuses conversations avec le Professeur J.Tits m'ont été très utiles; je lui en suis vivement reconnaissant.

§1: Définitions

Soit F un corps commutatif (de caractéristique différente de 2); soit $F^{\times} = F-\{0\}$ le groupe multiplicatif de F et $\hat{F} = F^{\times}/F^{\times 2}$. Soit encore F_s une clôture séparable de F. La théorie de Kummer établit une correspondance biunivoque entre les sous-groupes finis de $\hat{F}$ et les extensions abéliennes finies d'exposant 2 de F contenues dans F_s. Le sous-groupe $(M)_F$ de $\hat{F}$ associé par cette correspondance à une extension M est le noyau de l'homomorphisme canonique de $\hat{F}$ dans $\hat{M}$; de sorte que si l'extension M est obtenue en adjoignant à F les racines carrées (dans F_s)

d'éléments $a_1, \ldots, a_n$ de $F^\times$, alors $(M)_F$ est le sous-groupe de $\hat{F}$ engendré par les images de $a_1, \ldots, a_n$. De plus, la théorie de Kummer met en dualité le groupe $(M)_F$ et le groupe de Galois G de M sur F; en désignant par X(G) le groupe des caractères de G, on a un isomorphisme:

$$\kappa: X(G) \to (M)_F$$

défini de la manière suivante: si χ est un caractère non trivial de G, alors le sous-corps M_χ des éléments de M invariants par ker χ est une extension quadratique de F et le sous-groupe $(M_\chi)_F$ de $\hat{F}$ contient donc deux éléments; par définition, $\kappa\chi$ est l'élément non trivial de $(M_\chi)_F$.

Soit
$$\sigma: X(G) \otimes \hat{F} \to Br_2(F)$$
l'homomorphisme défini par:
$$\sigma(\Sigma\, \chi \otimes f_\chi) = \otimes(\kappa\chi, f_\chi).$$
Comme les éléments $\kappa\chi$ sont tous triviaux dans $\hat{M}$, l'image de σ est contenue dans le noyau de l'homomorphisme ρ d'extension des scalaires:
$$\rho: Br_2(F) \to Br_2(M).$$

Par ailleurs, les éléments décomposables de ker σ sont de la forme $\chi \otimes f$, où f appartient à l'image de l'application déduite de la norme, de $\hat{M}_\chi$ dans $\hat{F}$.

Soit
$$\nu: \bigoplus'_{\chi\in X(G)} \hat{M}_\chi \to X(G) \otimes \hat{F}$$
(le ' indiquant que le caractère trivial est exclu de la sommation)
l'homomorphisme défini par:
$$\nu((m_\chi)_{\chi\in X(G)}) = \Sigma\, \chi \otimes N_{M_\chi/F}(m_\chi);$$
l'image de ν est donc le sous-groupe de ker σ engendré par les éléments décomposables qu'il contient et le complexe suivant est analogue à la suite (0.1):
$$\bigoplus'_\chi \hat{M}_\chi \xrightarrow{\nu} X(G) \otimes \hat{F} \xrightarrow{\sigma} Br_2(F) \xrightarrow{\rho} Br_2(M).$$

Pour le calcul du groupe d'homologie de ce complexe en $Br_2(F)$, il est utile de lui ajouter un terme à droite, semblable au dernier terme d'un complexe défini par R.Elman, T.Y.Lam et A.R.Wadsworth [8]: pour $g \in G$, soit M^g le corps des éléments de M invariants par g et soit

$$\gamma : Br_2(M) \to \bigoplus_{g\in G}{}' Br_2(M^g)$$

(le ' indiquant que la sommation porte sur les éléments non triviaux de G) la somme des homomorphismes de corestriction, de $Br_2(M)$ dans $Br_2(M^g)$. Comme, pour $g \in G-\{1\}$, le corps M^g est de codimension 2 dans M, le composé des homomorphismes d'extension des scalaires, de Br(F) dans Br(M), et de corestriction, de Br(M) dans $Br(M^g)$, est l'homothétie de rapport 2 composée avec l'homomorphisme d'extension des scalaires, de Br(F) dans $Br(M^g)$; par conséquent, l'image de ρ est incluse au noyau de γ. (Cela peut également se voir à l'aide d'un résultat de C.Riehm [11,p.93] et W.Scharlau [13] (voir aussi [2,Th.10.16]) d'après lequel une algèbre simple de rang fini sur son centre L possède une involution de seconde espèce sur un sous-corps K (de codimension 2 dans L) si et seulement si sa corestriction, de L à K, est triviale; en effet, il est clair qu'une algèbre de centre M provenant, par extension des scalaires, d'une algèbre simple à involution de centre F, possède des involutions de seconde espèce sur tous les sous-corps de codimension 2 de M contenant F.)

Soit C(M/F) le complexe:

$$\bigoplus_{\chi\in X(G)}{}' \hat{M}_\chi \xrightarrow{\nu} X(G) \otimes \hat{F} \xrightarrow{\sigma} Br_2(F) \xrightarrow{\rho} Br_2(M) \xrightarrow{\gamma} \bigoplus_{g\in G}{}' Br_2(M^g).$$

Les groupes d'homologie de ce complexe en $X(G) \otimes \hat{F}$, $Br_2(F)$ et $Br_2(M)$ sont notés respectivement: $N_1(M/F)$, $N_2(M/F)$ et $N_3(M/F)$. Si n est un entier positif et i = 1,2 ou 3, on dit que le corps F *possède la propriété* $P_i(n)$ si, pour toute extension abélienne M d'exposant 2, de rang $[M:F] \leq 2^n$, le groupe $N_i(M/F)$ est trivial; on dit qu'il *possède la propriété* P_i s'il possède la propriété $P_i(n)$ pour tout entier positif n.

Exemples: 1) Il est immédiat que tout corps possède les propriétés $P_1(1)$ et $P_2(1)$. Le fait que tout corps possède la propriété $P_3(1)$ résulte, comme cas particulier, d'un théorème de J.K.Arason [4,Cor.4.6]. Dans la suite (corollaires 1.3, 2.5 et 2.8), on montre que tout corps possède les propriétés $P_1(2)$, $P_2(2)$ et $P_3(2)$.

2) Il n'est pas difficile de voir que tout corps local (localement compact, de caractéristique différente de 2) possède les propriétés P_i pour $i = 1$, 2 et 3; R.Elman, T.Y.Lam, A.R.Wadsworth et l'auteur ont montré que les corps globaux (de caractéristique différente de 2) possèdent également ces propriétés.

3) Soit $\mathbb{Q}(X)$ le corps des fractions rationnelles en une indéterminée sur le corps des nombres rationnels. D.B.Shapiro, A.R.Wadsworth et l'auteur ont prouvé que $\mathbb{Q}(X)$ ne possède pas la propriété $P_1(3)$ et, d'autre part, S.A.Amitsur, L.H.Rowen et l'auteur [3], [18], ont montré qu'il ne possède pas la propriété $P_2(3)$ (une démonstration en est donnée à la fin du §4).

Remarques: 1) Si un corps F possède les propriétés P_1 et P_2, alors il est clair que l'homomorphisme de k_2F dans $Br_2(F)$ défini par le symbole quaternionien est injectif et que son image est le sous-groupe de $Br_2(F)$ neutralisé par l'extension abélienne maximale d'exposant 2 de F (contenue dans une clôture séparable de F), mais l'exemple 3 ci-dessus montre que la réciproque n'est pas vraie: en effet, d'après un théorème de S.Bloch [5], l'homomorphisme de $k_2\mathbb{Q}(X)$ dans $Br_2(\mathbb{Q}(X))$ est bijectif.

2) Je ne connais pas d'exemple de corps dont on ait prouvé qu'il ne possède pas la propriété P_3.

Pour terminer ce §, on se propose de donner des propriétés $P_1(n)$ et $P_2(n)$ des formulations plus explicites.

Pour toute extension abélienne finie M/F d'exposant 2, on désigne par $Br_2(M/F)$ le sous-groupe de $Br_2(F)$ neutralisé par M et par $\mathcal{D}\acute{e}c(M/F)$ l'image dans $Br_2(F)$ de l'homomorphisme σ défini dans le complexe $C(M/F)$. Les éléments de $\mathcal{D}\acute{e}c(M/F)$ sont des produits tensoriels de la forme:

$$\underset{i}{\otimes}(m_i, x_i)$$

où, pour tout i, m_i est un élément de $F^{\times}$ qui est un carré dans M. On dit que les éléments de $\mathcal{D}\acute{e}c(M/F)$ *se décomposent (en produit tensoriel d'algèbres de quaternions) suivant* M (voir [18]). Par définition,

$$\mathcal{N}_2(M/F) = Br_2(M/F)/\mathcal{D}\acute{e}c(M/F).$$

Si $a_1, \ldots, a_n$ est une famille d'éléments de $F^{\times}$ telle que

$$M = F(\sqrt{a_1}, \ldots, \sqrt{a_n})$$

(ce qui entraîne la relation: $[M:F] \leqslant 2^n$, l'égalité ayant lieu si et seulement si les images de $a_1, \ldots, a_n$ dans $\hat{F}$ sont linéairement indépendantes, $\hat{F}$ étant considéré comme espace vectoriel sur le corps à deux éléments), alors

$$\mathcal{D}\acute{e}c(M/F) = \{ \underset{i}{\otimes}(a_i, x_i) \mid x_i \in F^{\times} \}.$$

Par conséquent, pour que le corps F possède la propriété $P_2(n)$, il faut et il suffit que, pour toute famille $a_1, \ldots, a_n$ d'éléments de $F^{\times}$, tout élément α de $Br_2(F)$ neutralisé par $F(\sqrt{a_1}, \ldots, \sqrt{a_n})$ soit de la forme:

$$\alpha = (a_1, x_1) \otimes \ldots \otimes (a_n, x_n),$$

avec $x_1, \ldots, x_n \in F^{\times}$.

En ce qui concerne la propriété $P_1(n)$, on a la proposition suivante:

1.1 PROPOSITION : *Soient* n *un entier strictement positif et* $\underline{n} = \{1,\ldots,n\}$. *Pour qu'un corps* F *possède la propriété* $P_1(n)$, *il faut et il suffit que, pour toute famille* $(a_i)_{i\in\underline{n}}$ *et toute famille* $(b_i)_{i\in\underline{n}}$ *d'éléments de* $F^{\times}$ *telles que:*

$$\underset{i}{\otimes}(a_i, b_i) = 1 \tag{1.1}$$

il y ait une famille $(x_P)_{P\subset\underline{n}}$ *d'éléments de* $F^{\times}$ *telle que:*

pour $P \subset \underline{n}$, $$\underset{i\in P}{\otimes}(a_i, x_P) = 1 \tag{1.2}$$

et pour $i \in \underline{n}$, $$\underset{P\ni i}{\otimes}(a_i, x_P) = (a_i, b_i) \tag{1.3}$$

Remarque: Pour tout $i \in \underline{n}$, la relation (1.2) indique que $(a_i, x_{\{i\}}) = 1$; on peut donc supprimer le facteur $(a_i, x_{\{i\}})$ dans le premier membre de (1.3) et modifier l'énoncé en ne tenant compte que des parties de $\underline{n}$ contenant au moins deux éléments.

Démonstration de la proposition: Soit $M = F(\sqrt{a_1}, \ldots, \sqrt{a_n})$. Pour $P \subset \underline{n}$, on pose:

$$a_P = \prod_{i \in P} a_i$$

(en particulier, $a_\emptyset = 1$) et l'on définit:

$$\hat{N}(a_P) = \{ x \in \hat{F} \mid (a_P, x) = 1 \}.$$

Comme l'ensemble $P(\underline{n})$ des parties de $\underline{n}$ est un groupe abélien (d'exposant 2) pour l'opération de différence symétrique, on peut former le produit tensoriel $P(\underline{n}) \otimes \hat{F}$; on considère alors la suite:

$$\bigoplus_{P \subset \underline{n}} \hat{N}(a_P) \xrightarrow{\phi'} P(\underline{n}) \otimes \hat{F} \xrightarrow{\sigma'} Br_2(F)$$

où ϕ' envoie la famille $(x_P)_{P \subset \underline{n}}$ sur $\Sigma\, P \otimes x_P$ et σ' envoie $\Sigma\, P \otimes x_P$ sur $\otimes(a_P, x_P)$.

En désignant par $\hat{b}_i$ l'image de b_i dans $\hat{F}$, la relation (1.1) est équivalente à:

$$\sum_i \{i\} \otimes \hat{b}_i \in \ker \sigma'$$

et l'existence d'une famille (x_P) satisfaisant les relations (1.2) et (1.3) est équivalente à: $\sum_i \{i\} \otimes \hat{b}_i \in \operatorname{im} \phi'$.

Pour démontrer la proposition, il suffit donc d'établir le lemme suivant:

1.2 LEMME : *Avec les notations précédentes, le groupe* $N_1(M/F)$ *est isomorphe à* $\ker \sigma'/\operatorname{im} \phi'$.

Soit G le groupe de Galois de M sur F. Pour tout caractère $\chi \in X(G)$, on définit:

$$\hat{N}(\kappa\chi) = \{ x \in \hat{F} \mid (\kappa\chi, x) = 1 \}.$$

Soit

$$\phi: \bigoplus_{\chi \in X(G)} \hat{N}(\kappa\chi) \to X(G) \otimes \hat{F}$$

l'homomorphisme qui envoie la famille (m_χ) sur $\Sigma\, \chi \otimes m_\chi$; comme pour tout caractère χ non trivial le groupe $\hat{N}(\kappa\chi)$ est l'image de $\hat{M}_\chi$ dans $\hat{F}$ par l'application déduite de la norme, il est clair que l'image de ϕ est identique à l'image de l'homomorphisme ν défini dans le complexe C(M/F).

L'application ε qui envoie une partie P de $\underline{n}$ sur le caractère $\kappa^{-1}(\hat{a}_P)$ est un épimorphisme de $P(\underline{n})$ sur X(G), qui induit des épimorphismes:

$$\varepsilon_1 : \bigoplus_{P \subset \underline{n}} \hat{N}(a_P) \to \bigoplus_{\chi} \hat{N}(\kappa\chi)$$

et

$$\varepsilon_2 : P(\underline{n}) \otimes \hat{F} \to X(G) \otimes \hat{F}$$

de la manière suivante: $\varepsilon_1(x_P) = (y_\chi)$, avec $y_\chi = \prod_{P\in\varepsilon^{-1}\chi} x_P$

$$\varepsilon_2(\Sigma\, P \otimes x_P) = \Sigma\, \varepsilon P \otimes x_P.$$

On a alors un diagramme commutatif, dont les lignes sont exactes:

$$\begin{array}{ccccccccc}
1 & \to & \ker \varepsilon_1 & \longrightarrow & \bigoplus_P \hat{N}(a_P) & \longrightarrow & \bigoplus_\chi \hat{N}(\kappa\chi) & \to & 1 \\
 & & \downarrow & & \downarrow \phi' & & \downarrow \phi & & \\
1 & \to & \ker \varepsilon_2 & \longrightarrow & P(\underline{n}) \otimes \hat{F} & \longrightarrow & X(G) \otimes \hat{F} & \to & 1 \\
 & & \downarrow & & \downarrow \sigma' & & \downarrow \sigma & & \\
 & & 1 & \longrightarrow & Br_2(F) & = & Br_2(F) & \to & 1
\end{array}$$

Pour achever la démonstration, il suffit de prouver que ϕ' induit un épimorphisme de $\ker \varepsilon_1$ sur $\ker \varepsilon_2$. Soit $x \in \ker \varepsilon_2$. Comme $\ker \varepsilon_2$ s'identifie à $(\ker \varepsilon) \otimes \hat{F}$, on peut écrire:

$$x = \sum_{P\in\ker\varepsilon} P \otimes x_P.$$

Soit $\quad y_P = x_P \quad$ pour $P \in \ker \varepsilon - \{\emptyset\}$

$\quad\quad y_P = 1 \quad$ pour $P \notin \ker \varepsilon$

et $\quad y_\emptyset = \prod_{\substack{P\in\ker\varepsilon \\ P\neq\emptyset}} x_P.$

Comme $\hat{N}(a_P) = \hat{F}$ pour $P \in \ker \varepsilon$, la famille (y_P) est dans $\bigoplus_P \hat{N}(a_P)$. Une simple vérification montre alors que cette famille est dans le noyau de ε_1 et que son image par ϕ' est x. □

1.3 COROLLAIRE : *Tout corps possède la propriété* $P_1(2)$.

Dans le cas où $n = 2$, la proposition 1.1, modifiée en ne tenant compte que des parties de $\underline{n}$ contenant au moins deux éléments, indique qu'un corps F possède la propriété $P_1(2)$ si et seulement si, pour a_1, a_2, b_1, $b_2 \in F^\times$ tels que: $(a_1,b_1) = (a_2,b_2)$,

il y a un élément $x_{\{1,2\}} \in F^\times$ tel que:

$$(a_1,b_1) = (a_1,x_{\{1,2\}}) = (a_2,x_{\{1,2\}}) = (a_2,b_2).$$

L'existence d'un tel élément $x_{\{1,2\}}$ est bien connue (et facile à établir): voir par exemple [9,p.69], [16,p.267] ou [3,Lem.6.3]. □

§2: Une suite exacte associée à une tour d'extensions

Dans ce §, M désigne une extension abélienne finie d'un corps (commutatif) F, de caractéristique différente de 2, de groupe de Galois G_F^M d'exposant 2. Soit K un sous-corps de M contenant F. On se propose de relier les groupes $N_i(K/F)$ et $N_i(M/F)$, pour i = 1, 2, 3, par une longue suite exacte, et d'en déduire plusieurs résultats connus.

Soit G_K^M (resp. G_F^K) le groupe de Galois de M sur K (resp. de K sur F) et soit ε l'épimorphisme canonique de G_F^M sur G_F^K. La théorie de Galois donne une suite exacte:

$$1 \to G_K^M \to G_F^M \xrightarrow{\varepsilon} G_F^K \to 1 \tag{2.1}$$

Si K ≠ F, soit g un élément de G_F^K, autre que l'identité. On a un diagramme commutatif:

$$\begin{array}{ccc} Br_2(K) & \longrightarrow & Br_2(M) \\ {\scriptstyle Cor}\big\downarrow & & \big\downarrow{\scriptstyle \oplus Cor} \\ Br_2(K^g) & \longrightarrow & \bigoplus_{h\in\varepsilon^{-1}g} Br_2(M^h) \end{array} \tag{2.2}$$

les flèches verticales étant les corestrictions et les flèches horizontales les homomorphismes d'extension des scalaires. (Comme précédemment, K^g (resp. M^h) désigne le corps des éléments de K (resp. de M) invariants par g (resp. par h).)

Soit Ξ la somme des homomorphismes d'extension des scalaires:

$$\Xi : \bigoplus{}'_{g\in G_F^K} Br_2(K^g) \to \bigoplus_{h\in G_F^M - G_K^M} Br_2(M^h)$$

(Le ' indique que la sommation porte sur les éléments de G_F^K autres que l'identité.) On a donc:

$$\ker \Xi = \bigoplus{}'_{g\in G_F^K} \Big(\bigcap_{h\in\varepsilon^{-1}g} Br_2(M^h/K^g) \Big),$$

où $Br_2(M^h/K^g)$ est le sous-groupe de $Br_2(K^g)$ neutralisé par M^h. Comme le diagramme (2.2) est commutatif, la somme des corestrictions induit un homomorphisme γ' de $Br_2(M/K)$ sur ker Ξ.

Soit $\mathcal{D}éc(M/F)$ l'ensemble des éléments de $Br_2(F)$ décomposés suivant M, c'est-à-dire l'image de l'homomorphisme σ défini dans le complexe $C(M/F)$, et soit $\mathcal{M}_2(M/K/F)$ le groupe d'homologie en $Br_2(M/K)$ de la suite: $\mathcal{D}\acute{e}c(M/F) \to Br_2(M/K) \xrightarrow{\gamma'} \ker \Xi$,

l'homomorphisme de $\mathcal{D}\acute{e}c(M/F)$ dans $Br_2(M/K)$ étant induit par l'extension des scalaires. Soit encore $\mathcal{M}_3(M/K/F) = \operatorname{coker} \gamma'$.

2.1 LEMME : *Il y a une suite exacte naturelle:*

$$\mathcal{N}_2(K/F) \to \mathcal{N}_2(M/F) \to \mathcal{M}_2(M/K/F) \to \mathcal{N}_3(K/F) \to \mathcal{N}_3(M/F).$$

Si $\mathcal{N}_3(M/K) = 1$, *alors cette suite se prolonge d'un terme:*

$$\ldots \quad \mathcal{N}_3(M/F) \to \mathcal{M}_3(M/K/F).$$

Comme $\mathcal{N}_2(M/F) = Br_2(M/F)/\mathcal{D}\acute{e}c(M/F)$, la suite indiquée dans l'énoncé est une partie de la suite d'homologie du diagramme commutatif suivant, dont les lignes sont exactes:

$$\begin{array}{ccccccccc}
1 & \to & \mathcal{D}\acute{e}c(M/F) & \longrightarrow & Br_2(F)/\mathcal{D}\acute{e}c(K/F) & \longrightarrow & Br_2(F)/\mathcal{D}\acute{e}c(M/F) & \to & 1 \\
 & & \downarrow & & \downarrow & & \downarrow & & \\
1 & \to & Br_2(M/K) & \longrightarrow & Br_2(K) & \longrightarrow & Br_2(M) & & \\
 & & \downarrow{\scriptstyle \gamma'} & & \downarrow{\scriptstyle \gamma_K} & & \downarrow{\scriptstyle \gamma_M} & & \\
1 & \to & \ker \Xi & \longrightarrow & \bigoplus'_{g \in G_F^K} Br_2(K^g) & \longrightarrow & \bigoplus'_{h \in G_F^M} Br_2(M^h) & &
\end{array}$$

Si $\mathcal{N}_3(M/K)$ est trivial, alors l'image de $Br_2(K)$ dans $Br_2(M)$ contient le noyau de γ_M; dans ce cas, la suite d'homologie peut donc être prolongée. □

Le lemme 2.1 relie les groupes $\mathcal{N}_i(K/F)$ et $\mathcal{N}_i(M/F)$ pour $i = 2, 3$; il s'agit à présent de comparer les premiers termes des complexes $C(K/F)$ et $C(M/F)$.

De la suite (2.1) on déduit par dualité la suite exacte:

$$1 \to X_F^K \to X_F^M \xrightarrow{\tau} X_K^M \to 1$$

(où l'on a posé, pour abréger, $X_F^K = X(G_F^K)$, etc.), d'où la suite exacte:

$$1 \to X_F^K \otimes \hat{F} \to X_F^M \otimes \hat{F} \to X_K^M \otimes \hat{F} \to 1.$$

Si $\chi \in X_F^K$, alors les corps K_χ et M_χ des éléments de K et de M (respectivement) invariants par ker χ sont identiques. La somme des groupes $\hat{K}_\chi$ s'injecte donc canoniquement dans la somme des groupes $\hat{M}_\chi$, et l'on a un diagramme commutatif, dont les lignes sont exactes:

$$\begin{array}{ccccccccc}
1 & \to & \bigoplus'_{X_F^K} \hat{K}_\chi & \longrightarrow & \bigoplus'_{X_F^M} \hat{M}_\chi & \longrightarrow & \bigoplus_{X_F^M - X_F^K} \hat{M}_\chi & \to & 1 \\
 & & \downarrow{\scriptstyle \nu_K} & & \downarrow{\scriptstyle \nu_M} & & \downarrow{\scriptstyle \nu'} & & \\
1 & \to & X_F^K \otimes \hat{F} & \longrightarrow & X_F^M \otimes \hat{F} & \longrightarrow & X_K^M \otimes \hat{F} & \to & 1 \\
 & & \downarrow{\scriptstyle \sigma_K} & & \downarrow{\scriptstyle \sigma_M} & & \downarrow{\scriptstyle \sigma'} & & \\
1 & \to & Br_2(K/F) & \longrightarrow & Br_2(M/F) & \longrightarrow & Br_2(M/K) & &
\end{array}$$

Les homomorphismes ν' et σ' sont définis par:

$$\nu'(m_\chi) = \Sigma\, \tau\chi \otimes N_{M_\chi/F}(m_\chi)$$

$$\sigma'(\Sigma\, \chi \otimes f) = \otimes(\kappa\chi, f/K).$$

On en déduit la suite exacte d'homologie:

$$\ker \nu' \to N_1(K/F) \to N_1(M/F) \to \ker \sigma'/\mathrm{im}\, \nu' \to N_2(K/F) \to N_2(M/F) \qquad (2.3)$$

2.2 PROPOSITION : *Soit* M *une extension abélienne finie d'exposant 2 d'un corps* F *et soit* K *un sous-corps de* M *contenant* F. *Pour* i = 0, 1, 2, 3, *on désigne par* $M_i(M/K/F)$ *le groupe d'homologie de la suite:*

$$1 \to \bigoplus_{X_F^M - X_F^K} \hat{M}_\chi \xrightarrow{\nu'} X(G_K^M) \otimes \hat{F} \xrightarrow{\sigma'} Br_2(M/K) \xrightarrow{\gamma'} \bigoplus'_{g \in G_F^K} [\bigcap_{h \in \varepsilon^{-1} g} Br_2(M^h/K^g)] \to 1$$

en $\oplus \hat{M}_\chi$, *en* $X(G_K^M) \otimes \hat{F}$, *en* $Br_2(M/K)$ *et en* $\oplus[\cap Br_2(M^h/K^g)]$ *respectivement. Il y a alors une suite exacte naturelle:*

$$M_0(M/K/F) \to N_1(K/F) \to N_1(M/F) \to M_1(M/K/F) \to$$
$$\to N_2(K/F) \to N_2(M/F) \to M_2(M/K/F) \to$$
$$\to N_3(K/F) \to N_3(M/F).$$

Si $N_3(M/K) = 1$, *alors cette suite se prolonge d'un terme:*

$$\ldots \to N_3(M/F) \to M_3(M/K/F).$$

Comme l'image de σ' dans $Br_2(M/K)$ est identique à l'image de $\mathcal{D}éc(M/F)$ par extension des scalaires, la définition de $M_2(M/K/F)$ donnée précédemment est équivalente à celle de l'énoncé. Pour démontrer la proposition, il suffit de réunir la suite (2.3) à celle du lemme 2.1. □

La fin de ce § est consacrée à l'examen de quelques cas particuliers.

a) Si $M = K$:

alors $M_i(M/K/F) = 1$ pour $i = 0, 1, 2$ et 3; la proposition 2.2 n'apporte donc aucune information.

b) Si $K = F$:

alors $M_i(M/K/F) = N_i(M/F)$ pour $i = 1, 2$ et $M_3(M/K/F) = 1$. Encore une fois, la proposition 2.2 n'apporte pas d'information.

c) Si K est une extension quadratique de F :

alors, comme tout corps possède les propriétés $P_i(1)$, pour $i = 1, 2$ et 3, les groupes $N_i(K/F)$ sont triviaux et la proposition 2.2 indique que les groupes $N_i(M/F)$ sont isomorphes aux groupes $M_i(M/K/F)$ pour $i = 1, 2$ et aussi pour $i = 3$ si $N_3(M/K) = 1$.

Pour mettre les groupes $M_i(M/K/F)$ sous une forme plus explicite, supposons que $M = F(\sqrt{a_1}, \ldots, \sqrt{a_n}, \sqrt{b})$ et $K = F(\sqrt{b})$, où $a_1, \ldots, a_n$, b sont des éléments de $F^\times$ dont les images dans $\hat{F}$ sont linéairement indépendantes. Soit $\underline{n} = \{1, \ldots, n\}$. Le groupe $X(G_K^M) \otimes \hat{F}$ s'identifie alors au groupe des familles $(f_i)_{i\in\underline{n}}$ d'éléments de $\hat{F}$, par l'isomorphisme qui envoie la famille (f_i) sur $\Sigma\, \tau\kappa^{-1}(\hat{a}_i) \otimes f_i$, où κ est l'application de $X(G_F^M)$ dans $\hat{F}$ défini par la théorie de Kummer (§1).

Par ce même isomorphisme, $M_1(M/K/F) \cong A_1/B_1$,

où A_1 est le groupe des familles $(f_i)_{i\in\underline{n}}$ d'éléments de $\hat{F}$ telles que:

$$\bigotimes_i (a_i, f_i/F) \otimes K = 1$$

et B_1 est le groupe des familles (f_i) pour lesquelles il existe des familles $(x_P)_{P\subset\underline{n}}$, $(y_P)_{P\subset\underline{n}}$ d'éléments de $\hat{F}$ (ou de $F^\times$) telles que:

pour $P \subset \underline{n}$, $\quad \bigotimes_{i\in P} (a_i, x_P/F) = 1$

et $\quad \bigotimes_{i\in P} (a_i, y_P/F) = (b, y_P/F)$

et pour $i \in \underline{n}$, $\quad \bigotimes_{P\ni i} (a_i, x_P y_P/F) = (a_i, f_i/F)$.

On a aussi $$M_2(M/K/F) = A_2/B_2,$$

où A_2 est le groupe des éléments de $Br_2(K)$ neutralisés par M et dont la corestriction sur F est triviale ou, ce qui revient au même d'après le résultat de Riehm et Scharlau cité au §1, le groupe des classes d'algèbres simples sur K neutralisées par M, possédant une involution de première espèce et une involution de seconde espèce sur F,

et B_2 est le groupe des éléments α de $Br_2(K)$ pour lesquels il existe une famille $(f_i)_{i\in \underline{n}}$ d'éléments de $F^\times$ telle que:

$$\alpha = (a_1,f_1/F) \otimes \ldots \otimes (a_n,f_n/F) \otimes K.$$

Enfin, $$M_3(M/K/F) = A_3/B_3,$$

où A_3 est l'intersection des groupes $Br_2(E/F)$, E parcourant l'ensemble des extensions de F de codimension 2 dans M, ne contenant pas K,

et B_3 est l'image de $Br_2(M/K)$ par la corestriction, de Br(K) dans Br(F).

En particulier, supposons n = 1, de sorte que $M = F(\sqrt{a}, \sqrt{b})$ et $K = F(\sqrt{b})$. Comme, d'après le corollaire 1.3, tout corps possède la propriété $P_1(2)$, on a $N_1(M/F) = 1$, d'où $M_1(M/K/F) = 1$ et $A_1 = B_1$, ce que l'on peut formuler comme suit:

2.3 COROLLAIRE : *Soient* a, b, f *des éléments non nuls d'un corps commutatif* F. *Si l'élément* (a,f/F) *de* $Br_2(F)$ *est neutralisé par* $F(\sqrt{b})$, *alors il y a un élément* y *de* $F^\times$ *tel que:*

$$(a,y/F) = (b,y/F)$$

et $$(a,y/F) = (a,f/F);$$

en d'autres termes, tout élément de $Br(F(\sqrt{a})/F) \cap Br(F(\sqrt{b})/F)$ *est de la forme* (a,y/F), *avec* (ab,y/F) = 1.

2.4 COROLLAIRE : *Si* M *est une extension abélienne élémentaire de rang 4 d'un corps* F *et si* K *est une extension quadratique de* F *contenue dans* M, *alors* $M_3(M/K/F) = 1$.

En effet, si $M = F(\sqrt{a}, \sqrt{b})$ et $K = F(\sqrt{b})$, alors, avec les mêmes notations que précédemment, on a : $A_3 = Br(F(\sqrt{a})/F) \cap Br(F(\sqrt{ab})/F)$. D'après le corollaire 2.3, tout élément α de A_3 s'écrit sous la forme:

$$\alpha = (a,x/F),$$

avec $(b,x/F) = 1$. D'après cette dernière relation, il y a un élément z de K tel que $N_{K/F}(z) = x$; alors

$$\alpha = \mathrm{Cor}(a,z/K),$$

ce qui montre que $A_3 = B_3$. □

2.5 COROLLAIRE : *Tout corps commutatif possède la propriété* $P_3(2)$.

Cela résulte immédiatement de la proposition 2.2 et du corollaire 2.4, car tout corps possède la propriété $P_3(1)$ (voir aussi [18,Lem.3.2]). □

Pour établir, de manière analogue, que tout corps possède la propriété $P_2(2)$, on utilise le lemme suivant (qui se trouve sous une forme légèrement différente dans [2,Th.10.21]):

2.6 LEMME : *Soit* K *une extension quadratique d'un corps* F *et soit* a *un élément non nul de* F. *Si* $x \in K^\times$ *est tel que:*

$$(a,N_{K/F}(x)/F) = 1 \qquad (2.4)$$

alors il existe un élément $y \in F^\times$ *tel que:*

$$(a,x/K) = (a,y/F) \otimes K. \qquad (2.5)$$

Soit Q l'algèbre de quaternions sur K dont les éléments i, j de la base usuelle satisfont les relations: $i^2 = a$, $j^2 = x$. (La classe de Q dans Br(K) est donc $(a,x/K)$.) Si a est un carré dans K, le lemme est trivial; on peut donc supposer que la sous-algèbre K(i) de Q est un sous-corps. Ce sous-corps, que l'on note M, est une extension abélienne élémentaire de F, de rang 4.

Soit g (resp. h) l'automorphisme non trivial de M qui laisse invariants les éléments de K (resp. de F(i)). L'automorphisme g est donc la restriction à M de l'automorphisme intérieur de Q associé à j.

La condition (2.4) indique qu'il existe dans F(i) un élément u tel que:

$$u.g(u) = x.h(x). \qquad (2.6)$$

(Si $x \in F$, on peut choisir $u = x$; ainsi, on a de toutes façons $u \neq -x$.)

Soit alors $$j' = j(1 + x^{-1}u).$$

On a $$j'^2 = x(1 + g(x^{-1}u))(1 + x^{-1}u),$$

d'où, par la relation 2.6 :

$$j'^2 = (x + h(x)) + (u + g(u)).$$

Comme le carré de j' est dans $F^\times$, la sous-F-algèbre de Q engendrée par i et j' est une algèbre de quaternions Q' de centre F et l'on a:

$$Q = Q' \otimes K,$$

d'où $$(a,x/K) = (a,y/F) \otimes K$$

en posant $$y = (x + h(x)) + (u + g(u)).$$ □

2.7 COROLLAIRE : *Si* M *est une extension abélienne élémentaire de rang 4 d'un corps* F *et si* K *est une extension quadratique de* F *contenue dans* M, *alors* $M_2(M/K/F) = 1$.

Soit, comme précédemment, $M = F(\sqrt{a}, \sqrt{b})$ et $K = F(\sqrt{b})$. Tout élément α du groupe A_2 défini ci-dessus est de la forme:

$$\alpha = (a,x/K),$$

où x est un élément de $K^\times$ qui satisfait la relation (2.4). L'existence d'un élément y satisfaisant la relation (2.5) montre que α est dans B_2. □

2.8 COROLLAIRE : *Tout corps commutatif possède la propriété* $P_2(2)$.

Cela résulte immédiatement de la proposition 2.2 et du corollaire 2.7, car tout corps possède la propriété $P_2(1)$. □

Le corollaire 2.8 est contenu implicitement dans une démonstration de M.Racine [10] du résultat suivant, dû à A.A.Albert [1]:

2.9 COROLLAIRE : *Tout corps à involution de première espèce, de degré 4, se décompose en produit tensoriel de deux corps de quaternions.*

A.A.Albert a démontré [2,Th.11.9] que tout corps de degré 4 contient un sous-corps commutatif maximal qui est une extension abélienne élémentaire du centre. (La démonstration de cette assertion est particulièrement simple lorsque le corps possède une involution de première espèce.) L'énoncé résulte donc du corollaire 2.8. □

Le corollaire 2.8 apporte d'ailleurs la précision suivante: si D est un corps à involution de première espèce de degré 4, de centre F, qui contient un sous-corps commutatif de la forme $K_1 \otimes K_2$, où K_1 et K_2 sont des extensions quadratiques de F, alors il y a dans D des corps de quaternions Q_1, Q_2 de centre F, tels que Q_i contient K_i pour i = 1, 2 et $D = Q_1 \otimes Q_2$.

Comme conséquence du lemme 2.6, on trouve également un cas particulier d'un résultat de R.Elman et T.Y.Lam [7,n°2.13]:

2.10 COROLLAIRE : *Soit* K *une extension quadratique d'un corps* F *et* a *un élément de* $F^\times$. *Soit* $N_F(a)$ *(resp.* $N_K(a)$*) l'ensemble des éléments non nuls de* F *(resp. de* K*) représentés par la forme quadratique* $X^2 - aY^2$. *Soit* $N_{K/F}$ *l'application "norme" de* $K^\times$ *dans* $F^\times$. *Alors*

$$N_{K/F}^{-1}(N_F(a)) = F^\times.N_K(a)$$

et

$$N_F(a) \cap N_{K/F}(K^\times) = F^{\times 2}.N_{K/F}(N_K(a)).$$

Il suffit de démontrer la première égalité, car la deuxième en est une conséquence directe. Comme

$$N_F(a) = \{ z \in F^\times \mid (a,z/F) = 1 \},$$

un élément x de $K^\times$ est dans $N_{K/F}^{-1}(N_F(a))$ si et seulement si la relation (2.4) est satisfaite. Par ailleurs, $x \in F^\times.N_K(a)$ si et seulement si $F^\times$ contient un élément y satisfaisant la relation (2.5); le corollaire est donc une autre formulation du lemme 2.6. □

§3: La conjecture principale pour les corps à involution de degré 8

3.1 PROPOSITION : *Soit* n *un entier positif et* M *une extension abélienne élémentaire de rang* 2^{n+1} *d'un corps commutatif* F *(de caractéristique différente de 2). Si* F *possède la propriété* $P_1(n)$ *et si* M *contient une extension quadratique* K *de* F *telle que* $N_2(M/K) = 1$, *alors tout élément de* $Br_2(M/F)$ *est la classe de similitude d'un produit tensoriel de* 2^n *algèbres de quaternions; de plus, tout élément de* $N_2(M/F)$ *peut être représenté par la classe d'un produit tensoriel de* $2^n - n - 1$ *algèbres de quaternions.*

Si $n = 0$, la proposition est triviale; on peut donc supposer $n \geqslant 1$. Soit $a_1, \ldots, a_n, b$ une famille d'éléments de $F^\times$ qui sont des carrés dans M mais non dans F, et telle que b soit un carré dans K, de sorte que:

$$M = F(\sqrt{a_1}, \ldots, \sqrt{a_n}, \sqrt{b}) \quad \text{et} \quad K = F(\sqrt{b}).$$

Soit $\alpha \in Br_2(M/F)$. Alors $\alpha \otimes K$ est un élément de $Br_2(M/K)$ qui est dans le noyau de la corestriction de Br(K) dans Br(F), puisque le composé des homomorphismes d'extension des scalaires, de Br(F) dans Br(K), et de corestriction, de Br(K) dans Br(F), est l'homothétie de rapport 2. Comme, par hypothèse, $Br_2(M/K) = \mathcal{D}\acute{e}c(M/K)$, on peut trouver des éléments $u_1, \ldots, u_n$ de $K^\times$ tels que:

$$\alpha \otimes K = (a_1, u_1/K) \otimes \ldots \otimes (a_n, u_n/K),$$

et puisque $Cor(\alpha \otimes K) = 1$, on a:

$$(a_1, N_{K/F}(u_1)/F) \otimes \ldots \otimes (a_n, N_{K/F}(u_n)/F) = 1.$$

Soit $\underline{n} = \{1, \ldots, n\}$. Comme le corps F possède la propriété $P_1(n)$, il existe, d'après la proposition 1.1, une famille $(x_P)_{P \subset \underline{n}}$ d'éléments de $F^\times$ telle que:

$$\text{pour } P \subset \underline{n}: \quad (\prod_{i \in P} N_{K/F}(u_i), x_P/F) = 1 \tag{3.1}$$

$$\text{et pour } i \in \underline{n}: \quad (N_{K/F}(u_i), \prod_{P \ni i} x_P/F) = (a_i, N_{K/F}(u_i)/F) \tag{3.2}$$

D'après la relation (3.1), on a, pour $i \in \underline{n}$:

$$(N_{K/F}(u_i), x_{\{i\}}/F) = 1,$$

par conséquent il suffit, dans le premier membre de la relation (3.2), de former le produit tensoriel sur les parties P de $\underline{n}$ de cardinal au moins égal à 2. A l'aide du lemme 2.6, on déduit des relations (3.1) et (3.2) l'existence de familles $(v_i)_{i\in\underline{n}}$, $(w_P)_{P\subset\underline{n}}$ d'éléments de $F^\times$ telles que :

$$\text{pour } P \subset \underline{n}, \qquad (x_P, \prod_{i\in P} u_i/K) = (x_P, w_P/F) \otimes K \tag{3.3}$$

$$\text{et pour } i \in \underline{n}, \quad (a_i.\prod_{P\ni i}{}' x_P, u_i/K) = (a_i.\prod_{P\ni i}{}' x_P, v_i/F) \otimes K \tag{3.4}$$

le ' indiquant que le produit porte sur les parties P contenant au moins deux éléments.

La relation (3.4) peut aussi s'écrire:

$$(a_i, u_i/K) = [(a_i, v_i/F) \otimes (\bigotimes_{P\ni i}{}' (x_P, v_i/F)) \otimes K] \otimes [\bigotimes_{P\ni i}{}' (x_P, u_i/K)]$$

et l'on en déduit:

$$\alpha \otimes K = [\bigotimes_i (a_i, v_i/F) \otimes (\bigotimes_{P\subset\underline{n}}{}' \bigotimes_{i\in P} (x_P, v_i/F)) \otimes K] \otimes [\bigotimes_{P\subset\underline{n}}{}' \bigotimes_{i\in P} (x_P, u_i/K)].$$

Comme, par la relation (3.3),

$$\bigotimes_{i\in P} (x_P, u_i/K) = (x_P, w_P/F) \otimes K,$$

on en déduit aussi:

$$\alpha \otimes K = [\bigotimes_i (a_i, v_i/F)] \otimes [\bigotimes_{P\subset\underline{n}}{}' (x_P, y_P/F)] \otimes K,$$

où

$$y_P = w_P.\prod_{i\in P} v_i.$$

Comme K est une extension quadratique de F, il y a dans $F^\times$ un élément z tel que: $\alpha = [\bigotimes_i (a_i, v_i/F)] \otimes (b, z/F) \otimes [\bigotimes_{P\subset\underline{n}}{}' (x_P, y_P/F)]$,

d'où la proposition, car le nombre de parties de $\underline{n}$ de cardinal au moins égal à 2 est $2^n - n - 1$ et car

$$[\bigotimes_i (a_i, v_i/F)] \otimes (b, z/F) \in \mathcal{D}ec(M/F). \qquad \square$$

3.2 COROLLAIRE : *Soit* M *une extension abélienne élémentaire de rang 8 d'un corps commutatif* F. *Tout élément de* $Br_2(M/F)$ *est la classe de similitude d'un produit tensoriel de quatre algèbres de quaternions et tout élément de* $N_2(M/F)$ *peut être représenté par une algèbre de quaternions.*

Cela résulte, comme cas particulier, de la proposition 3.1, en vertu des corollaires 1.3 et 2.8. □

L.H.Rowen a démontré [12,Th.6.2] que tout corps à involution de première espèce de degré 8 (et de caractéristique différente de 2) contient un sous-corps commutatif maximal qui est une extension abélienne élémentaire du centre. A l'aide du corollaire 3.2, on en tire immédiatement la conclusion suivante:

3.3 COROLLAIRE : *Tout corps à involution de première espèce, de degré 8 (et de caractéristique différente de 2), est semblable à un produit tensoriel de quatre algèbres de quaternions.* □

Dans l'énoncé des corollaires 3.2 et 3.3, on ne peut pas, en général, réduire à 3 le nombre de facteurs du produit tensoriel, car S.A.Amitsur, L.H.Rowen et l'auteur [3] ont construit des corps à involution de première espèce, de degré 8, qui ne se décomposent pas en produit tensoriel de trois corps de quaternions.

§4: La propriété $P_2(3)$

L'objet de ce § est de mettre en évidence un isomorphisme entre le groupe N_2 d'une extension abélienne élémentaire de rang 8 d'un corps F (de caractéristique différente de 2) et un quotient de sous-groupes du groupe multiplicatif $F^{\times}$. Cet isomorphisme, qui est implicite dans [3,§4], rend plus calculable le groupe N_2 et permet d'interpréter de manière géométrique la propriété $P_2(3)$.

4.1 PROPOSITION : *Soit* M *une extension abélienne finie d'exposant 2 d'un corps* F. *Soient* L_1, L_2 *deux sous-corps de* M *contenant* F, *tels que*

$M = L_1.L_2$, *et soit* $K = L_1 \cap L_2$. *Soit* ε_1 *(resp.* ε_2*) l'épimorphisme canonique du groupe de Galois de* L_1 *sur* F *(resp. de* L_2 *sur* F*) sur le groupe de Galois* G_F^K *de* K *sur* F. *Pour* $g \in G_F^K$, $g \neq 1$, *et* $i = 1, 2$, *on pose:*

$$\Gamma_i(g) = \bigcap_{h \in \varepsilon_i^{-1} g} Br_2(L_i^h/K^g).$$

Soit encore γ_3 *la somme (sur* $g \in G_F^K - \{1\}$*) des corestrictions, de* $Br_2(L_1/K) \cap Br_2(L_2/K)$ *dans* $\bigoplus_g{}'[\Gamma_1(g) \cap \Gamma_2(g)]$. *Si* $N_3(L_1/K) = 1$, *alors il y a une suite exacte naturelle:*

$$M_2(L_2/K/F) \to M_2(M/L_1/F) \to \operatorname{coker} \gamma_3 \to M_3(L_2/K/F).$$

Soit
$$T: \bigoplus_{g \in G_F^K}{}' \Gamma_2(g) \to \bigoplus_{h \in G_F^{L_1}}{}' Br_2(L_1^h)$$

la somme des homomorphismes d'extension des scalaires, de $Br_2(K^g)$ dans $Br_2(L_1^h)$, pour $h \in \varepsilon_1^{-1} g$. On a donc: $\ker T = \bigoplus_g{}'[\Gamma_1(g) \cap \Gamma_2(g)]$. Les lignes du diagramme commutatif suivant sont exactes:

$$\begin{array}{ccccccc}
 & & 1 & \longrightarrow & \mathcal{D}\acute{e}c(L_2/F) & \longrightarrow & \mathcal{D}\acute{e}c(M/F) \\
 & & \downarrow & & \downarrow & & \downarrow \\
1 & \to & \bigcap_i Br_2(L_i/K) & \longrightarrow & Br_2(L_2/K) & \longrightarrow & Br_2(M/L_1) \\
 & & \downarrow \gamma_3 & & \downarrow \gamma_2 & & \downarrow \gamma_1 \\
1 & \to & \ker T & \longrightarrow & \bigoplus_g{}' \Gamma_2(g) & \xrightarrow{T} & \bigoplus_{h \in G_F^{L_1}}{}' Br_2(L_1^h)
\end{array} \tag{4.1}$$

où γ_1, γ_2 sont des sommes de corestrictions.

Comme $N_3(L_1/K) = 1$, l'image de $Br_2(L_2/K)$ dans $Br_2(M/L_1)$ contient le noyau de γ_1. De plus, comme $M = L_1.L_2$, tout élément de $Br_2(M/L_1)$ qui est l'image d'un élément de $\mathcal{D}\acute{e}c(M/F)$ est aussi l'image d'un élément de $\mathcal{D}\acute{e}c(L_2/F)$. La proposition s'en déduit par une chasse au diagramme (4.1), car les groupes d'homologie des colonnes en $Br_2(L_2/K)$ et en $Br_2(M/L_1)$ respectivement, sont $M_2(L_2/K/F)$ et $M_2(M/L_1/F)$, et car $\operatorname{coker} \gamma_2$ est égal à $M_3(L_2/K/F)$. □

4.2 COROLLAIRE : *Soit* M *une extension abélienne élémentaire de rang 8 d'un corps* F. *Soient* L_1, L_2 *deux sous-corps de* M, *de rang 4 sur* F, *dont l'intersection, que l'on note* K, *est de rang 2 sur* F. *Le groupe* $N_2(M/F)$

est alors naturellement isomorphe au conoyau de l'homomorphisme de corestriction, de $Br(L_1/K) \cap Br(L_2/K)$ *dans* $\bigcap_{E\in\mathcal{E}} Br(E/F)$, *où* $\mathcal{E}$ *est l'ensemble des extensions quadratiques de* F *contenues dans* $L_1 \cup L_2$, *autres que* K.

D'après la proposition 2.2 (ou le lemme 2.1) et le corollaire 2.8, le groupe $N_2(M/F)$ est isomorphe, par extension des scalaires, à $M_2(M/L_1/F)$. D'après les corollaires 2.4 et 2.7, les groupes $M_2(L_2/K/F)$ et $M_3(L_2/K/F)$ sont triviaux; enfin, si g est l'automorphisme non trivial de K sur F, alors le groupe $\Gamma_1(g)$ (resp. $\Gamma_2(g)$) est l'intersection des groupes Br(E/F), où E parcourt l'ensemble des extensions quadratiques de F contenues dans L_1 (resp. L_2), autres que K. Le corollaire découle donc de la proposition 4.1. □

Pour donner à ce résultat une forme plus explicite, il est commode d'utiliser la notation suivante (déjà introduite dans l'énoncé du corollaire 2.10): si a est un élément non nul d'un corps commutatif F, on désigne par $N_F(a)$ l'ensemble des éléments de $F^\times$ représentés par la forme quadratique $X^2 - aY^2$; on a donc:

$$N_F(a) = \{ x \in F^\times \mid (a,x/F) = 1 \},$$

ce qui montre que $N_F(a)$ est un sous-groupe de $F^\times$.

4.3 COROLLAIRE : *Soit* F *un corps commutatif et* M *l'extension de* F *obtenue en adjoignant des racines carrées de trois éléments non nuls* a_1, a_2, a_3 *de* F, *de sorte que* $M = F(\sqrt{a_1}, \sqrt{a_2}, \sqrt{a_3})$. *Le groupe* $N_2(M/F)$ *est isomorphe au quotient de*

$$Br(F(\sqrt{a_1})/F) \cap Br(F(\sqrt{a_2})/F) \cap Br(F(\sqrt{a_3})/F) \cap Br(F(\sqrt{a_1 a_2 a_3})/F)$$

par le sous-groupe des éléments $\alpha \in Br(F)$ *pour lesquels il existe* $x \in F^\times$ *tel que:*

$$\alpha = (a_1,x) = (a_2,x) = (a_3,x). \quad (4.2)$$

De plus, il est isomorphe au quotient $n(a_1,a_2;a_3)/d(a_1,a_2;a_3)$,

où $n(a_1,a_2;a_3) = [N_F(a_1).N_F(a_3)] \cap [N_F(a_2).N_F(a_3)] \cap [N_F(a_1 a_2).N_F(a_3)]$

et $d(a_1,a_2;a_3) = [N_F(a_1) \cap N_F(a_2) \cap N_F(a_1 a_2)].N_F(a_3)$.

Si les images de a_1, a_2, a_3 dans $\hat{F}$ ne sont pas linéairement indépendantes ($\hat{F}$ étant considéré comme espace vectoriel sur le corps à deux éléments), alors le corps M est de rang strictement inférieur à 8 et, d'après le corollaire 2.8, le groupe $N_2(M/F)$ est trivial; par ailleurs, une simple vérification montre que dans ce cas les quotients ci-dessus sont triviaux; il suffit donc de considérer le cas où les images de a_1, a_2 et a_3 sont linéairement indépendantes.

Soient $L_1 = F(\sqrt{a_1}, \sqrt{a_1 a_2 a_3})$, $L_2 = F(\sqrt{a_2}, \sqrt{a_3})$ et $K = L_1 \cap L_2$. L'ensemble $\mathcal{E}$ décrit dans l'énoncé du corollaire 4.2 est alors:

$$\mathcal{E} = \{ F(\sqrt{a_1}), F(\sqrt{a_2}), F(\sqrt{a_3}), F(\sqrt{a_1 a_2 a_3}) \}.$$

Pour démontrer la première partie de l'énoncé, il reste à montrer que le groupe des éléments $\alpha \in Br(F)$ satisfaisant les relations (4.2) est l'image de $Br(L_1/K) \cap Br(L_2/K)$ dans $Br(F)$ par la corestriction. D'après le corollaire 2.3, tout élément de $Br(L_1/K) \cap Br(L_2/K)$ peut s'écrire:

$$\beta = (a_1 a_2 a_3, y/K),$$

avec $y \in N_K(a_1 a_2)$. La corestriction de cet élément dans $Br(F)$ est:

$$\text{Cor } \beta = (a_1 a_2 a_3, N_{K/F}(y)/F);$$

or, d'après le corollaire 2.10,

$$F^{\times 2}.N_{K/F}(N_K(a_1 a_2)) = N_F(a_1 a_2) \cap N_F(a_2 a_3) \cap N_F(a_3 a_1);$$

on peut donc écrire:

$$\text{Cor } \beta = (a_1 a_2 a_3, x/F),$$

avec $x \in N_F(a_1 a_2) \cap N_F(a_2 a_3) \cap N_F(a_3 a_1)$.

En conclusion, un élément $\alpha \in Br(F)$ est la corestriction d'un élément de $Br(L_1/K) \cap Br(L_2/K)$ si et seulement si $F^\times$ contient un élément x tel que:

$$\alpha = (a_1 a_2 a_3, x/F)$$

et

$$(a_1 a_2, x/F) = (a_2 a_3, x/F) = (a_3 a_1, x/F) = 1.$$

La première partie de l'énoncé est ainsi établie, car ces relations sont clairement équivalentes aux relations (4.2).

Comme on peut écrire $M = F(\sqrt{a_1 a_3}, \sqrt{a_2 a_3}, \sqrt{a_3})$, il résulte de la partie de l'énoncé déjà démontrée que le groupe $N_2(M/F)$ est isomorphe au quotient du groupe des éléments $\alpha \in Br(F)$ pour lesquels il existe x_1, x_2,

$x_3, x_4 \in F^\times$ tels que:

$$\alpha = (a_1 a_3, x_1) = (a_2 a_3, x_2) = (a_3, x_3) = (a_1 a_2 a_3, x_4) \qquad (4.3)$$

par le sous-groupe des éléments $\alpha \in Br(F)$ pour lesquels il existe $x \in F^\times$ tel que

$$\alpha = (a_1 a_3, x) = (a_2 a_3, x) = (a_3, x). \qquad (4.4)$$

L'application qui envoie un élément (a_3, y) de $Br(F)$ sur la classe de y dans le quotient $F^\times/N_F(a_3)$ est un isomorphisme de $Br(F(\sqrt{a_3})/F)$ sur $F^\times/N_F(a_3)$. Par cet isomorphisme, le groupe des éléments α satisfaisant les relations (4.3) est envoyé sur $n(a_1, a_2; a_3)/N_F(a_3)$, d'après le corollaire 2.3, et le groupe des éléments α satisfaisant les relations (4.4) est envoyé sur $d(a_1, a_2; a_3)/N_F(a_3)$. □

A l'aide de ce corollaire, il est possible d'interpréter de manière géométrique la trivialité de $\mathcal{N}_2(M/F)$, pour $M = F(\sqrt{a_1}, \sqrt{a_2}, \sqrt{a_3})$; en effet, il est clair qu'un élément u de $F^\times$ est dans $n(a_1, a_2; a_3)$ si et seulement si chacune des équations:

$$\begin{aligned} X_0^2 - a_1 X_1^2 &= u(X_2^2 - a_3 X_3^2) \\ Y_0^2 - a_2 Y_1^2 &= u(Y_2^2 - a_3 Y_3^2) \qquad (4.5) \\ Z_0^2 - a_1 a_2 Z_1^2 &= u(Z_2^2 - a_3 Z_3^2) \end{aligned}$$

possède une solution non triviale.

D'autre part, l'élément u est dans $d(a_1, a_2; a_3)$ si et seulement si le système:

$$X_0^2 - a_1 X_1^2 = Y_0^2 - a_2 Y_1^2 = u(Z_0^2 - a_3 Z_1^2) \qquad (4.6)$$

(auquel on peut ajouter l'équation $= T_0^2 - a_1 a_2 T_1^2$, puisque

$$N_F(a_1) \cap N_F(a_2) \subset N_F(a_1 a_2)\ )$$

possède une solution non triviale.

Ainsi, pour $M = F(\sqrt{a_1}, \sqrt{a_2}, \sqrt{a_3})$, la condition $\mathcal{N}_2(M/F) = 1$ est équivalente à celle-ci: pour tout $u \in F$, si chacune des équations (4.5) possède une solution non triviale, alors le système (4.6) en possède également.

Les résultats précédents peuvent être utilisés pour prouver que certains corps ne possèdent pas la propriété $P_2(3)$; voici un exemple tiré de l'article de S.A.Amitsur, L.H.Rowen et l'auteur [3,§5]:

Soit $F = \mathbb{Q}(X)$ le corps des fractions rationnelles en une indéterminée sur le corps des nombres rationnels. On choisit:

$$a_1 = X \text{ , } a_2 = -1 \text{ , } a_3 = X^2 + 1 \text{ et } u = X.$$

Les relations

$$0^2 - X.1^2 = X(X^2 - (1 + X^2).1^2)$$

$$X^2 - (-1).X^2 = X((1 + X)^2 - (1 + X^2).1^2)$$

$$0^2 - (-X).1^2 = X(1^2 - (1 + X^2).0^2)$$

montrent que les équations (4.5) possèdent des solutions non triviales. Cependant on peut voir, par réduction modulo 2, que le système (4.6) n'a que la solution triviale. Par conséquent, si $F = \mathbb{Q}(X)$ et $M = F(\sqrt{X}, \sqrt{-1}, \sqrt{X^2+1})$, le groupe $N_2(M/F)$ n'est pas réduit à l'élément neutre. (On peut calculer que ce groupe possède deux éléments.)

§5: Extensions de rang impair

D'après un théorème de A.Brumer [6], si un système d'équations tel que (4.6) possède une solution non triviale dans une extension de rang impair du corps de base, alors il en possède aussi dans le corps de base. Il en résulte que, si M et K sont des extensions d'un corps F (de caractéristique différente de 2) contenues dans une même clôture algébrique de F, si M/F est une extension abélienne élémentaire de rang 8 et K/F une extension de rang impair, alors l'homomorphisme d'extension des scalaires induit une injection de $N_2(M/F)$ dans $N_2(M.K/K)$, en désignant par M.K le corps composé de M et de K. L'objet de ce § est de prouver que ce résultat est également vrai pour une extension abélienne finie d'exposant 2 et de rang arbitraire.

a) Le cas d'une extension radicielle

5.1 PROPOSITION : *Soit* F *un corps commutatif de caractéristique impaire. Soient* M *et* K *deux extensions de* F *contenues dans une clôture algébrique de* F. *On suppose que* M/F *est une extension abélienne finie d'exposant 2 et que* K/F *est radicielle. Alors, pour* i = 1, 2, 3, *le groupe* $N_i(M/F)$ *est isomorphe au groupe* $N_i(M.K/K)$, *par extension des scalaires.*

Comme K/F est radicielle, l'homomorphisme canonique de $\hat{F}$ dans $\hat{K}$ est un isomorphisme. De plus, le groupe de Galois G de M sur F est canoniquement isomorphe au groupe de Galois de M.K sur K et, pour tout caractère χ de X(G), les groupes $\hat{M}_\chi$ et $\widehat{M.K}_\chi$ sont isomorphes, par extension des scalaires. Enfin, d'après le théorème d'Albert-Hochschild [14,p.91], les groupes $Br_2(F)$ et $Br_2(K)$ sont isomorphes, par extension des scalaires. Ainsi, l'extension des scalaires induit un isomorphisme du complexe C(M/F) vers le complexe C(M.K/K). □

b) Le cas d'une extension séparable

5.2 PROPOSITION : *Soit* F *un corps commutatif (de caractéristique différente de 2). Soient* M *et* K *deux extensions de* F *contenues dans une clôture séparable de* F. *On suppose que* M/F *est une extension abélienne finie d'exposant 2 et que* K/F *est de rang impair. Alors, pour* i = 1, 2, 3, *l'extension des scalaires induit une injection de* $N_i(M/F)$ *dans* $N_i(M.K/K)$.

Comme dans le cas précédent, le groupe de Galois de M sur F s'identifie au groupe de Galois de M.K sur K. L'injection naturelle de $\hat{F}$ dans $\hat{K}$ est scindée par l'homomorphisme de $\hat{K}$ dans $\hat{F}$ induit par la norme; de même, l'injection naturelle de $Br_2(F)$ dans $Br_2(K)$ est scindée par la corestriction. L'extension des scalaires induit par conséquent un monomorphisme scindé du complexe C(M/F) vers le complexe C(M.K/K). □

5.3 COROLLAIRE : *Soit* F *un corps commutatif (de caractéristique différente de 2). Soient* M *et* K *deux extensions de* F *contenues dans une même clôture algébrique de* F. *On suppose que* M/F *est une extension abélienne finie d'exposant 2 et que* K/F *est de rang impair. Alors, pour* i = 1, 2, 3, *l'extension des scalaires induit une injection de* $N_i(M/F)$ *dans* $N_i(M.K/K)$. □

§6: Corps complets pour une valuation discrète

Le principal résultat de ce § s'énonce comme suit: soit M une extension abélienne finie d'exposant 2 d'un corps F complet pour une valuation discrète, dont le corps résiduel $\underline{F}$ est de caractéristique différente de 2. Soit $\underline{M}/\underline{F}$ l'extension résiduelle. Pour i = 1, 2, 3, les groupes $N_i(M/F)$ et $N_i(\underline{M}/\underline{F})$ sont naturellement isomorphes. A la fin du §, on utilise ce résultat pour construire, à partir d'une extension M/F abélienne élémentaire de rang 8 telle que $N_i(M/F) \neq 1$, une extension abélienne élémentaire M'/F' de rang 2^n, où n est un entier arbitraire supérieur à 3, telle que $N_i(M'/F') \neq 1$.

a) Le cas d'une extension non ramifiée

6.1 PROPOSITION : *Soit* M *une extension abélienne finie d'exposant 2 d'un corps* F *complet pour une valuation discrète. Soit* $\underline{M}/\underline{F}$ *l'extension résiduelle. On suppose que le corps résiduel* $\underline{F}$ *n'est pas de caractéristique 2 et que l'extension* M/F *n'est pas ramifiée. Alors, pour* i = 1, 2, 3, *les groupes* $N_i(\underline{M}/\underline{F})$ *et* $N_i(M/F)$ *sont naturellement isomorphes.*

Soit G le groupe de Galois de M sur F, qui s'identifie naturellement au groupe de Galois de $\underline{M}$ sur $\underline{F}$. Les groupes $\hat{\underline{F}}$ et $\hat{F}$ sont liés par une suite exacte:

$$1 \to \hat{\underline{F}} \to \hat{F} \to \mathbb{Z}/2\mathbb{Z} \to 0 \qquad (6.1)$$

par laquelle $\hat{\underline{F}}$ est identifié à un sous-groupe de $\hat{F}$ (voir [9,p.144]). Si π est une uniformisante de F, la suite ci-dessus est scindée par l'homomorphisme de $\mathbb{Z}/2\mathbb{Z}$ dans $\hat{F}$ qui envoie l'élément non trivial sur la classe de π.

Comme $\underline{F}$ n'est pas de caractéristique 2, tout élément de $Br_2(F)$ est neutralisé par une extension non ramifiée de F et l'on a une suite exacte, due à E.Witt (voir [15,p.195]):

$$1 \to Br_2(\underline{F}) \to Br_2(F) \to \mathrm{Hom}(G_s, \mathbb{Z}/2\mathbb{Z}) \to 0 \qquad (6.2)$$

où G_s est le groupe de Galois d'une clôture séparable $\underline{F}_s$ de $\underline{F}$ et $\mathrm{Hom}(G_s, \mathbb{Z}/2\mathbb{Z})$ est le groupe des homomorphismes continus de G_s dans $\mathbb{Z}/2\mathbb{Z}$.

La suite de cohomologie associée à la suite exacte:

$$0 \to \mathbf{Z}/2\mathbf{Z} \to \underline{F}_s^\times \xrightarrow{2} \underline{F}_s^\times \to 1,$$

où ^{2} est l'homomorphisme qui envoie tout élément sur son carré, permet d'identifier $\mathrm{Hom}(G_s, \mathbf{Z}/2\mathbf{Z})$ à $\hat{\underline{F}}$, car, d'après le théorème 90 de Hilbert, $H^1(G_s, \underline{F}_s^\times) = 1$; la suite (6.2) peut donc aussi s'écrire:

$$1 \to \mathrm{Br}_2(\underline{F}) \to \mathrm{Br}_2(F) \to \hat{\underline{F}} \to 1. \qquad (6.3)$$

Si π est une uniformisante de F, cette suite exacte est scindée par l'homomorphisme de $\hat{\underline{F}}$ dans $\mathrm{Br}_2(F)$ qui envoie tout élément f de $\hat{\underline{F}}$ sur $(f,\pi/F)$ (en identifiant f à un élément de $\hat{F}$, à l'aide de la suite (6.1)).

Les lignes du diagramme commutatif suivant sont des suites exactes analogues à (6.1) et (6.3):

$$\begin{array}{ccccccccc}
1 & \longrightarrow & \bigoplus'_{\chi} \hat{\underline{M}}_\chi & \longrightarrow & \bigoplus'_{\chi} \hat{M}_\chi & \overset{\frown}{\longrightarrow} & \bigoplus'_{\chi} \mathbf{Z}/2\mathbf{Z} & \longrightarrow & 0 \\
 & & \downarrow{\scriptstyle \underline{\nu}} & & \downarrow{\scriptstyle \nu} & & \downarrow{\scriptstyle \nu'} & & \\
1 & \longrightarrow & X(G)\otimes\hat{\underline{F}} & \longrightarrow & X(G)\otimes\hat{F} & \overset{\frown}{\longrightarrow} & X(G) & \longrightarrow & 1 \\
 & & \downarrow{\scriptstyle \underline{\sigma}} & & \downarrow{\scriptstyle \sigma} & & \downarrow{\scriptstyle \sigma'} & & \\
1 & \longrightarrow & \mathrm{Br}_2(\underline{F}) & \longrightarrow & \mathrm{Br}_2(F) & \overset{\frown}{\longrightarrow} & \hat{\underline{F}} & \longrightarrow & 1 \\
 & & \downarrow{\scriptstyle \underline{\rho}} & & \downarrow{\scriptstyle \rho} & & \downarrow{\scriptstyle \rho'} & & \\
1 & \longrightarrow & \mathrm{Br}_2(\underline{M}) & \longrightarrow & \mathrm{Br}_2(M) & \overset{\frown}{\longrightarrow} & \hat{\underline{M}} & \longrightarrow & 1 \\
 & & \downarrow{\scriptstyle \underline{\gamma}} & & \downarrow{\scriptstyle \gamma} & & \downarrow{\scriptstyle \gamma'} & & \\
1 & \longrightarrow & \bigoplus'_{g} \mathrm{Br}_2(\underline{M}^g) & \longrightarrow & \bigoplus'_{g} \mathrm{Br}_2(M^g) & \overset{\frown}{\longrightarrow} & \bigoplus'_{g} \hat{\underline{M}}^g & \longrightarrow & 1
\end{array} \qquad (6.4)$$

Les lignes sont scindées par le choix d'une uniformisante π de F, qui est aussi une uniformisante de M, ainsi que de M_χ et de M^g pour $\chi \in X(G)$ et $g \in G$, puisque l'extension M/F n'est pas ramifiée.

Pour achever la démonstration, il suffit de prouver que la colonne de droite de ce diagramme est une suite exacte; comme les lignes sont scindées, la suite exacte d'homologie donne alors les isomorphismes cherchés.

Pour $\chi \in X(G)$, la norme $N_{M_\chi/F}(\pi)$ est un carré dans F, par conséquent l'homomorphisme ν' est trivial. Si l'on identifie X(G) à $(\underline{M})_{\underline{F}}$, par l'isomorphisme κ de la théorie de Kummer, alors l'homomorphisme σ' est l'injection naturelle de $(\underline{M})_{\underline{F}}$ dans $\hat{\underline{F}}$; d'autre part, l'homomorphisme γ' est la somme des homomorphismes induits par les normes, de $\hat{\underline{M}}$ dans $\hat{\underline{M}}^g$. L'exactitude de la colonne de droite du diagramme (6.4) résulte alors d'un théorème de R.Elman, T.Y.Lam et A.R.Wadsworth [8,Th.2.1]. □

b) Le cas d'une extension ramifiée

6.2 PROPOSITION : *Soit* M *une extension abélienne finie d'exposant 2 d'un corps* F *complet pour une valuation discrète. Soit* $\underline{M}/\underline{F}$ *l'extension résiduelle. On suppose que le corps résiduel* $\underline{F}$ *n'est pas de caractéristique 2 et que l'extension* M/F *est ramifiée. Alors, pour* i = 1, 2, 3, *les groupes* $N_i(\underline{M}/\underline{F})$ *et* $N_i(M/F)$ *sont naturellement isomorphes.*

Le principe de la démonstration est le même que dans le cas précédent, mais il faut tenir compte de ce que le groupe de Galois $\underline{G}$ de l'extension résiduelle $\underline{M}/\underline{F}$ est un quotient du groupe de Galois G de M/F: on a une suite exacte:

$$0 \to \mathbb{Z}/2\mathbb{Z} \to G \xrightarrow{\varepsilon} \underline{G} \to 1. \tag{6.5}$$

Pour $g \in G$, soit $\underline{M}^g$ le corps résiduel de M^g; c'est aussi le sous-corps de $\underline{M}$ des éléments invariants par εg, ce que l'on exprime par l'égalité:

$$\underline{M}^g = \underline{M}^{\varepsilon g}.$$

Des suites exactes analogues à (6.3) lient les groupes $Br_2(\underline{M})$ et $Br_2(M)$, ainsi que les groupes $Br_2(\underline{M}^{\varepsilon g})$ et $Br_2(M^g)$, et forment les lignes d'un diagramme commutatif, pour $g \neq 1$:

$$\begin{array}{ccccccccc} 1 & \longrightarrow & Br_2(\underline{M}) & \longrightarrow & Br_2(M) & \longrightarrow & \hat{\underline{M}} & \longrightarrow & 1 \\ & & \downarrow & & \downarrow \mathrm{Cor} & & \downarrow & & \\ 1 & \longrightarrow & Br_2(\underline{M}^{\varepsilon g}) & \xrightarrow{\theta_g} & Br_2(M^g) & \longrightarrow & \hat{\underline{M}}^{\varepsilon g} & \longrightarrow & 1 \end{array} \tag{6.6}$$

Si $\varepsilon g \neq 1$, la flèche verticale de gauche est la corestriction et la flèche de droite est induite par la norme; si $\varepsilon g = 1$, la flèche verticale de gauche est l'homomorphisme trivial et la flèche de droite est l'identité. Soit

$$\theta : \bigoplus_{g \in \underline{G}}' Br_2(\underline{M}^g) \to \bigoplus_{g \in G}' Br_2(M^g)$$

l'homomorphisme qui envoie toute famille (α_g) sur la famille (β_g) définie par:

$$\beta_g = \theta_g(\alpha_{\varepsilon g}), \qquad \text{pour } g \in G \text{ et } \varepsilon g \neq 1$$

$$\beta_g = 1 \qquad \text{pour l'(unique) élément } g \in G$$
$$\text{tel que } g \neq 1 \text{ et } \varepsilon g = 1.$$

On déduit du diagramme (6.6) un diagramme commutatif:

$$\begin{array}{ccccccccc}
1 & \longrightarrow & Br_2(\underline{M}) & \longrightarrow & Br_2(M) & \longrightarrow & \hat{\underline{M}} & \longrightarrow & 1 \\
 & & \downarrow{\scriptstyle \underline{\gamma}} & & \downarrow{\scriptstyle \gamma} & & \downarrow{\scriptstyle \gamma'} & & \\
1 & \longrightarrow & \bigoplus'_{g\in\underline{G}} Br_2(\underline{M}^g) & \overset{\theta}{\longrightarrow} & \bigoplus'_{g\in G} Br_2(M^g) & \longrightarrow & \operatorname{coker}\theta & \longrightarrow & 1
\end{array} \qquad (6.7)$$

où $\underline{\gamma}$ et γ sont les sommes de corestrictions définies dans les complexes $C(\underline{M}/\underline{F})$ et $C(M/F)$ respectivement. Il est facile de vérifier que l'homomorphisme γ' est injectif.

Par ailleurs, la suite (6.5) donne, par dualité, une suite exacte:

$$1 \to \underline{X} \to X \overset{\nu'}{\to} \mathbb{Z}/2\mathbb{Z} \to 0,$$

où l'on a posé $X = X(G)$ et $\underline{X} = X(\underline{G})$. Cette suite exacte permet d'identifier $\underline{X}$ à un sous-groupe de X. Soit π une uniformisante de F qui est un carré dans M. En comparant la suite ci-dessus à la suite (6.1):

$$1 \to \hat{\underline{F}} \to \hat{F} \overset{\nu}{\to} \mathbb{Z}/2\mathbb{Z} \to 0,$$

par laquelle $\hat{\underline{F}}$ est identifié à un sous-groupe de $\hat{F}$, on obtient la suite exacte:

$$1 \to \underline{X} \otimes \hat{\underline{F}} \to X \otimes \hat{F} \overset{\psi}{\to} X \oplus \hat{\underline{F}} \to 1,$$

où ψ envoie tout élément $\Sigma\, \chi \otimes f$ sur le couple: $(\Sigma\, v(f)\chi, \Pi(f\pi^{v(f)})^{v'(\chi)})$.

Le diagramme (6.7) peut alors être complété de la manière suivante:

$$\begin{array}{ccccccccc}
1 & \longrightarrow & \bigoplus'_{\underline{\chi}} \hat{M}_{\underline{\chi}} & \longrightarrow & \bigoplus'_{\chi} \hat{M}_\chi & \longrightarrow & (\bigoplus'_{\underline{X}} \mathbb{Z}/2\mathbb{Z}) \oplus (\bigoplus_{X-\underline{X}} \hat{M}_\chi) & \longrightarrow & 1 \\
 & & \downarrow{\scriptstyle \underline{\nu}} & & \downarrow{\scriptstyle \nu} & & \downarrow{\scriptstyle \nu'} & & \\
1 & \longrightarrow & \underline{X} \otimes \hat{\underline{F}} & \longrightarrow & X \otimes \hat{F} & \longrightarrow & X \oplus \hat{\underline{F}} & \longrightarrow & 1 \\
 & & \downarrow{\scriptstyle \underline{\sigma}} & & \downarrow{\scriptstyle \sigma} & & \downarrow{\scriptstyle \sigma'} & & \\
1 & \longrightarrow & Br_2(\underline{F}) & \longrightarrow & Br_2(F) & \longrightarrow & \hat{\underline{F}} & \longrightarrow & 1 \\
 & & \downarrow{\scriptstyle \underline{\rho}} & & \downarrow{\scriptstyle \rho} & & \downarrow{\scriptstyle \rho'} & & \\
1 & \longrightarrow & Br_2(\underline{M}) & \longrightarrow & Br_2(M) & \longrightarrow & \hat{\underline{M}} & \longrightarrow & 1 \\
 & & \downarrow{\scriptstyle \underline{\gamma}} & & \downarrow{\scriptstyle \gamma} & & \downarrow{\scriptstyle \gamma'} & & \\
1 & \longrightarrow & \bigoplus'_{\underline{G}} Br_2(\underline{M}^g) & \longrightarrow & \bigoplus'_{G} Br_2(M^g) & \longrightarrow & \operatorname{coker}\theta & \longrightarrow & 1
\end{array}$$

Dans ce diagramme commutatif, dont les lignes sont exactes, ρ' est l'homomorphisme trivial et les homomorphismes ν' et σ' sont définis par:

$$\nu'(z_{\underline{\chi}}, m_\chi) = (\sum_{\chi-\underline{X}} v(m'_\chi)\chi, \prod_{\chi-\underline{X}} m'_\chi . \pi^{v(m'_\chi)})$$

où $m' = N_{M_\chi/F}(m_\chi)$

et
$$\sigma'(\chi, f) = (-\pi)^{v'(\chi)} . (\kappa\chi) . f,$$

où κ est l'isomorphisme de $X(G)$ sur $(M)_F$ défini par la théorie de Kummer.

Il est clair que σ' est surjectif. Pour achever la démonstration, il ne reste donc plus qu'à prouver que la colonne de droite est exacte en $X \oplus \hat{\underline{F}}$ et que l'image de $\ker \nu'$ dans $\ker \underline{\sigma}/\mathrm{im}\ \underline{\nu}$ (= $N_1(\underline{M}/\underline{F})$) est triviale.

A cet effet, il convient de remarquer que, pour $\chi \in X - \underline{X}$, l'image de $\hat{M}_\chi$ par l'application "norme" dans $\hat{F}$ est un sous-groupe à deux éléments:

$$N_{M_\chi/F}(\hat{M}_\chi) = \{ 1, -\kappa\chi \}.$$

(C'est une simple application d'un théorème de T.A.Springer: voir [9, p.147].) Dès lors, l'image de ν' est le sous-groupe de $X \oplus \hat{\underline{F}}$ engendré par les éléments de la forme: $(\chi, (-\kappa\chi).\pi)$, où $\chi \in X - \underline{X}$, et il est facile de vérifier que c'est aussi le sous-groupe des éléments de la forme $(\chi, (-\pi)^{v'(\chi)}.\kappa\chi)$, donc $\mathrm{im}\ \nu' = \ker \sigma'$.

Pour prouver que l'image de $\ker \nu'$ dans $\ker \underline{\sigma}/\mathrm{im}\ \underline{\nu}$ est triviale, on considère un élément $a = ((z_{\underline{\chi}}), (m_\chi))$ du noyau de ν'. Soit

$$A = \{ \chi \in X - \underline{X} \mid N_{M_\chi/F}(m_\chi) = -\kappa\chi \}.$$

Comme $a \in \ker \nu'$, on a:

$$\sum_{\chi\in A} \chi = 0 \quad \text{et} \quad \prod_{\chi\in A} (-\kappa\chi).\pi = 1,$$

ce qui implique que le nombre d'éléments de A est pair. Soit $\chi_\pi \in X$ le caractère tel que $\kappa\chi_\pi = \pi$. L'image de a dans $\ker \underline{\sigma}/\mathrm{im}\ \underline{\nu}$ est la classe de

$$\sum_{\chi\in A} (\chi + \chi_\pi) \otimes (-\kappa\chi).\pi.$$

Si l'on pose $\underline{\chi} = \chi + \chi_\pi$, il est clair que $(-\kappa\chi).\pi$ est un élément de $\hat{\underline{F}}$ contenu dans l'image de $\hat{\underline{M}}_{\underline{\chi}}$ par l'application "norme"; par conséquent, l'image de a dans $\ker \underline{\sigma}/\mathrm{im}\ \underline{\nu}$ est triviale. □

c) Application

6.3 COROLLAIRE : *Soit* M *une extension abélienne finie d'exposant 2 d'un corps commutatif* F. *Pour tout entier positif* n, *on désigne par* M_n *le corps des séries de Laurent formelles itérées:*

$$M_n = M((t_1)) \dots ((t_n))$$

et par F_n *le sous-corps:* $F_n = F((t_1^2)) \dots ((t_n^2))$.

L'extension M_n/F_n *est abélienne d'exposant 2 et de rang*

$$[M_n:F_n] = 2^n.[M:F]$$

et, pour i = 1, 2, 3, *les groupes* $N_i(M/F)$ *et* $N_i(M_n/F_n)$ *sont naturellement isomorphes.*

Pour n ≥ 1, le corps F_n est complet pour la valuation t_n^2-adique et l'extension M_n/F_n est ramifiée; l'extension résiduelle est M_{n-1}/F_{n-1} (en convenant que M_0 = M et F_0 = F). Par conséquent, le corollaire se déduit de la proposition 6.2 par induction sur n. □

A partir de l'exemple de S.A.Amitsur, L.H.Rowen et l'auteur, indiqué au §4, il est donc possible de construire, pour tout entier n ≥ 3, une extension M/F abélienne d'exposant 2 et de rang 2^n, telle que $N_2(M/F)$ ne soit pas trivial; il existe par conséquent, pour tout entier n ≥ 3, des corps à involution de première espèce, de degré 2^n, qui ne se décomposent pas en produit tensoriel de corps de quaternions (voir [18]). Cependant, puisque le centre d'un tel corps, construit à partir de l'exemple de [3] par application du corollaire 6.3, est une extension transcendante pure d'un corps de séries formelles itérées sur un corps de fractions rationnelles sur $\mathbb{Q}$, il résulte du théorème de E.Witt déjà cité (voir la suite (6.3)) et du théorème de S.Bloch [5] que ces corps à involution sont semblables à des produits tensoriels de corps de quaternions.

Références

1 ALBERT,A.A. Normal Division Algebras of Degree Four over an Algebraic Field, *Trans. Amer. Math. Soc.* *34* (1932) 363-372.

2 ALBERT,A.A. *Structure of Algebras*, Amer. Math. Soc. Coll. Pub. 24, Providence, 1968.

3 AMITSUR,S.A., L.H.ROWEN et J.-P.TIGNOL Division algebras of degree 4 and 8 with involution, *Israel J. Math.* *33* (1979) 133-148.

4 ARASON,J.K. Cohomologische Invarianten Quadratischer Formen, *J.Algebra* *36* (1975) 448-491.

5 BLOCH,S. Torsion Algebraic Cycles, K_2, and Brauer Groups of Function Fields, *Bull. Amer. Math. Soc.* *80* (1974) 941-945.

6 BRUMER,A. Remarques sur les couples de formes quadratiques, *C.R. Acad. Sc. Paris, Sér. A* *286* (1978) 679-681.

7 ELMAN,R. et T.Y.LAM Quadratic Forms under Algebraic Extensions, *Math. Ann.* *219* (1976) 21-42.

8 ELMAN,R., T.Y.LAM et A.R.WADSWORTH Quadratic Forms under Multiquadratic Extensions, *Nederl. Akad. Wetensch. Proc. Ser A* *83* (1980) 131-145.

9 LAM,T.Y. *The Algebraic Theory of Quadratic Forms*, Benjamin, Reading, 1973.

10 RACINE,M.L. A Simple Proof of a Theorem of Albert, *Proc. Amer. Math. Soc.* *43* (1974) 487-488.

11 RIEHM,C. The Corestriction of Algebraic Structures, *Invent. Math.* *11* (1970) 73-98.

12 ROWEN,L.H. Central Simple Algebras, *Israel J. Math.* *29*(1978) 285-301.

13 SCHARLAU,W. Zur Existenz von Involutionen auf einfachen Algebren, *Math. Z.* *145* (1975) 29-32.

14 SERRE,J.-P. *Cohomologie Galoisienne*, Lecture Notes in Math. 5, Springer, Berlin, 1973.

15 SERRE,J.-P. *Corps Locaux*, Hermann, Paris, 1968.

16 TATE,J. Relations between K_2 and Galois Cohomology, *Invent. Math.* *36* (1976) 257-274.

17 TIGNOL,J.-P. Sur les classes de similitude de corps à involution de degré 8, *C.R.Acad.Sc.Paris, Sér. A* *286* (1978) 875-876.

18 TIGNOL,J.-P. Produits croisés abéliens (à paraître dans *J.Algebra*).

THE GOLDIE RANK OF VIRTUALLY POLYCYCLIC GROUPS

Shmuel ROSSET

A polycyclic group is a group which has a finite normal series descending to {1} with all quotients cyclic. If all quotients are infinite cyclic. The group is poly - $\mathbb{Z}$ or "poly-infinite-cyclic". A group is "virtually (*)" if it has a (*) subgroup of finite index. (This is also called "(*) by finite") it is easy to see that the class of virtually polycyclic groups is the same as the class of poly - ($\mathbb{Z}$ or finite) groups i.e. groups with a normal series with all quotients finite or infinite cyclic. It is also known and easy to prove that a virtually polycyclic group has normal poly - $\mathbb{Z}$ subgroups of finite index.

Throughout this note we abbreviate "Virtually Polycylic" to <u>VP</u>.

It is known [6, p. 129] that if k is a field, or more generally a (commutative) integral domain, and Γ is a group then the group ring $k\Gamma$ is a <u>prime</u> ring iff Γ has no non-trivial finite normal subgroup. Thus it is reasonable to call a group <u>prime</u> when it has this property.

For example if Γ is virtually free abelian (the case of [7]) i.e. if Γ fits into an exact sequence $1 \to A \to \Gamma \to G \to 1$ where A is a finitely generated free abelian normal subgroup and G is finite, then : if the action of G on A is faithful then Γ is prime.

It is easy to see that, over a Noetherian ring of coefficients k, the group ring of a VP group Γ is Noetherian. Assume from now on that k is a commutative <u>Noetherian domain</u>. If Γ is a prime VP group then the group ring $k\Gamma$ is a prime Noetherian ring. A theorem of Goldie [5] implies that $k\Gamma$ has a total "classical" (i.e. öre) ring of fractions (left and right) which is simple artinian, hence isomorphic to $M_r(D)$ for some unique positive integer r and skew field D. We call r the Goldie rank of Γ over k. Obviously the Goldie rank over a domain of coefficients is the same as the rank over its field of fractions. We denote it by $r_k(\Gamma)$ although we conjecture below that it is independent of k.

As a non-trivial example of this concept it is a theorem ([2], [1]) that the Goldie rank of a torsion-free VP group is 1 (irrespective of k). It is now easy to observe that $r(\Gamma) = 1 \Leftrightarrow \Gamma$ is torsion free. Thus we can

Conjecture (A) : the Goldie rank of a prime VP group is equal to the least common multiple of the orders of finite subgroups (independently of k).

Here we prove the easy half of the conjecture, namely that over any k the least common multiple of the orders of finite subgroups divides the Goldie rank. We do this by defining an Euler characteristic for modules over $k\Gamma$, which is a rational number. Then the Goldie rank is easily seen to be the least common multiple of the denominators appearing in Euler characteristics. On the other hand we show that if H is a finite subgroup then $1/|H|$ is the characteristic of some module.

There is another, stronger, conjecture :

Strong Conjecture (B). Assume Γ is a VP group and k a Noetherian domain. Then the Grothendieck group of the category of finitely generated $k\Gamma$ module, $G_0(k\Gamma)$, is the sum of the inductions from the finite subgroups of Γ.

We show below how (B) $\Rightarrow$ (A). It should be remarked that the strong conjecture (B) has been proved [4] for crystallographic Γ and $k = \mathbb{Z}$.

Finally the contrast between the Goldie rank in the case of group rings and the general situation of crossed product should be stressed. The groups which "lead" to crossed products and division algebras are the virtually free abelian groups i.e. those fitting into an exact sequence

$$(\underline{\alpha})\, 1 \to A \to \Gamma \to G \to 1$$

with G faithfully represented on A. If k is a field let $L = k(A)$, the field of fractions of the group ring kA; G acts on L. Let $K = L^G$. If $i : A \to L^*$ is the natural inclusion let $\beta = i_*(\underline{\alpha}) \in H^2(G,L^*)$. It is easy to show [7] that the crossed product (L,G,β) is isomorphic to the total ring of fractions of $k\Gamma$. Now, it is true that Γ is torsion free <u>iff</u> the restrictions of $\underline{\alpha}$ to all cyclic subgroups of prime order of G is non-trivial. Also the injectivity of i_* is proved in [7]. Thus one might be tempted to generalize conjecture (A) as follows : suppose L is a galois extension of K with group G and $\beta \in H^2(G,L^*)$. Define the <u>Goldie</u> <u>rank</u> in this situation to be r if $(L,G,\beta) \approx M_r(D)$ with D a division algebra. Is r equal to the l.c.m. of the numbers $|H|$, H running over the subgroups of G such that $\beta/H = 0$? It turns out that in this generality conjecture (A) is hopelessly wrong. Examples over global fields can be constructed where $\beta/H \neq 0$ for all non-trivial H, but (L,G,β) is not a division algebra.

This shows that, as it stands, our conjecture is specific to division rings originating from group rings !

I am grateful to O. Gabber for some helpful conversation.

1. The Euler Characteristic

Let k be a Noetherian integral domain which is regular (in the sense of homological algebra). Let K be its field of fractions. If Γ is a poly-$\mathbb{Z}$ group then $k\Gamma$ is also regular and over K projective modules are stably free [3]. Thus every finitely generated $k\Gamma$ module M has a finite projective resolution

$$0 \to P_n \to \dots \to P_0 \to M \to 0$$

and we can assume, by adding free summands if necessary, that $K\otimes_k P_i$ is a finitely generated free K module for each i. Following [9, p. 81] we define the Euler characteristic of M to be

$$\chi(M) = \Sigma(-1)^j \mathrm{rk}_{K\Gamma}(K\otimes_k P_j).$$

Since this rank can be computed by

$$\mathrm{rk}_{K\Gamma}(K\otimes_k P_j) = \dim_K(K\otimes_{k\Gamma} P_j)$$

we see, using the fact that the Euler characteristic of a complex equals that of its homology, that

$$\chi(M) = \Sigma(-1)^j \dim_K \mathrm{Tor}_j^{k\Gamma}(K,M).$$

This invariant way to get $\chi(M)$, which will be useful below, also shows that it is well defined, independently of the resolution chosen.

We now give the general definition. Let Γ be a virtually polycyclic (=VP) group, and let M be a k module. Let Γ' be a poly-$\mathbb{Z}$ subgroup of finite index and let M' be M considered as Γ' module. We define

$$\chi(M) = \chi_\Gamma(M) := (\Gamma : \Gamma')^{-1}\chi_{\Gamma'}(M') .$$

We note immediately that this definition is independent of the choice of Γ'. For suppose $\Gamma'' \subset \Gamma'$ is another choice. If $F_{\cdot} \to M'$ is a free resolution over Γ' then it is also over Γ'' and clearly for F free over $k\Gamma'$

$$rk_{k\Gamma''}F = (\Gamma' : \Gamma'')rk_{k\Gamma'}F .$$

Thus the Euler characteristic is, in general, a rational number.

Example. Let H be a finite group and M be k with trivial H action. It is immediate the $\chi_H(M) = |H|^{-1}$. This trivial example will be used below.

We now tie the Euler characteristic and the Goldie rank. Assume Γ is prime and the Goldie rank is r i.e. that $K(\Gamma) \approx M_r(D)$ with D a division ring.

Lemma 1 : If $\Gamma' \subset \Gamma$ is a prime subgroup of finite index in Γ, then $K(\Gamma')\cdot k\Gamma = K\Gamma \cdot K(\Gamma') = k(\Gamma)$. (Here $K(\Gamma')$ is imbedded in $K(\Gamma)$ in the obvious way.)

Note that every non-zero-divisor in $k\Gamma'$ also does not divide zero in $k\Gamma$, so that indeed $K(\Gamma') \subseteq K(\Gamma)$.

Proof. By passing to a smaller subgroup, if necessary, we can assume that Γ' is poly-$\mathbb{Z}$ and is normal in Γ. Then $k\Gamma'$ is a domain, $D' = K(\Gamma')$ is a division ring, and, because of Γ'-s normality, $K(\Gamma') \cdot k\Gamma = k\Gamma \cdot K(\Gamma') = E$. Clearly $(E : K(\Gamma')) = (\Gamma : \Gamma')$ as left or right vector space. If u is a non-zero-divisor in $k\Gamma$ then right multiplication by u is a (left-) $k(\Gamma')$ linear injective map of E to itself, so u has a left inverse. Similarly on the right. By the universal property of $K(\Gamma)$, $E = K(\Gamma)$, q.e.d.

<u>Proposition 2</u>. With the above notation let U be a simple left $K(\Gamma)$ module and assume that $K(\Gamma)\otimes_{k\Gamma}M \approx U^s$. Then

$$\chi(M) = \frac{s}{r} .$$

<u>Proof</u>. Let Γ' be a poly-$\mathbb{Z}$ subgroup of finite index, say $(\Gamma : \Gamma') = t$. It is clear that $k\Gamma'$ is a domain so that $D' = K(\Gamma')$ is a division ring naturally imbedded in $K(\Gamma)$. We know from lemma 1 that $K(\Gamma) = D' \cdot k\Gamma$. By <u>flatness</u> $D' \cdot k\Gamma \approx D'\otimes_{k\Gamma'}k\Gamma$ as D' vector spaces. It follows that the dimension $(K(\Gamma) : K(\Gamma')) = (\Gamma : \Gamma') = t$. As $K(\Gamma)$ module $K(\Gamma) \approx U^r$. It follows from

$$\dim_{D'} U^r = r \dim_{D'} U \neq t$$

that $\dim_{D'} U = t/r$. Now, to compute $\chi(M)$ assume, for simplicity, that k is a field i.e. $k = K$. Let $F_{\bullet} \to M \to 0$ be a finite free resolution over $k\Gamma'$. We can compute the rank of a free $k\Gamma'$ module F by $\dim_{D'} D'\otimes_{k\Gamma'}F$. By flatness

$$\sum_j (-1)^j \dim_{D'}(D'\otimes_{k\Gamma'}F_j) = \dim_{D'}(D'\otimes_{k\Gamma'}M).$$

Thus

$$\chi(M) = \frac{1}{t}\dim_{D'}(D'\otimes_{k\Gamma'}M) = \frac{1}{t}\dim_{D'}(U^s) = \frac{1}{t}\cdot\frac{ts}{r} = \frac{s}{r}\text{, q.e.d.}$$

Identifying U as a minimal left non-zero ideal let $I = k\Gamma \cap U$. Clearly $K(\Gamma) \cdot I = U$ so, by the proposition

$$\chi(I) = \frac{1}{r} .$$

The following are easy consequences.

<u>Corollary 1</u>. The Goldie rank is the least common multiple of the denominator of Euler characteristics, in reduced form, of finitely

generated $k\Gamma$ modules.

Corollary 2. The group ring $k\Gamma$ is a domain iff, for every finitely generated $k\Gamma$ module M, $\chi(\Gamma)$ is an integer.

Thus the idea is to compute Euler characteristics of $k\Gamma$ modules by some other means, if possible, and deduce results on the Goldie rank and zero divisors.

As a non-trivial illustration let k be a Dedekind domain of algebraic integers. A theorem of Swan [10] states that if G is a finite group and Q is a projective kG module then $K\otimes_k Q$ is free over KG. We show now that this implies that when Γ is a torsionfree VP group the Euler characteristic of a $k\Gamma$ module is always integral and hence that $k\Gamma$ is a domain. This, incidentally, suffices to show that $K\Gamma$ is domain for all fields K of characteristic 0. (See also [2]).

Lemma 3. Let Γ be a torsion from VP group and k a Dedekind domain of algebraic integers. If M is a finitely generated $k\Gamma$ module which is torsion free (hence k-projective) then M has a finite projective resolution.

Proof. We assume, for simplicity, that M is free over k. Serre 9 proved that $\mathbb{Z}$ has a finite projective resolution over $\mathbb{Z}\Gamma$. Tensoring over $\mathbb{Z}$ with M, using that $P\otimes M$ (with diagonal action) is $k\Gamma$ projective and the Noetherianity of $k\Gamma$ the lemma is proved.

Let $P_{\cdot} \to M \to 0$ be such a resolution and let Γ' be a normal poly-$\mathbb{Z}$ subgroup of finite index with $\Gamma/\Gamma' = G$. By adding free summands to the P_j-s we can assume that $K\otimes P_j$ is $K\Gamma'$ free for all j. By Swan's theorem $K\otimes_{k\Gamma'} P$ is a complex of free KG modules. Hence $\chi(M)$ is an integer.

2. Induction

Let H be a subgroup of a prime VP group Γ. Then it is easily seen that H is VP too. Let N be a finitely generated kH module and let $M = k\Gamma \otimes_{kH} N$ be the $k\Gamma$ module induced from N.

Proposition 4. The Euler characteristic is compatible with induction, i.e., in the notation above, $\chi_\Gamma(M) = \chi_H(N)$.

Proof. Let Γ' be a normal poly-$\mathbb{Z}$ subgroup of Γ and let $H' = H \cap \Gamma'$. Then H' is also poly-$\mathbb{Z}$. We use the formula

$$\chi(M) = \frac{1}{(\Gamma : \Gamma')} \Sigma_j (-1)^j \dim_K \mathrm{Tor}_j^{k\Gamma'}(K,M) .$$

Let $Q_. \to N \to 0$ be a projective resolution of N over kH. Then $R\Gamma \otimes_{RH} Q_. \to M \to 0$ is a resolution of M over $k\Gamma$, so certainly over $k\Gamma'$ as well. Tensoring with K over $k\Gamma'$ we get

$$K \otimes_{k\Gamma'} k\Gamma \otimes Q \cong K[\Gamma/\Gamma'] \otimes_{kH} Q.$$

so we need to understand the group ring $K[\Gamma/\Gamma']$ as an H module. Let $\ell = (\Gamma/\Gamma' : H/H')$. Obviously as a $K[H/H']$ module, and hence also as a KH module, it is isomorphic to $(K[H/H'])$. It follows that

$$\begin{aligned} K \otimes_{k\Gamma'} k\Gamma \otimes_{kH} Q_. &\cong K[H/H']^\ell \otimes_{kH} Q_. \\ &\cong (K)^\ell \otimes_{KH'} KH \otimes_{kH} Q_. \\ &\cong (K)^\ell \otimes_{kH'} Q_. \end{aligned}$$

It follows that

$$\dim_K \mathrm{Tor}_.^{k\Gamma'}(K,N) = \ell \cdot \dim_K \mathrm{Tor}_.^{kH'}(K,N).$$

Since $(\Gamma : \Gamma') = \ell \cdot (H:H')$ the proposition follows.

In particular assume that Γ has a finite subgroup H of order h. Let N be k with trivial H action. Then $\chi_H(N) = 1/h$. Let $M = k\Gamma \otimes_{kH} N \cong k[\Gamma/H]$. By the above $\chi_\Gamma(M) = 1/h$. It follows that the order of every finite subgroup of Γ divides the Goldie rank of Γ (computed over k.)

<u>Corollary</u>. The least common multiple of the orders of finite subgroups of Γ divides the Goldie rank $r_k(\Gamma)$.

It is a simple matter now to show that the strong conjecture (B) implies conjecture (A). Since, by cor. 1 to Prop. 2, $r_k(\Gamma)$ is the least common multiple of the denominators of Euler characteristics and since the denominator of the characteristic of a module induced from a finitely generated kH module, with finite H, has denominator dividing $|H|$ we see that conjecture (B) implies that $r_k(\Gamma)$ divides the least common multiple of the orders of finite subgroups. By the corollary above this is an equality.

REFERENCES

1. G.H. Cliff, Zero divisors and idempotents in group rings, preprint (?).

2. D.R. Farkas and R.L. Snider, K_0 and Noetherian group rings, J. of Algebra 42 (1976) 192-198.

3. F.T. Farrel and W.C. Hsiang, A formula for $K_1R_\alpha[T]$, in "Application of categorical algebra", Proc. Symposia in Pure Math. XVIII, AMS 1970.

4. F.T. Farrel and W.C. Hsiang, The topological Euclidean space form, Inventiones Math. 45 (1978) 181-192.

5. I.N. Herstein, "Non-commutative rings", Carus Math. Monographs # 15, MAA 1968.

6. D.S. Passman, "The algebraic structure of group rings", J. Wiley * Sons, N.Y. 1977.

7. S. Rosset, Group extensions and division algebras, J. of Algebra 53 (1978) 297-304.

8. J.P. Serre, Cohomologie des groupes discrets, in Prospects in Math. Annals of Math. Studies 70, Princeton 1971.

9. R.G. Swan, Induced representations and projective modules, Ann. of Math. 71 (1960) 552-578.

Tel-Aviv University

Israel

BRAUER GROUPS OF RATIONAL FUNCTION FIELDS OVER GLOBAL FIELDS

B. FEIN and M. SCHACHER

1. INTRODUCTION

Let B(K) denote the Brauer group of a field K. The structure of B(K) for K a global field (that is, either an algebraic number field or an algebraic function field of transcendence degree one over a finite constant field) was completely determined in the 1930's by the work of Albert, Brauer, Hasse, and Noether [7, Chapter 7]. In this article we are concerned primarily with the case when K is a rational function field over a global field. For such fields K, the structure (as an abelian group) of B(K) was completely determined in [11] by the authors and J. Sonn. Our goal here is to make the proof of this result more accessible; towards that end we will prove several preliminary results that are basic for the argument and we will explain some of the essential ideas involved in the proof of the results announced in [11].

B(K), being an abelian torsion group, is determined by its various p-primary components. We denote, in general, the p-primary component of an abelian torsion group G by G_p. For K a field of characteristic $q > 0$, the basic result used in determining $B(K)_q$ is a theorem of Witt which asserts that $B(K)_q$ is divisible; we prove this in § 2. For p different from the characteristic of K, and $K = E(t)$ where t is transcendental over E, the basic result used in determining $B(K)_p$ is a theorem due originally to Faddeev, but generalized and simplified by Auslander and Brumer, which shows how $B(K)_p$ is built up from $B(E)_p$ and the p-primary components of various character groups of finite separable extensions of E. We also prove the Auslander-Brumer-Faddeev Theorem in § 2.

We digress from our main theme in § 3. We show there how an important theorem of Bloch (the original proof of which appears in Bloch's article in these proceedings) follows from the Auslander-Brumer-Faddeev Theorem together with a result of Tate (proved in Rosset's article (CMH 52 (1977) 519-523)). We also discuss briefly the relationship of this result to the conjecture that the Brauer group is, in general, generated by cyclic algebras. Finally, in § 4 through § 8, we discuss the structure of the Brauer group of a rational function field over a global field.

We fix now some notation that will be maintained throughout this article. We denote a separable closure of a field K by K_s and we denote the multiplicative group of K by K^x. p will always denote a rational prime. If E is Galois over K, then Gal(E/K) is the Galois group of E over K. If K has characteristic different from p, then $\varepsilon(p^n)$ will always denote a primitive p^n-th root of unity over K. We will abbreviate the phrase "finite dimensional simple algebra with center K" by f.d.c.s. We denote elements of B(K) by [A]; here A is a f.d.c.s. Our Brauer groups are written additively, so $[A \otimes_K B] = [A] + [B]$.

If G is a group and α is a cardinal, then $\oplus_\alpha G$ denotes the direct sum of α copies of G. We let Q and Z represent, respectively, the rationals and the integers. We denote the (continuous) character group of a field K by X(K); $X(K) = \mathrm{Hom}(\mathrm{Gal}(K_s/K), Q/Z)$ where we consider only the (continuous) homomorphisms whose kernels have finite index in $Gal(K_S/K)$. We will introduce additional notation as needed.

2. WITT'S THEOREM AND THE AUSLANDER-BRUMER-FADDEEV THEOREM

We assume that the reader has a basic familiarity with Galois cohomology for finite and profinite groups; as basic references we refer the reader to [17], [20], [21], and to the relevant articles in [6]. We recall, in particular, that if E is a Galois extension of F, then $H^q(Gal(E/F),E^x) \cong \varinjlim H^q(Gal(E_i/F),E_i^x)$ where E_i runs through the family of all finite Galois extensions in E of F.

We turn first to Witt's Theorem. The proof we present is an adaption of one implicit in [20].

Theorem (Witt) : Let K be a field of characteristic $p > 0$. Then $B(K)_p$ is divisible.

Proof. We must show that the map $B(K)_p \overset{p}{\to} B(K)_p$, where $\alpha \in B(K)_p$ is sent to $p\alpha$, is surjective. Let $G = Gal(K_S/K)$. Then $B(K)_p \cong H^2(G,K_S^x)_p$ where K_S^x is a G-module under Galois action. Let $K_S^x \overset{p}{\to} K_S^x$ denote the map sending $a \in K_S$ to a^p. Since K has characteristic p, the sequence

$$1 \to K_S^x \overset{p}{\to} K_S^x \to K_S^x/(K_S^x)^p \to 1$$

is exact. This sequence gives rise to the cohomology sequence :

$$\ldots \to H^2(G,K_S^x)_p \overset{p}{\to} H^2(G,K_S^x)_p \to H^2(G,K_S^x/(K_S^x)^p)_p \to \ldots$$

Thus we have to show that $H^2(G,K_S^x/(K_S^x)^p)_p = 0$. Since each $a \in K_S^x$ has only finitely many conjugates under the action of G, it follows that we can decompose $K_S^x/(K_S^x)^p$ as direct sum $\oplus V_j$, $j \in J$, where each V_j is a finite dimensional Z/pZ - vector space and is generated (not necessarily freely) by the conjugates under G of some coset of $(K_S^x)^p$ in K_S^x. V_j is then a G-module under Galois action and we have $H^2(G,K_S^x/(K_S^x)^p)_p \cong \oplus H^2(G,V_j)_p$, $j \in J$. Fix $j \in J$ and set $V = V_j$. We are reduced to showing that $H^2(G,V)_p = 0$. Let H be a Sylow p-subgroup of G. ($H = Gal(K_S/L)$ where $K_S \supset L \supset K$ and L is maximal subject to being the union of p'-extensions of K.) Then $res : H^2(G,V)_p \to H^2(H,V)_p$ is a monomorphism [20,I-11] and so, without loss of generality, we may assume that G is a pro-p-group. Assume inductively that $H^2(G,W)_p = 0$ if W is a G-submodule or G-factor module of V of lower Z/pZ-dimension. If V is not an irreducible G-module, then we have $0 \to W_1 \to V \to W_2 \to 0$ is exact, where W_1 and W_2 are G-modules of lower dimension. By induction, $H^2(G,W_1)_p = H^2(G,W_2)_p = 0$. But we have an exact cohomology sequence $\ldots \to H^2(G,W_1)_p \to H^2(G,V)_p \to H^2(G,W_2)_p \to \ldots$ and so $H^2(G,V)_p = 0$. Thus, we may assume that V is an irreducible G-module. Let $v \in V$, $v \neq 0$. Then $v = a(K_S^x)^p$ for some $a \in K_S^x$. Let $K_S \supset E \supset K$ where E is the normal closure of K(a) in K_S. Let $U = Gal(K_S/E)$. Since G is a pro-p-group, G/U is a finite p-group. Since V is an irreducible G-module, every element of V is of the form ngv for some $n \in Z/pZ$ and some $g \in G$. But U is normal in G so if $u \in U$ we have $g^{-1}ug \in U$. Since U fixes a, U fixes v so $u(ngv) = n(ugv) = ng(g^{-1}ugv) = ngv$. Thus U acts trivially on V so V is a G/U-module. Since G/U is a p-group and $|V|$ is a power of p, there must be at least one other orbit of length one of V under the action of G/U other than $\{0\}$. Let $\{v'\}$ be an orbit of length one, $v' \neq 0$. Then $gv' = v'$ for all $g \in G$ and so we conclude that $V = (Z/pZ)v'$. Thus

$V \cong Z/pZ$ where G acts trivially. Now consider the map $K_S \xrightarrow{f} K_S$ given by $f(a) = a^p - a$. Since $x^p - x - b$ is separable in $K_S[x]$ for any $b \in K_S$ and since K_S is separably closed, we see that f is surjective. Since Z/pZ is the kernel, we have $0 \to Z/pZ \to K_S \xrightarrow{f} K_S \to 0$. This gives rise to the cohomology sequence :

$$\ldots \to H^1(G,K_S^+) \to H^2(G,Z/pZ) \to H^2(G,K_S^+) \to \ldots .$$

where K_S^+ denotes the additive group of K_S. By the normal basis theorem, $H^1(G,K_S^+) = H^2(G,K_S^+) = 0$ [6, page 124] and so $H^2(G,V) = 0$ as was to be proved.

Before stating and proving the Auslander-Brumer-Faddeev Theorem, we give two preliminary results. The first of these is simply a very special case of Shapiro's Lemma and is implicit in various places in the literature (see, for example, [17, Satz 4.19] and [21, page 128, exercise 3]).

<u>Lemma 1</u> : Let G be a profinite group, H a subgroup of G of finite index, and let G\H denote a complete set of left coset representatives of H in G. Let M be a trivial H-module and let M' denote the direct sum of $|G\backslash H|$ copies of M made into a G-module by permutation action in the natural way. Let i denote the inclusion of M in M' by embedding M as the first factor in M' and let π denote the projection of M' onto its first factor. Then $\pi \circ \text{res} : H^r(G,M') \to H^r(H,M)$ and $\text{cor} \circ i : H^r(H,M) \to H^r(G,M')$ are inverse isomorphisms.

We will not prove this result completely. One proceeds, as is standard, by dimension shifting. For $r = 0$, $(M')^G = \{(m,\ldots,m) \mid m \in M\}$ and so $\pi \circ \text{res} : (m,\ldots,m) \to (m,\ldots,m) \to m$ while $\text{cor} \circ i : m \to (m,0,\ldots,0) \to (m,m,\ldots,m)$ as was to be proved.

Lemma 2 : Let K be a field of characteristic $q \geq 0$, let t be transcendental over K, and let $p \neq q$. Then $B(K_S(t))_p = 0$.

Proof. Suppose not. Then there exists a division ring D central over $K_S(t)$ such that the class [D] of D in $B(K_S(t))$ has order p. But then D has index a power of p [19, Theorem 29.22]. Let $\tilde{K}$ be an algebraic closure of K_S. By Tsen's Theorem [14, Chapter 3], D is split by $\tilde{K}(t)$. Since $\tilde{K}(t)$ is algebraic over $K_S(t)$, there exists a field E, $\tilde{K} \supset E \supset K_S$ with $[E : K_S]$ finite such that E splits D. If $q = 0$, then $\tilde{K} = K_S$ so $E = K_S$ and we have a contradiction. If $q > 0$, then E is purely inseparable over K_S so $[E : K_S]$ is a power of q. This contradicts [19, Theorem 28.5].

The proof we shall give of the Auslander-Brumer-Faddeev Theorem is essentially the one appearing in [4].

Theorem (Auslander-Brumer [4], Faddeev [8]) : Let K be an infinite field of characteristic $q \geq 0$, let t be transcendental over K, and let $p \neq q$. Then :

(2.1) $$B(K(t))_p \cong B(K)_p \oplus \oplus_{|K|} (\oplus_F X(F)_p)$$

where F ranges over all finite extensions of K in K_S. Moreover, every class of order p^n in $B(K(t))$ can be represented in the form $[A \otimes_K K(t)] + \mathrm{cor}[L_1(t)/F_1(t), \sigma_1, b_1] + \ldots + \mathrm{cor}[L_r(t)/F_r(t), \sigma_r, b_r]$. Here [A] is an element of B(K) of order dividing p^n, F_i is a finite separable extension of K, L_i is cyclic over F_i of degree dividing p^n, $\langle\sigma_i\rangle = \mathrm{Gal}(L_i/F_i)$, $b_i \in F_i[t]$ and cor denotes the corestriction map from $B(F_i(t))$ to $B(K(t))$.

Proof. By Lemma 2, $B(K(t))_p = B(K_s(t)/K(t))_p \cong H^2(G,K_s(t)^\times)_p$, where $G = Gal(K_s(t)/K(t)) \cong Gal(K_s/K)$. Let N denote the set of monic irreducible polynomials in $K_s[t]$ and, for $P(t) \in N$, let $<P(t)> = \{P(t)^n | n \in Z\}$. Then $K_s(t)^\times \cong K^\times \oplus \oplus_P <P(t)>$ where the sum is taken over all $P(t) \in N$. Let M be a subset of N such that $\{<P(t)> | P(t) \in M\}$ is a complete set of representatives of the distinct orbits of $\{<P(t)> | P(t) \in N\}$ under the action of G. Let $G_P = \{\sigma \in G | P(t)^\sigma = P(t)\}$. G_P is a subgroup of finite index in G; we let $G \backslash G_P$ denote a complete set of left coset representatives of G_P in G. Then $V_P = \oplus_\sigma <P(t)^\sigma>$, $\sigma \in G\backslash G_P$, is a G-module and $K_s(t)^\times = K_s^\times \oplus \oplus_P V_P$, the sum over all $P \in M$. Thus $B(K(t))_p \cong H^2(G,K_s^\times)_p \oplus \oplus_P H^2(G,V_P)_p$, the sum being over all $P \in M$. Since $<P(t)>$ is a trivial G_P- module we have $H^2(G,V_P)_p \cong H^2(G_P,<P(t)>)_p$ by Lemma 1. Let $K_P = (K_s)^{G_P}$, the fixed field of G_P. Since $<P(t)> \cong Z$, $H^2(G_P,<P(t)>)_p \cong H^2(G_P,Z)_p \cong H^1(G_P,Q/Z)_p = Hom(G_P,Q/Z)_p = X(K_P)$ [21, page 194]. Since $P(t)$ is monic irreducible and K_s is separably closed, $P(t) = t - a$ or $t^{q^r} - a$ if $q > 0$, for some $a \in K_s$. It follows that $K_P = K(a)$. Thus each finite extension of K in K_s appears as a K_P and since $K(a) = K(a+k)$ for any $k \in K$, each $K(a)$ appears $|K|$ times. Since $H^2(G,K_s^\times)_p \cong B(K)_p$, the first assertion of the theorem follows. Next suppose $\alpha \in B(K(t))_p$ has order p^n. Then $\alpha = \beta + \gamma_1 + \ldots + \gamma_r$ where $\beta \in H^2(G,K_s^\times)_p$, $\gamma_i \in H^2(G,V_{P_i})_p$, $i = 1,\ldots,r$, and $\beta, \gamma_1,\ldots,\gamma_r$ have orders dividing p^n. It is clear that β has the form $[A \otimes_K K(t)]$ for some [A] of order dividing p^m in B(K). Let $P = P_i$ and set $\gamma = \gamma_i$. By the above argument and Lemma 1, $\pi \circ res(\gamma)$ corresponds to a character ψ of order p^m, $m \le n$, in $X(K_P)$. Let L be the fixed field of ker ψ. Then $L(t)$ is a cyclic extension of $K_P(t)$ of degree p^m. Also, as shown above, $P(t) \in K_P(t)$. For a suitable choice of generator σ of $Gal(L(t)/K_P(t))$ we have $\pi \circ res(\gamma) = [(L(t)/K_P(t),\sigma,P(t))]$. By Lemma 1, $\gamma = cor \circ i([L(t)/K_P(t),\sigma,P(t))])$, proving the theorem.

3. BLOCH'S THEOREM

Let K be an arbitrary infinite field. The following is perhaps the major unsolved problem in the theory of B(K) :

Question 1 : Let $[A] \in B(K)$. Is A similar to a tensor product of cyclic K-algebras ?

If K is a local or global field, the answer to Question 1 is yes and A is already a cyclic algebra [19, § 14 and Theorem 32.20]. In general, however, A might be similar to a tensor product of cyclic algebras without being itself cyclic or even similar to a cyclic algebra. For example, let $K = R((t_1,t_2))$ be the field of formal Laurent series in t_1 and t_2 over the reals. Let $D = (K(\varepsilon(4))/K,\sigma,-1) \otimes_K (K(t_1^{\frac{1}{2}})/K,\tau,t_2)$. Using the fact that K has no cyclic extensions of degree 4 and only 8 square classes, it is not difficult to see that D is a division ring which is not similar to any cyclic K-algebra.

The answer to Question 1 is yes if A is a K-division ring of index 2 or 3 [1, pages 146 and 177] or if A is a K-division ring of index 4 where K has characteristic different from 2 and $\varepsilon(4) \in K$ [22]. The answer is also yes if K has characteristic p and [A] has order a power of p in B(K) [1, page 109]; in that case A is already similar to a cyclic K-algebra. If A is a K-division ring of index 8 with [A] of order 2 in B(K), K of characteristic different from 2, then A is similar to a tensor product of four quaternion algebras over K [24] so the answer is also yes in this case. No examples are known where Question 1 has a negative answer.

Let ζ_1 and ζ_2 be two $n \times n$ generic matrices over K; that is, the entries of ζ_1 and ζ_2 are $2n^2$ commuting indeterminates over K. The ring generated over K by ζ_1 and ζ_2 is an integral domain having an Ore quotient ring $K<n;\zeta_1,\zeta_2>$. $K<n;\zeta_1,\zeta_2>$ is a division ring of index n over its center F_n; $K<n;\zeta_1,\zeta_2>$ is referred to as a generic division ring over K [18, Chapter 3, § 1]. The field F_n is a subfield of index n! in a purely transcendental extension of K of transcendence degree n^2+1 over K [18, Theorem 6.3, page 95]. This leads to :

Question 2 : Is F_n purely transcendental over K ?

The answer to Question 2 is yes if n = 2 [18, Theorem 6.5, page 99], if n = 3 [12], or if n = 4 [13]. No examples are known where F_n is not purely transcendental over K.

There turns out to be an interesting relationship between Questions 1 and 2. Suppose that $K<n:\zeta_1,\zeta_2>$ is similar to a tensor product of cyclic F-algebras and that L is a field, $L \supset K$. Then, using a specialization argument similar to that employed in [2, § 4], one can show that every L-division ring of index n is similar to a tensor product of cyclic L-algebras. This reduces Question 1 to the corresponding question for $K<n;\zeta_1,\zeta_2>$. The relevance of Question 2 to this question is now clear from the following result of Bloch [5] :

Theorem (Bloch) : Let K be a field such that if [A] has order dividing n in B(K), then A is similar to a tensor product of cyclic K-algebras (e.g. K local, global or algebraically closed). Assume K has characteristic not dividing n and that $\varepsilon(n) \in K$. Let E be a finitely generated purely transcendental extension of K and let D be a f.d.c.s. E-algebra, [D] of order dividing n in B(E). Then D is similar to a tensor product of cyclic E-algebras.

Proof. Bloch's original proof appears in his article in these proceedings. The proof we present is based on the Auslander-Brumer-Faddeev Theorem. We proceed by induction on the transcendence degree of E over K and so can assume that $E = K(t)$. Let F be a finite separable extension of K and let L be a cyclic extension of F of degree dividing n. Let $b \in F(t)$, let $\langle\sigma\rangle = \mathrm{Gal}(L/F)$, and let cor denote the corestriction map from $B(F(t))$ to $B(K(t))$. By the Auslander-Brumer-Faddeev Theorem it is enough to prove that $\mathrm{cor}((L(t)/F(t),\sigma,b)$ is similar to a tensor product of cyclics. Since $\varepsilon(n) \in K$, there are elements $x,y \in (L(t)/F(t),\sigma,b)$ which generate this algebra such that $x^m \in F(t)$, $y^m = b$, and $xy = \delta yx$ where m divides n, $m = [L:F]$, and δ is a primitive m-th root of unity. Let $B_n(M)$ denote the subgroup of the Brauer group of a field M annihilated by n. According to the article of Rosset in CMH 52 (1977), 519-523, we have a commutative diagram :

$$\begin{array}{ccc} K_2(F(t)) & \xrightarrow{N} & B_n(F(t)) \\ \Big\downarrow{\scriptstyle tr} & & \Big\downarrow{\scriptstyle cor} \\ K_2(K(t)) & \xrightarrow{N} & B_n(K(t)) \end{array}$$

Since $[(L(t)/F(t),\sigma,b)] = N(x^m,b)$, we have $\mathrm{cor}((L(t)/F(t),\sigma,b)) = N \circ \mathrm{tr}(x^m,b)$ which is an element of $B_n(K(t))$ represented as a tensor product of cyclic algebras. This proves Bloch's Theorem.

4. THE MAIN THEOREM

Throughout this section F will be a global field of characteristic $q \geq 0$. By a rational function field over F we mean a purely transcendental extension K of F with $1 \leq$ t.d. $K/F < \infty$ (t.d. K/F is the transcendence degree of K over F). Since a description of $B(K)$ will eventually hinge on a determination of its Ulm invariants, we describe first the relevant concepts in that theory. In what follows p will always denote a prime integer, ω the first infinite ordinal (which we regard also as a cardinal when convenient), and $\omega 2$ the least limit ordinal greater than ω.

Suppose G is an abelian torsion group. The analogue of the Sylow decomposition says that G can be recovered from its primary components because $G \cong \oplus G_p$, p prime. Hence it is enough to devise a set of invariants for the p-primary group G_p. Since DG_p is injective, one has $G_p \cong DG_p \oplus RG_p$ for some suitable complement RG_p of DG_p in G_p. RG_p is not unique, but it is unique up to isomorphism since $RG_p \cong G_p/DG_p$. RG_p is called a (or the) reduced subgroup of G_p; it has no divisible subgroups. Structure theory says that DG_p is just a direct sum of copies of $Z(p^\infty) = (Q/Z)_p$, and if $DG_p \cong \oplus_r Z(p^\infty)$ then r is a complete invariant for DG_p. The Ulm invariants of G_p, which we describe next, are invariants of RG_p.

The Ulm subgroups $G_p(\lambda)$ are defined inductively for any ordinal λ by : $G_p(0) = G_p$, $G_p(\lambda+1) = pG(\lambda)$, and $G_p(\lambda) = \bigcap_{\beta<\lambda} G_p(\beta)$ for λ a limit ordinal. The least ordinal λ with $G_p(\lambda) = G_p(\lambda+1)$ is called the Ulm length of G; for that λ we have $G_p(\lambda) = DG_p$. We write $\ell_p(G)$ for the Ulm length of G_p. We set $P_\lambda(G_p) = \{\alpha \in G_p \mid p\alpha = 0 \text{ and } \alpha \in G_p(\lambda)\}$. The dimension of $P_\lambda(G_p)/P_{\lambda+1}(G_p)$ over the field of p elements is called the λ-th Ulm invariant of G_p and denoted $U_p(\lambda,G)$. One checks easily that $U_p(\lambda,G) = U_p(\lambda,RG_p)$. The Ulm invariants are not in general a complete set of invariants for RG_p, but they are known to be so if G_p is countable. Since the rational function fields K under consideration are countable, the groups B(K) will be countable as well, and so the Ulm invariants will be a complete set of invariants for the reduced part of its p-primary components.

Let $K = F(t_1,\dots,t_n)$ be a rational function field in n variables over F. To describe B(K) we will need a cyclotomic constant depending only on F. Suppose $p \neq q = \mathrm{char}(F)$. We define $\phi(F,p)$ to be the largest integer r so that :

(i) $\varepsilon(p^r) \in F(\varepsilon(p))$ for p odd

or

(ii) $\varepsilon(2^r) \in F(\varepsilon(4))$ if $p = 2$.

For any integer $n \geq 0$ let H(F,p,n) be the abelian group with generators x, y_i $(i = \phi(F,p), \phi(F,p)+1,\dots,k,\dots)$ and relations $p^{n+1}x = 0$, $p^i y_i = x$ $(i = \phi(F,p),\dots)$. Let

$$G(F,p,n) = \oplus_\omega [Z(p^\infty) \oplus \bigoplus_{n=0}^{\infty} H(F,p,n)].$$

We set $T_i = \oplus_\omega Z/2^i Z$ where $Z/2^i Z$ is the cyclic group of order 2^i. Our main result is :

<u>Theorem A</u>. Let $K = F(t_1,\dots,t_n)$ be a rational function field over the global field F of characteristic $q \geq 0$. We have :

(1) if $q > 0$, then $B(K)_q \cong \oplus_\omega Z(q^\infty)$

(2) for $p \neq q$, if either p is odd or $p = 2$ and $\varepsilon(4) \in F$, then $B(K)_p \cong G(F,p)$

(3) if $p = 2$, $p \neq q$ and $\varepsilon(4) \notin F$, then $B(K)_2 \cong G(F,2) \oplus T_1$.

A few remarks on Theorem A are in order. One is the surprising fact that the answer for B(K) is independent of n (as long as $n \geq 1$), so we get the :

Corollary. If E and K are rational function fields over the global field F, then $B(E) \cong B(K)$.

Of course Theorem A and the Corollary deal only with the structure of B(K) as an abstract abelian torsion group; we make no claim to have discovered the nature of all central simple algebras over K. Thus all questions as to whether they are similar to products of cyclic algebras (discussed earlier), have abelian splitting fields, or have solvable splitting fields are open even in this fairly simple case (unless we have the requisite roots of unity to apply Bloch's theorem).

We do not prove Theorem A directly, but rather a reformulation of it in terms of Ulm invariants. This amounts to :

Theorem B. Set $K = F(t_1,\ldots,t_n)$ as before with $\mathrm{char}(F) = q \geq 0$. Then :

(1) If $q > 0$, $B(K)_q \cong \oplus_\omega Z(q^\infty)$

(2) For $p \neq q$, $DB(K)_p \cong \oplus_\omega Z(p^\infty)$

(3) For all $p \neq q$, $\ell_p(B(K)) = \omega 2$

(4) For $p \neq q$, p odd, $U_p(\lambda, B(K)) = 0$ for $0 \leq \lambda < \phi(F,p) - 1$ and $U_p(\lambda, B(K)) = \omega$ for $\phi(F,p) - 1 \leq \lambda < \omega 2$.

(5) If $p \neq q$ and $p = 2$, then :

(a) If $\varepsilon(4) \in F$, $U_2(\lambda, B(K)) = 0$ for $0 \leq \lambda < \phi(F,2) - 1$ and $U_2(\lambda, B(K)) = \omega$ for $\phi(F,2) - 1 \leq \lambda < \omega 2$.

(b) If $\varepsilon(4) \notin F$, $U_2(0,B(K)) = \omega$, $U_2(\lambda',B(K)) = 0$ for $0 < \lambda < \phi(F,2) - 1$ and $U_2(\lambda,B(K)) = \omega$ for $\phi(F,2) - 1 \leq \lambda < \omega 2$.

We will not work out here the exercise of showing that the invariants listed in Theorem B are exactly those attached to the groups in Theorem A. By the completeness property of Ulm invariants in the countable case, Theorems A and B are equivalent. The rest of the notes is dedicated to a discussion of the proof of Theorem B.

Part (i) of Theorem B is not difficult using the Theorem of Witt proved in Section 2. For Witt's theorem says that $B(K)_q$ is divisible, and so a direct sum of copies of $Z(q^\infty)$. That one gets as many as ω copies of $Z(q^\infty)$ will follow from the analysis of Brauer groups of global fields which we come to later; this theory will also provide (ii) of Theorem B.

The rest of Theorem B comes from using the exact sequence of Auslander-Brumer-Faddeev proved in Section 2. Since Ulm invariants are additive on direct sums, this sequence reduces our problem to a study of the divisibility properties and Ulm invariants of character groups. We indicate now how our analysis of character groups is conducted.

Suppose E is a field, E_S the separable closure of E, $G = \mathrm{Gal}(E_S/E)$, and $X(E) = \mathrm{Hom}(G,Q/Z)$. If $\sigma \in X(E)$, we make correspond to σ a cyclic extension M_σ of E via : $M_\sigma = E_S^H$, H = kernel σ. It is easy to show that $[M_\sigma : E]$ = order of σ and M_σ/E is Galois with cyclic Galois group. Similarly, every finite cyclic extension of E arises from a character in X(E) in this way. The passage is not unique because finitely many characters σ can describe the same cyclic extension M_σ; nevertheless, we would like to think of characters and cyclic extensions as identified. We need to give a meaning to the equation $p\tau = \sigma$ for $\sigma, \tau \in X(E)$. The answer is : this happens if and only if $M_\tau \supset M_\sigma \supset E$ with $[M_\tau : M_\sigma] = p$.

For G an abelian torsion group, an element $x \in G_p$ is said to be of infinite height $\Longleftrightarrow x \in G_p(\omega)$, i.e. the equation $p^n y_n = x$ has a solution for all n. Then x lies in a divisible subgroup of G_p if and only if some set of these p^n-th roots can be lined up in coherent fashion so that $py_{n+1} = y_n$ for all n; otherwise it is called reduced. The elements x in the groups H(F,p,n) of Theorem A are reduced but of infinite height. To interpret this at the level of characters : if $\sigma \in X(E)_p$, then σ is of infinite height $\Longleftrightarrow$ for every $n \geq 1$ in Z there is a field $M_n \supset M_\sigma \supset E$ with $[M_n : M_\sigma] = p^n$ and M_n/E still cyclic. σ will lie in a divisible subgroup $\Longleftrightarrow$ the M_n can be arranged so that $\ldots M_{n+1} \supset M_n \supset M_{n-1} \supset \ldots \supset M_\sigma \supset E$; when this happens the field $M = \bigcup_n M_n$ is called a Z_p-extension (see [15]).

5. BRAUER GROUPS OF GLOBAL FIELDS

If F is a global field, we let $\{F_\alpha\}$ be the set of completions of F at the primes α (both finite and infinite). For each α, one knows by local class field theory (see [21]) that :

(i) $B(F_\alpha) \cong Q/Z$ if α finite

(ii) $B(F_\alpha) \cong Z/2Z \subset Q/Z$ if α is real infinite

(iii) $B(F_\alpha) = \{0\}$ if α is complex.

These isomorphisms are accomplished by a local invariant map $\mathrm{inv}_\alpha : B(F_\alpha) \to Q/Z$.

For $[A] \in B(F)$, we set $\phi : B(F) \to \bigoplus_\alpha B(F_\alpha)$ to be the sum of the natural restriction maps $B(F) \to B(F_\alpha)$; global class field theory (see [3]) says the image of ϕ is in the direct sum (i.e. $\mathrm{inv}_\alpha[A] = 0$ for almost all α). We define $\psi : \oplus B(F_\alpha) \to Q/Z$ by $\psi((x_\alpha)) = \sum_\alpha \mathrm{inv}_\alpha(x_\alpha)$. The principal result we need (see [3, Chapter 7]) is : The sequence $0 \to B(F) \overset{\phi}{\to} \bigoplus_\alpha B(F_\alpha) \overset{\psi}{\to} Q/Z \to 0$ is exact. From this one easily derives :

Theorem 5.1. For any global field F;

$$B(F) \cong \oplus_s Z/2Z \oplus_\omega Q/Z$$

where s is the number of real primes of F (and s = 0 if F has no real primes).

We return to (i) and (ii) of Theorem B. The exact sequence (2.1) says that for any rational function field K over F, $B(F)_p$ is a direct summand of $B(K)_p$ for $p \neq q$. Since $B(F)_p$ contains ω copies of $Z(p^\infty)$ for any prime p, the same is true of $B(K)_p$. This, together with the theorem of Witt in Section 2, allows us to read off (i) and (ii) of Theorem B.

6. CHARACTERS OF INFINITE HEIGHT

Suppose E is any field finitely generated over its prime field (either Q or Z/pZ). We derive here a fundamental tool for controlling the cyclic extensions which can correspond to characters in $X(E)_p(\omega)$. This is accomplished in :

Lemma 6.1. Suppose $\sigma \in X(E)_p(\omega)$ and $M = M_\sigma$. Then M/E is unramified for every discrete rank one valuation of E whose residue class field has characteristic prime to p.

We indicate the proof of the Lemma briefly; full details can be found in [10]. Suppose v is such a valuation, E_v the completion, and $\overline{E_v}$ the residue field. Since E is finitely generated over its prime field, both E_v and $\overline{E_v}$ contain finitely many roots of unity. Writing v again for any extension of v to M, we conclude likewise that M_v and $\overline{M_v}$ have finitely many roots of unity. Since $\sigma \in X(E)_p(\omega)$, there is a cyclic extension M_N of E so that $M_N \supset M \supset E$ and $[M_N : M] = p^N$. Write v again for an extension of v to M_N, and letting T = the maximal unramified extension of E_v in M_v, we have : $(M_N)_v \supset M_v \supset T \supset E_v$. If $T \neq M_v$, then (by our characteristic assumption), M_v/T is totally and tamely ramified and cyclic, say of dimension p^m. It follows from Hilbert theory (see [25,4-10]) that $(M_N)_v/T$ is totally and tamely ramified of dimension p^{m+N}, and still cyclic. Tamely ramified extensions of dimension p^{m+N} are obtained by adjoining a p^{m+N}-th root of a prime element; they are cyclic $\Longleftrightarrow \varepsilon(p^{m+N}) \in T$. As N is arbitrary, we would get all $\varepsilon(p^{m+N}) \in T \subset M_v$, contradicting the fact that M_v contains finitely many roots of 1. Thus $T = M_v$ is unramified over E_v.

When E is a global field, Lemma 6.1 says the following : $\sigma \in X(E)_p(\omega) \Longleftrightarrow M_\sigma/E$ is unramified at all primes of E except perhaps the finite number which lie over the rational prime p. By [15] there can only be a finite number of such M_σ of a given dimension; this will be one of the key ingredients involved in the derivation of the Ulm invariants of X(E).

7. THE CASE n = 1.

The aim of this section is to complete the proof of Theorem B when $n = 1$. We set $t = t_1$, and so by (2.1) :

(7.1) $$B(F(t))_p \cong B(F)_p \oplus \oplus_{|E|} X(V)_p$$

where the V are global fields containing E. The Ulm invariants of $B(F)_p$ are evident since it is essentially divisible, and is in fact divisible when p is odd. Thus it is enough to find the Ulm invariants and Ulm length of the $X(V)_p$. One friendly consequence of the exact sequence above is the following : since Ulm invariants are additive on direct sums, and since each X(V) occurs ω times, then for any ordinal λ :

Lemma 7.1. If $U_p(\lambda, B(F(t))) \neq 0$, then $U_p(\lambda, B(F(t)) = \omega$.

Thus for $K = F(t)$, all our Ulm invariants will be 0 or ω, and it is only a question of deciding when each occurs. We first settle the matter of the Ulm length.

Theorem 7.2. If V is a global field of characteristic q and $p \neq q$, then $\ell_p(X(V)) < \omega 2$.

Proof. We fit together the pieces from the last section. Suppose σ is a reduced character in $X(V)_p(\omega 2)$. If M_σ is the corresponding cyclic extension of V, then :

(1) $M_\sigma \supset V$ lies in no Z_p-extension of V.

(2) For any $N \geq 0$, there is a cyclic extension M_N of V with $M_n \supset M_\sigma \supset V$ and $[M_N : M_\sigma] = p^N$.

(3) Each extension M_N/V is reduced of infinite height, but each M_N lies in no Z_p-extension of V.

By Lemma 6.1, each M_N/V is unramified outside of the finite number of primes dividing p. By [15], there are a finite number of possibilities for each M_N. Now let Γ be the directed graph whose vertices are reduced elements of $X(V)_p(\omega)$, and such that δ is connected to τ if $p\tau = \delta$. Lemma 6.1 translates to say : each vertex of Γ has a finite number of edges emanating from it. It follows from the "Konig infinity lemma" that Γ is finite; if Γ were infinite then it would contain an infinite ascending path corresponding to a Z_p-extension of V, and this would contradict the fact that all characters considered are reduced. This says that $\ell_p(X(V)) \leq \omega + r$ for some integer r, completing the proof of Theorem 7.2.

The derivation of the Ulm invariants for B(K) when $K = F(t)$ will proceed in two steps.

Step 1 : Compute $U_p(\lambda, B(F(t)))$ when $0 \leq \lambda < \omega$

Step 2 : Compute $U_p(\lambda, B(F(t)))$ when $\omega \leq \lambda < \omega 2$.

We begin with Step 1. In general we will have to know when a cyclic extension $M \supset V$ of dimension p^n can be enlarged to a cyclic extension (of V) of dimension p^{n+1}. There are well-known number theoretic criteria for this, beginning with :

<u>Lemma 7.3</u>. Suppose $K \supset E$ are fields with $[K : E] = p^n$, K/E cyclic. Suppose also $\varepsilon(p) \in F$. Then K can be enlarged to an extension $M \supset K \supset E$ where M/E is cyclic of dimension $p^{n+1} \Longleftrightarrow \varepsilon(p)$ is a norm from K to E. This will always happen when $\varepsilon(p^{n+1}) \in E$.

The last remark is clear since $N_{K/E}(\varepsilon(p^{n+1})) = \varepsilon(p)$ when $\varepsilon(p^{n+1}) \in E$. For a proof of Lemma 7.3 see [6, p. 124]. Lemma 7.3 is true for all fields, but depends on the presence of the root of unity $\varepsilon(p)$. When K and E are global fields, this can be sharpened considerably and $\varepsilon(p)$ is not needed.

Lemma 7.4. Suppose $K \supset E$ are global fields with K/E cyclic of dimension p^n, $p \neq \text{char } E$. If $p = 2$ we assume $\sqrt{-1} \in E$. Then K can be enlarged to a cyclic extension M/E of dimension p^{n+m} for $m \geq 1 \iff$ any p^m-th root of unity is a local norm from K_α to E_α for each completion E_α of E (and there is no condition if 1 is the only p^m-th root of 1 in E_α).

For a proof of Lemma 7.4 see [3, Chapter 10].

Suppose now $K = F(t)$, p is odd, $p \neq \text{char}(F)$, and $[V : F] < \infty$. If $\sigma \in P_\lambda(X(V)_p)$ (see Section 4) then for $0 \leq \lambda < \phi(F,p)-1$ we have : $M \supset M_\sigma \supset V$ where M/V is cyclic, $[M : M_\sigma] = p^\lambda$ and $[M_\sigma : V] = p$. By definition of $\phi(F,p)$, if α is any prime of V with $\varepsilon(p) \in V_\alpha$, then $\varepsilon(p^{\lambda+1}) \in V_\alpha$. Lemma 7.4 then says that $\varepsilon(p)$ is a local norm from M to V everywhere, and so M can be enlarged to $M' \supset M \supset M_\sigma \supset V$ with M'/M_σ cyclic of dimension $p^{\lambda+1}$. We conclude $\sigma \in P_{\lambda+1}(X(V)_p)$. Thus $U_p(\lambda, X(V)_p) = 0$ when $0 \leq \lambda < \phi(F,p)-1$.

Assume then $\phi(F,p)-1 \leq \lambda < \omega$. By standard density arguments, there is a finite prime π of F, $\pi \nmid 2$ if $q = 0$, and such that $\varepsilon(p^{\lambda+1}) \in F_\pi$ but $\varepsilon(p^{\lambda+2}) \notin F_\pi$. We set $W = F_\pi(a)$ where $a^{p^{\lambda+1}}$ is a prime of F_π. Then W/F_π is cyclic and totally ramified. By the Grunwald-Wang Theorem [3, Chapter 10] there is a cyclic extension V of F so that $[V : F] = p^{n+1}$ and $V_\pi = W$. Let S be the subfield of V with $V \supset S \supset F$ and

$[S:F] = p$, and suppose τ is in $X(F)_p$ with $S = M_\tau$. Since $S \subset V$ we have $\tau \in P_\lambda(X(F)_p)$. W cannot be extended to a cyclic extension of F_π of dimension $p^{\lambda+2}$ since any such would be totally and tamely ramified and $\varepsilon(p^{\lambda+2}) \notin F_\pi$. It follows that V cannot be extended to a cyclic extension of F of dimension $p^{\lambda+2}$, since any such would be totally ramified at π. Thus $\tau \in P_\lambda(X(F)_p)$ but $\tau \notin P_{\lambda+1}(X(F)_p)$. We conclude $U_p(\lambda, X(F)_p) \neq 0$. By Lemma 7.1, $U_p(\lambda, B(K)) = \omega$.

The argument that the Ulm invariants $U_2(\lambda, B(K))$ for finite λ are as required in Theorem B is a little more technical than the above; we refer the reader to [9] rather than reproduce it here.

We proceed to Step 2 : to evaluate the $U_p(\lambda, B(K))$ when $\omega < \lambda \leq \omega 2$. We will need the following Lemma, which is a handy way of constructing characters of infinite height.

Lemma 7.5. Suppose $E \supset F$ are global fields with E/F cyclic of dimension p^n and everywhere unramified, $p \neq \text{char } F$. Assume $\sqrt{-1} \in F$ if $p = 2$. Then E can be extended to a cyclic extension M of F with $[M:F] = p^{n+m}$ for any $m \geq 1$ (i.e. any character τ corresponding to M is of infinite height).

Since E/F is everywhere unramified, units are local norms. Then any p^m-th root of 1 in any completion F_π is a local norm, and Lemma 7.4 then shows E can be so extended.

To manufacture characters of infinite height which are not divisible, we need the following incisive theorem of J. Sonn. Sonn's theorem is proved in [23].

Theorem 7.6. Let F be a global field, $p \neq \text{char } F$, and $n \geq 1$ an integer. Then there is a finite solvable extension E of F having the property : there is a cyclic extension L of E of dimension p^n which is unramified at all primes of E. If M is any cyclic extension of E of dimension p^{n+m} with $m \geq 1$ and $M \cap L \neq E$, then some finite prime of E not dividing p ramifies in M.

Theorem 7.6 has as a consequence : for any global field F, $p \neq \text{char } F$, there is an extension E of F which has $Z/p^n Z$ as a direct summand of its class group; we are not aware of any other way to prove this result.

We apply Theorem 7.6 when $n = \lambda + 1$. Let L_1 be the subfield of L with $L \supset L_1 \supset E$ and $[L_1 : E] = p$. We can choose characters σ and τ in $X(E)_p$ with $M_\sigma = L, M_\tau = L_1$ and $p^\lambda \sigma = \tau$. By Lemma 7.5, both σ and τ are of infinite height, so $\tau \in X(E)_p(\omega+\lambda)$. If $\delta \in X(E)_p$ with $p^{\lambda+1}\delta = \tau$, then by Theorem 7.6 M_δ is ramified at some prime of E not dividing p; Lemma 6.1 then says δ is not of inifite height. We conclude : $\tau \notin X(E)_p(\omega + \lambda + 1)$. It follows that $U_p(\omega + \lambda, B(K)) \neq 0$, and so $U_p(\omega + \lambda, B(K)) = \omega$. Putting this together with Theorem 7.2, we now have all the pieces to Theorem B when $K = F(t)$.

8. THE CASE OF GENERAL n.

Having proved Theorem B for $B(F(t_1,\ldots,t_n))$ when $n = 1$, we may assume Theorem B holds for B(K) with $K = F(t_1,\ldots,t_{n-1})$ and try to deduce the theorem for B(E), $E = K(t)$, $t = t_n$. From (2.1) we have :

$$0 \to B(K)_p \to B(E)_p \to \oplus \oplus_{|K|} (\oplus_L X(L)_p) \to 0$$

is split exact where $[L : K] < \infty$. Since Ulm invariants are additive over exact sequences, those which are ω will persist in passage from

$B(K)_p$ to $B(E)_p$. Thus, to finish the proof of Theorem B, we need only show :

(i) $U_p(\lambda, X(L)_p) = 0$ at the finite number of finite ordinals λ where $U_p(\lambda, B(K)_p) = 0$ and

(ii) $\ell_p(X(L)_p) < \omega 2$.

For (i), assume p is odd, $p \neq q = \mathrm{char}(F)$, and $[L : K] < \infty$. If $U_p(\lambda, B(K)_p) = 0$, then $0 \le \lambda < \phi(F,p)-1$, and so suppose $\sigma \in P_\lambda(X(L)_p)$. Then σ corresponds to a cyclic extension M of L so that $[M : L] = p$ and $L \subset M \subset N$ with N/L cyclic of dimension $p^{\lambda+1}$. We must show $\sigma \in P_{\lambda+1}(X(L)_p)$, and for this it will be enough to show N can be enlarged to $\bar{N}$ which is a field cyclic over L of dimension $p^{\lambda+2}$. Let $\varepsilon = \varepsilon(p)$ and set $L_1 = L(\varepsilon)$, $M_1 = M(\varepsilon)$, $N_1 = N(\varepsilon)$. Since $[L_1 : L] | p-1$ and is prime to p, we have $[M_1 : L_1] = p$, $[N_1 : M_1] = p^\lambda$, N_1/L_1 cyclic. By assumption, L_1 contains all p^m-th roots of 1 with $m = \phi(F,p)$, and $\lambda < m-1$. Since $\bar{N}_1/L_1$ is cyclic of degree $p^{\lambda+1}$, $\lambda+1 < m$, we conclude ε is a norm from N_1 to L_1 (in fact a norm of $\varepsilon(p^{\lambda+2}) \in N_1$). It follows from Lemma 7.3 that N_1 can be enlarged to $\bar{N}$ which is cyclic of dimension $p^{\lambda+2}$ over F; this says that $\mathrm{res}(\tau) = \tau_1 \in P_{\lambda+1}(X(L_1)_p)$, where res = restriction from L to L_1. Applying corestriction and remembering that cor∘res = multiplication by $a = [L_1 : L]$, we get : $a\sigma \in P_{\lambda+1}(X(L)_p)$. Since $(a,p) = 1$, $\sigma \in P_{\lambda+1}(X(L)_p)$ and so $U_p(\lambda, X(L)_p) = 0$. This finishes (i) when p is odd. (Admittedly, we could have used the same argument directly in the case n = 1, but it was still necessary to have the local-global criterion in Lemma 7.4 for Lemma 7.5).

We will make some remarks about the proof of (i) for p = 2 since there are some complications which have not appeared in print. We assume, as before, char $F = q \neq 2$. We must show, for $[L : K] < \infty$:

(a) If $\varepsilon(4) \in F$ and $\sigma \in P_\lambda(X(L)_2)$ with $0 < \lambda < \phi(F,2)-1$, then $\sigma \in P_{\lambda+1}(X(L)_2)$

and

(b) If $\varepsilon(4) \notin F$ and $\sigma \in P_\lambda(X(L)_2)$ for $0 < \lambda < \phi(F,2)-1$, then $\sigma \in P_{\lambda+1}(X(L)_2)$.

Part (a) is fairly easy; by hypothesis $\varepsilon(2^m) \in L$ with $m = \phi(F,2)$. If M is the associated quadratic extension for σ, then $L \subset M \subset N$ with N/L cyclic of dimension $2^{\lambda+1}$ and $\lambda+1 < m$. Thus -1 is a norm from N to L (of $\varepsilon(2^{\lambda+2})$ and Lemma 7.3 says N can be extended to $\bar{N}/L$ cyclic of dimension $2^{\lambda+2}$, so $\sigma \in P_{\lambda+1}(X(L)_2)$ as needed.

For part (b), let $L_1 = L(\varepsilon(4))$. We may assume $[L_1 : L] = 2$; otherwise we are back in case (a). The difficulty now is that cor∘res is multiplication by 2, which is no longer prime to 2 (in fact cor∘res$(\sigma) = 0$). To rectify the situation, we assume first that $\sigma \in P_\lambda(X(L)_2)$ with $\lambda \leq \phi(F,2)-3$ and $\phi(F,2) \geq 3$. We may also assume $\lambda \geq 1$, so $2\tau = \sigma$ with $\tau \in X(L)_2$. Let M be the cyclic 4 extension of L corresponding to τ, so $M_1 = M(\varepsilon(4)) = M(\sqrt{-1})$ is the cyclic 4 extension of L_1 corresponding to res τ. Since $\varepsilon(2^m)$ is in L_1, we get, again using Lemma 7.3, that M_1 can be extended to a cyclic extension N_1 of L_1 of dimension 2^m. Since $\lambda \leq m-3$, $\lambda+1 \leq m-2$ and $[N_1 : M_1] = 2^{m-2}$. Choose a character δ corresponding to N_1 so that $2^{m-2}\delta = \text{res}\,\tau$. Applying corestriction gives : $2^{m-2}\text{cor}\,\delta = \text{cor}\circ\text{res}\,\tau = 2\tau = \sigma$, and so $\sigma \in P_{m-2}(X(L)_2) \subset P_{\lambda+1}(X(L)_2)$, showing $U_2(\lambda, X(L)_2) = 0$.

There is only one case of Part (b) remaining - the one for which $1 \leq \lambda = m-2$. For this case we must fall back on a Theorem of J. Sonn, which appears in [23] :

Proposition 8.1. Suppose K is a field of characteristic not 2, $\sqrt{-1} \notin K$, and $K(\sqrt{-1}) = K(\varepsilon(2^m))$. If L is a quadratic extension of K which can be extended to a cyclic extension of degree 2^{m-1}, then L can also be extended to a cyclic extension of degree 2^m.

This finishes our proof of (i). Since (ii) does appear in the literature, we make only passing remarks about it here. In fact (ii) is equivalent to showing :

(*) If L is a field finitely generated over a global field F and $p \neq \operatorname{char} F$, then $\ell_p(X(L)) < \omega 2$.

In order to prove (*) it is enough to show :

(+) There are only finitely many elements in $X(L)_p(\omega)$ of a given order.

For, assuming (+) is known, we let Γ be the directed graph of reduced elements in $X(L)_p(\omega)$ as in the proof of Theorem 7.2. By (+) every vertex of Γ has finite order. It follows as in Theorem 7.2 that Γ is finite; otherwise we would have an infinite ascending chain in Γ corresponding to a divisible subgroup. If s is the maximum length of a chain in Γ, then $\ell_p(X(L)) = \omega + s < \omega 2$.

Let $r = \text{t.d.}(L/F)$. If $r = 0$ we are finished by Theorem 7.2, so we may assume $[L : K(t)] < \infty$ where t is transcendental over K, K is algebraically closed in L, and $\text{t.d.}(L/F) = r - 1$. Suppose $\sigma \in X(L)_p(\omega)$ of order p and M is the corresponding cyclic extension. There are two relevant cases.

Case 1 : M is a constant field extension, i.e. $M = LK_1$ where K_1/K is cyclic of dimension p.

In this circumstance we are able to prove in [10] that there are only finitely many choices for K_1. We are not able to prove that K_1 is of infinite height over K, nor can we resolve this point even when F is local, and this is one of the major obstructions in determining Brauer groups of function fields over local fields.

Case 2 : M is not a constant field extension.

In this case M is "geometric", i.e. if $\tilde{K}$ is the algebraic closure of K, $\tilde{L} = L\tilde{K}$ and $\tilde{M} = M\tilde{K}$ then $[\tilde{M} : \tilde{L}] = p$ and is cyclic. Furthermore, by Lemma 6.1 $\tilde{M}/\tilde{L}$ is unramified at all points of the natural theory of divisors for $\tilde{L}$ coming from linear polynomials in $\tilde{K}[t]$. One then knows from algebraic geometry that there are only finitely many choices for $\tilde{M}$. An intricate descent argument in [10] then shows that only finitely many M over L could satisfy $M\tilde{K} = \tilde{M}$, so there were only finitely many choices for M.

Cases 1 and 2 together show there were finitely many possibilities for M, so (+) follows. Thus (*) holds, and the inductive step in Theorem B is complete.

We end with a few remarks about Brauer groups of function fields over local fields. A local field for us is the completion of a global field at a finite prime. Easy arguments in local number theory give :

<u>Theorem 8.2</u>. If F is a local field, t and indeterminate and $p \neq \mathrm{char}(\bar{F})$, $\bar{F}$ = the residue class field, then :

> $B(F(t))_p \cong \oplus_{\omega} Z(p^{\infty}) \oplus A$, where A is a direct sum of cyclic p-groups.

A communication to us from J.-P. Serre indicates that the situation is quite different for $B(F(t_1,t_2))_p$. He constructs a local field F of equicharacteristic (i.e. char F = char $\bar{F} = q > 0$) and a function field L in one variable over F (in fact the function field of an elliptic curve) so that : for some $p \neq q$, $X(L)_p$ contains a reduced element of infinite height. This, together with (2.1) says that $B(F(t_1,t_2))_p$ has a reduced element of infinite height, and no such element occurs in Theorem 8.2. We conclude for this field F :

Theorem 8.3. $B(F(t_1)) \not\cong B(F(t_1,t_2))$.

We are not aware of whether Brauer groups of rational function fields over local fields of unequal characteristic will also depend on the number of variables.

REFERENCES

1. A.A. Albert, Structure of Algebras, Amer. Math. Soc. Coll. Publ. Vol. 24, Providence, R. I., 1961.

2. S.A. Amitsur, On central division algebras, Israel J. Math. 12 (1972), 408-420.

3. E. Artin and J. Tate, Class field theory, W.A. Benjamin, 1967.

4. M. Auslander and A. Brumer, Brauer groups of discrete valuation rings, Nederl. Akad. Wetensch. Proc. Ser A, 71 (1968), 286-296.

5. S. Bloch, Torsion algebraic cycles, K_2, and the Brauer group of function fields, Bull. A.M.S. 80 (1974), 941-945.

6. J.W.S. Cassels and A. Fröhlich, Algebraic Number Theory, Thompson, Washington, 1967.

7. M. Deuring, Algebren, Springer-Verlag, New York, 1966.

8. D.K. Faddeev, Simple algebras over a field of algebraic functions of one variable, Trudy Mat. Inst. Steklov 38 (1951), 321-344; Amer. Math. Soc. Transl. Ser. II 3 (1956), 15-38.

9. B. Fein and M. Schacher, Ulm invariants of the Brauer group of a field, Math. Z. 154 (1977), 41-50.

10. B. Fein and M. Schacher, Brauer groups and character groups of function fields, J. Algebra Vol. 61, No. 1, Nov. 1979, 249-255.

11. B. Fein, M. Schacher and J. Sonn, Brauer groups of rational function fields, Bull. A.M.S. (New Series) 1 (1979), 766-768.

12. E. Formanek, The center of the ring of 3×3 generic matrices, Lin. Mult. Alg. 7 (1979), 203-212.

13. E. Formanek, The center of the ring of 4×4 generic matrices, J. Alg. Vol. 62, No. 2 (1980), 304-320.

14. M. Greenberg, Lectures on Forms in Many Variables, Benjamin, New York, 1969.

15. K. Iwasawa, On Z_ℓ-extensions of algebraic number fields, Ann. of Math. 98 (1973) 187-326.

16. N. Jacobson, Lectures in Abstract Algebra, Vol. III, Van Nostrand, 1964.

17. J. Neukirch, Klassenkörpertheorie, B. I. Hochschulskripten, Mannheim, 1969.

18. C. Procesi, Rings with Polynomial Identities, Marcel Dekker, New York, 1973.

19. I. Reiner, Maximal Orders, Academic Press, New York, 1975.

20. J.-P. Serre, Cohomologie Galoisienne, Lecture Notes in Math. 9, Springer Verlag, Berlin, 1964.

21. J.-P. Serre, Corps Locaux, Hermann, Paris, 1962.

22. R. Snider, Is the Brauer group generated by cyclics ? Lect. Notes in Math. 734 (Ring Theory, Waterloo 1978), Springer Verlag, Berlin, 1979.

23. J. Sonn, Class groups and Brauer groups, Israel J. Math., Vol. 34 Nos 1-2 (1979), 97-106.

24. J.P. Tignol, Sur les classes de similitude de corps à involution de degré 8, C.R. Acad. Sc. Paris 286 (1978), 875-876.

25. E. Weiss, Algebraic Number Theory, New York, McGraw-Hill, 1963.

TORSION ALGEBRAIC CYCLES, K_2, AND BRAUER GROUPS OF FUNCTION FIELDS[1]

S. BLOCH

Introduction

Let F be a field and let $H^*(F,\mu_n)$ denote the Galois cohomology of $Gal(F_s/F)$ with coefficients in the group μ_n of n-th roots of 1, for some fixed n prime to char F. Bass and Tate have shown that the natural pairing

$$F^*/F^{*n} \times F^*/F^{*n} = H^1(F,\mu_n) \times H^1(F,\mu_n) \xrightarrow{\text{cup product}} H^2(F,\mu_n^{\otimes 2})$$

is a symbol on F. In other words there is an induced homomorphism (n-th power norm residue symbol) of the Milnor K_2 group [7]

$$R_{n,F}: K_2(F)/nK_2(F) \longrightarrow H^2(F,\mu_n^{\otimes 2}) .$$

Tate showed that $R_{n,F}$ is an isomorphism where F is a global field, and asked whether an analogous result held for arbitrary fields. The situation is particularly interesting when $\mu_n \subset F$, because in this case

$$H^2(F,\mu_n^{\otimes 2}) \cong H^2(F,\mu_n) \otimes \mu_n \cong {}_nBr(F) \otimes \mu_n \qquad (Br(F) = \text{Brauer group of } F).$$

(1) This paper was written in 1973 but never published. I am endebted to M. Kervaire for his efforts and interest in arranging publication at this late date, as well as for inviting me to speak on these matters at les Plans sur Bex in March, 1980. Travel support was provided by Troisième Cycle Romand de Mathématiques.

I have added footnotes at various points to try to bring the reader up to date on developments since 1973.

Surjectivity of R_n implies, for example, that every division algebra with exponent n and center F is split by an abelian extension field of F.

Theorem (0.1). Let F be a rational function field in r variables over a field k, and let n be an integer prime to char k. Then the maps

$$R_{n,k} : K_2(k)/nK_2(k) \longrightarrow H^2(k,\mu_n^{\otimes 2})$$

$$R_{n,F} : K_2(F)/nK_2(k) \longrightarrow H^2(F,\mu_n^{\otimes 2})$$

have isomorphic kernels and isomorphic cokernels.

Tate's question is related to an interesting problem in algebraic geometry. Let X be a smooth variety of finite type over k, $F = k(X)$. Let $\theta_n(F) = \mathrm{Coker}(R_{n,F})$ and let $CH^2(X)$ denote the Chow group of codimension two algebraic cycles modulo rational equivalence on X.

Theorem (0.2). There is an exact sequence

$$\theta_n(F) \longrightarrow {}_nCH^2(X) \longrightarrow G_n(X) \longrightarrow 0$$

where $G_n(X)$ is a finite group closely related to the group of n-torsion points on a hypothetical intermediate jacobian for codimension two algebraic cycles. Hence $\theta_n(F)$ finite implies ${}_nCH^2(X)$ finite.[2]

(2) It is now known that $\theta_n(F) = 0$ for F of transcendance degree 1 over a C_1-field, e.g. for F the function field of a surface over an algebraically closed field. This implies by a geometric argument, that ${}_nCH^d(X)$ is finite for $\dim X = d$, a theorem first proved by A. Roitman. For details, see S. Bloch, Lectures on Algebraic Cycles, available from Duke University department of mathematics, Durham, North Carolina 27706. Another proof of the vanishing of $\theta_n(F)$ for such F has been given by K. Kato.

When n is a power of 2, one can use the theory of quadratic forms to prove :

Theorem (0.3). Let F be a C_2-field (e.g. $F = k(X)$ where k is algebraically closed and $\dim X \leq 2$) of characteristic $\neq 2$. Then $R_{2^r,F}$ is injective for any r.

Fix a prime $\ell \neq \mathrm{char}(F)$, assume F contains all ℓ^{th}-power roots of 1, and let

$$R_{\infty,F} = \varinjlim_{\nu} R_{\ell^\nu,F} : K_2(F) \otimes \mathbb{Q}_\ell/\mathbb{Z}_\ell \longrightarrow Br(F) \otimes \mathbb{Z}_\ell(1) ,$$

where $\mathbb{Z}_\ell(1) = \varprojlim_{\nu} \mu_{\ell^\nu}$. In § 5 we assume $R_{\infty,F}$ is injective (e.g. $\ell = 2$, and F is C_2), and show in this case that surjectivity is equivalent to the vanishing of certain Galois cohomology groups

$$H^1(G, K_2'(F'))$$

where $G = Gal(F'/F)$ and $K_2' = K_2/\text{torsion}$. For a precise statement, see (5.1).

Note : At the time this was written I was unaware of the work of Lam and Elman [14]. Their results are sharper and predate mine.

1. Preliminaries on K-theory

Let K_n : (rings) $\longrightarrow$ (abelian groups) $n \geq 0$ be the algebraic K-functors of Grothendieck, Bass, Milnor, and Quillen. If X is a scheme, we define a sheaf $\underline{\underline{K}}_n$ on X for the Zariski topology by "sheafifying" the presheaf

$$U \longmapsto K_n(\Gamma(U, \mathcal{O}_X)) \ , \ U \xrightarrow[\text{open}]{\subset} X.$$

Some notation :

$$X^r = \{x \in X \mid \operatorname{codim}\overline{\{x\}} = r\} \quad ;$$

given $x \in X$, A an abelian group, $i_x A$ denotes the constant sheaf A on $\overline{\{x\}}$ extended by zero to X.

The following result was conjectured by Gersten [6], and recently proven by Quillen [9].

<u>Theorem</u> (1.1). Assume X is regular and of finite type over a field k. Then there exists a resolution of $\underline{\underline{K}}_n$ as follows :

$$(1.2)\quad 0 \longrightarrow \underline{\underline{K}}_n \longrightarrow \coprod_{x\in X^0} i_x K_n(k(x)) \longrightarrow \coprod_{x\in X^1} i_x K_{n-1}(k(x)) \longrightarrow \cdots$$

$$\longrightarrow \coprod_{x\in X^n} i_x K_0(k(x)) \longrightarrow 0 \ .$$

For example when $n = 2$ and X is irreducible with function field F, this resolution looks like

$$(1.3)\quad 0 \longrightarrow \underline{\underline{K}}_2 \longrightarrow K_2(F)_X \xrightarrow{T} \coprod_{x\in X^1} i_x(k(x)^*) \longrightarrow \coprod_{x\in X^2} i_x \mathbb{Z} \longrightarrow 0 \ .$$

The group $K_2(F)$ is known to be generated by "symbols" $\{f,g\}$, $f,g \in F^*$, and the <u>tame symbol</u> T is given by

$$T\{f,g\} = \coprod_{x\in X^1} (-1)^{\operatorname{ord}_x(f)\cdot\operatorname{ord}_x(g)} f^{\operatorname{ord}_x(g)} g^{-\operatorname{ord}_x(f)} \ .$$

We will need two variants of these ideas.

<u>Proposition</u> (1.4). Let $F = k(t)$ for t transcendental, $X = \operatorname{Spec} k[t]$. Then there are exact sequences,

(1.5) $$0 \longrightarrow K_n(k) \longrightarrow K_n(F) \longrightarrow \coprod_{x\varepsilon X^1} K_{n-1}(k(x)) \longrightarrow 0 .$$

Proof. The Brown-Gersten spectral sequence [5]

(1.6) $$H^p(X,\underline{\underline{K}}_q) \Longrightarrow K_{q-p}(X)$$ (Cohomology taken with respect to the Zariski topology),

together with Quillen's theorem $K_*(k) \cong K_*(X)$ [9], gives exact sequences

(1.7) $$0 \longrightarrow H^1(X,\underline{\underline{K}}_{q+1}) \longrightarrow K_q(k) \xrightarrow{\alpha} \Gamma(X,\underline{\underline{K}}_q) \longrightarrow 0 .$$

The map α is easily seen to be injective as well (define an inverse by choosing a k-rational point on X). Now (1.2) is a resolution of $\underline{\underline{K}}_n$ by acyclic sheaves, so there is an exact sequence

(1.8) $$0 \longrightarrow \Gamma(X,\underline{\underline{K}}_n) \longrightarrow K_n(F) \longrightarrow \coprod_{x\in X^1} K_{n-1}(k(x)) \longrightarrow H^1(X,\underline{\underline{K}}_n) \longrightarrow 0.$$

The sequence (1.5) follows from (1.7) and (1.8). Q.E.D.

Proposition (1.9). Let X be a regular algebraic k-scheme, and let n be an integer. Then the sequence

$$0 \longrightarrow \underline{\underline{K}}_2/n\underline{\underline{K}}_2 \longrightarrow (K_2/nK_2(F))_X \xrightarrow{T} \coprod i_x(k(x)^*/k(x)^{*n}) \longrightarrow \coprod i_x\mathbb{Z}/n\mathbb{Z} \rightarrow 0$$

obtained from (1.3) by reducing mod n, is exact.

Proof. Let $n = m\cdot p^e$ where $p = \operatorname{char} k$ and $(m,p) = 1$. Since for any abelian group A, one has $A/mp^e\cdot A \cong A/mA \times A/p^e\cdot A$ it suffices to consider the cases $n = p^e$ and $(n,p) = 1$. Breaking (1.3) into short exact sequences yields

$$0 \longrightarrow \underline{\underline{K}}_2 \longrightarrow K_2(F)_X \longrightarrow \mathrm{Im}(T) \longrightarrow 0$$

$$\circlearrowleft \quad \circlearrowleft \quad \circlearrowleft$$

$$\times n \quad \times n \quad \times n$$

(1.10)

$$0 \longrightarrow \mathrm{Im}(T) \longrightarrow \coprod i_x k(x)^* \longrightarrow \coprod i_x \mathbb{Z} \longrightarrow 0 .$$

$$\circlearrowleft \quad \circlearrowleft \quad \circlearrowleft$$

$$\times n \quad \times n \quad \times n$$

For an abelian sheaf or group A, ${}_nA$ will denote the kernel of multiplication by n. Using the serpent lemma on the diagrams (1.10), one sees that the proposition holds if and only if the map of sheaves

$${}_nK^2(F)_X \longrightarrow {}_n\mathrm{Im}(T)$$

is surjective.

Since ${}_n\mathrm{Im}(T) \subset \coprod i_x k(x)^*$ has no p-torsion, it suffices to consider the case $(n,p) = 1$. The question is local, so we may suppose given a point $\eta \in X$, an $x \in X^1$ such that $\eta \in \overline{\{x\}}$, and a $\zeta^n = 1$. We may choose a lifting $\zeta' \in \mathcal{O}^*_{X,x}$ such that $\zeta'^n = 1-\pi$ where $\mathrm{ord}_x(\pi) = 1$ and π has no other zeros or poles through η. The symbol $\{\pi,\zeta'\} \in K_2(F)$ satisfies

$$\{\pi,\zeta'\}^n = \{\pi,1-\pi\} = 1$$

$$(T\{\pi,\zeta'\})_{\text{stalk at }\eta} = \zeta \in i_x k(x)^* . \qquad \text{Q.E.D.}$$

<u>Remark</u> (1.10). The sequence (1.5) can be shown to be split, at least if $n \leq 3$ or $k = \bar{k}$ (cf. [7], thm. 2.3). Thus under those hypotheses,

the sequences

$$0 \longrightarrow K_n/rK_n(k) \longrightarrow K_n/rK_n(k(t)) \longrightarrow \coprod K_{n-1}/rK_{n-1}(k(x)) \longrightarrow 0$$

are exact for any integer r.(3)

<u>Corollary</u> (1.11). Let ${}_nCH^2(X)$ denote the kernel of multiplication by n on the Chow group $CH^2(X)$ of codimension two algebraic cycles modulo rational equivalence [10]. Then there is an exact sequence

$$(1.12)\quad H^1(X,\underline{\underline{K}}_2) \xrightarrow{xn} H^1(X,\underline{\underline{K}}_2) \longrightarrow H^1(X,\underline{\underline{K}}_2/n\underline{\underline{K}}_2) \longrightarrow {}_nCH^2(X) \longrightarrow 0 .$$

<u>Proof</u>. Using (1.9), one gets an exact sequence of sheaves

$$0 \longrightarrow {}_n\underline{\underline{K}}_2 \longrightarrow ({}_nK_2(F))_X \longrightarrow \coprod_{x\in X^1} i_x(\mu_n \cap k(x)^*) \longrightarrow 0 .$$

In particular, $H^r(X,{}_n\underline{\underline{K}}_2) = (0)$ for $r \geq 2$. Now take cohomology of the sequences

$$0 \longrightarrow {}_n\underline{\underline{K}}_2 \longrightarrow \underline{\underline{K}}_2 \longrightarrow n\underline{\underline{K}}_2 \longrightarrow 0$$

$$0 \longrightarrow n\underline{\underline{K}}_2 \longrightarrow \underline{\underline{K}}_2 \longrightarrow \underline{\underline{K}}_2/n\underline{\underline{K}}_2 \longrightarrow 0 ,$$

getting

$$(1.13)\quad H^1(\underline{\underline{K}}_2) \xrightarrow{xn} H^1(\underline{\underline{K}}_2) \longrightarrow H^1(\underline{\underline{K}}_2/n\underline{\underline{K}}_2) \longrightarrow {}_nH^2(\underline{\underline{K}}_2) \longrightarrow 0 .$$

An important consequence of (1.1) is the isomorphism [9]

$$(1.14)\qquad CH^r(X) \cong H^r(X,\underline{\underline{K}}_r)$$

for any r. Now (1.12) follows from (1.13) and (1.14). Q.E.D.

---

(3) Grayson has proven quite generally that the Gersten resolution (1.2) remains exact when reduced modulo r for any r.

2. Preliminaries from Etale Cohomology

Recently, A. Ogus and I proved an analogue of (1.1) for étale cohomology. We continue to assume X a regular algebraic k-scheme, and fix an integer n prime to char.(k). Let μ denote the étale sheaf of n-th roots of unity, and write $\mathcal{H}^*_X(r)$ for the Zariski sheaf associated to the presheaf

$$U \longmapsto H^*_{\acute{e}t}(U, \mu_U^{\otimes r}), \qquad U \xrightarrow[\text{open}]{\subset} X \ .$$

For a field L, let $H^*(L,r) = H^*_{\text{Galois}}(\text{Gal}(\bar{L}/L), \mu^{\otimes r})$

Theorem (2.1). With notation as above, there is a resolution of Zariski sheaves

$$0 \longrightarrow \mathcal{H}^q_X(r) \longrightarrow \coprod_{x\in X^0} i_x H^q(k(x),r) \longrightarrow \coprod_{x\in X^1} i_x H^{q-1}(k(x),r-1) \longrightarrow \cdots$$

$$\longrightarrow \coprod_{x\in X^q} i_x H^0(k(x),r-q) \longrightarrow 0 \ .$$

Notice that $H^0(k(x),0) = \mathbb{Z}/n\mathbb{Z}$, and $H^1(k(x),1) \cong k(x)^*/k(x)^{*n}$ (Kummer theory). When q = r = 2, and X is irreducible with quotient field F (2.1) yields

$$(2.2) \quad 0 \longrightarrow \mathcal{H}^2_X(2) \longrightarrow H^2(F,2)_X \longrightarrow \coprod_{x\in X^1} i_x(k(x)^*/k(x)^{*n})$$

$$\longrightarrow \coprod_{x\in X^2} i_x(\mathbb{Z}/n\mathbb{Z}) \longrightarrow 0 \ .$$

Theorem (2.3). The n-th power norm residue map R_n gives rise to a commutative diagram of exact sequences

$$\begin{array}{ccccccccc}
0 \longrightarrow & \underline{\underline{K}}_2/n\underline{\underline{K}}_2 & \longrightarrow & K_2/nK_2(F)_X & \xrightarrow{\;T\;} & \coprod_{X^1} i_x(k(x)^*/k(x)^{*n}) & \longrightarrow & \coprod_{X^2} i_x(\mathbb{Z}/n\mathbb{Z}) & \longrightarrow 0 \\
& \Big\downarrow{\scriptstyle R_{n,X}} & & \Big\downarrow{\scriptstyle R_{n,F}} & & \Big\| & & \Big\| & \\
0 \longrightarrow & \mathcal{H}^2_X(2) & \longrightarrow & H^2(F,2)_X & \xrightarrow{\;U\;} & \coprod_{X^1} i_x(k(x)^*/k(x)^{*n}) & \longrightarrow & \coprod_{X^2} i_x(\mathbb{Z}/n\mathbb{Z}) & \longrightarrow 0.
\end{array}$$

<u>Proof</u>. Notice that $R_{n,X}$ is defined once the rest of the diagram is checked to be commutative. The only problem in this regard is the center square. Given $x \in X^1$, let R be the corresponding discrete valuation ring. Let $\hat{R}$ be the completion of R, and let $\hat{F}$ denote the function field of $\hat{R}$. The left hand square in the diagram below obviously commutes,

$$\begin{array}{ccccc}
K_2/nK_2(F) & \longrightarrow & K_2/nK_2(\hat{F}) & \xrightarrow{\;T_x\;} & k(x)^*/k(x)^{*n} \\
\Big\downarrow{\scriptstyle R_{n,F}} & & \Big\downarrow{\scriptstyle R_{n,\hat{F}}} & (*) & \Big\downarrow\Big\| \\
H^2(F,2) & \longrightarrow & H^2(\hat{F},2) & \xrightarrow{\;U_x\;} & k(x)^*/k(x)^{*n}
\end{array} \quad ,$$

so we need only consider (*).

Since $\hat{R}$ is complete and equicharacteristic, we have $\hat{R} \cong k(x)[[t]]$ and there is a diagram

$$\begin{array}{ccccccc}
0 \longrightarrow & K_2/nK_2(\hat{R}) & \longrightarrow & K_2/nK_2(\hat{F}) & \underset{\sigma}{\overset{T_x}{\rightleftarrows}} & k(x)^*/k(x)^{*n} & \longrightarrow 0 \\
& \Big\downarrow & (**) & \Big\downarrow{\scriptstyle R_{n,\hat{F}}} & (*) & \Big\| & \\
0 \longrightarrow & H^2_{\text{ét}}(\mathrm{Spec}(\hat{R}),\mu_n^{\otimes 2}) & \longrightarrow & H^2(\hat{F},2) & \underset{\psi}{\overset{U_x}{\rightleftarrows}} & k(x)^*/k(x)^{*n} & \longrightarrow 0
\end{array}$$

The splittings σ and ψ are given by

$$\sigma(f) = \{f,t\}$$

$$\psi(f) = f \cup \bar{t} \qquad\qquad \bar{t} \in H^1(\hat{F},1) = \hat{F}^*/\hat{F}^{*n} \quad ,$$

where { } denotes the symbol ([8]). For the verification that $U_x \circ \psi = \text{Id}$, see [12]. Commutativity of (**) is clear, and that of (*) follows from $R_{n,\hat{F}}\{f,t\} = f \cup \bar{t}$. Q.E.D.

Corollary (2.4). $\text{Ker } R_{n,X} \cong \text{Ker } R_{n,F}$; and $\text{Coker } R_{n,X} \cong \text{Coker } R_{n,F}$. In particular, both $\text{Ker } R_{n,X}$ and $\text{Coker } R_{n,X}$ are constant sheaves.

Proposition (2.5). Let $k(t)$ be the function field of the affine line $A = \mathbb{A}^1_k$. Then for r,q there is an exact sequence

$$0 \longrightarrow H^q(k,r) \longrightarrow H^q(k(t),r) \longrightarrow \coprod_{x\varepsilon A^1} H^{q-1}(k(x),r-1) \longrightarrow 0 .$$

Proof. Let $\bar{k}$ be the algebraic closure of $k, \bar{A} = A \otimes_k \bar{k}$, $G = \text{Gal}(\bar{k}/k)$. One knows the étale cohomology of $\bar{A}$ (cf. [1])

$$H^q_{\text{ét}}(\bar{A},r) = \begin{cases} \mu_n^{\otimes r} & q = 0 \\ 0 & q > 0 . \end{cases}$$

The spectral sequence

$$E_2^{p,q} = H^p(G, H^q_{\text{ét}}(\bar{A},r)) \Longrightarrow H^{p+q}_{\text{ét}}(A,r)$$

yields

(2.6) $$H^q(A,r) \cong H^q(k,r) \text{ , all } q,r .$$

There is a spectral sequence (for any X)

$$E_2^{p,q} = H^p(X, \mathcal{H}^q_X(r)) \Longrightarrow H^{p+q}_{\text{ét}}(X,r) ,$$

which in the case X = A = curve gives exact sequences

$$0 \longrightarrow H^1(A, \mathcal{H}_A^{q-1}(r)) \longrightarrow H^q_{\text{ét}}(A,r) \longrightarrow \Gamma(A, \mathcal{H}_A^q(r)) \longrightarrow 0$$

$$\begin{array}{ccc} & & \Gamma(A,\mathcal{H}_A^q(r)) \\ \int\,\| & & \big\downarrow \text{k-rational point} \\ H^q(k,r) & = & H^q(k,r) \end{array} .$$

It follows that

(2.7) $\quad H^1(A, \mathcal{H}_A^{q-1}(r)) = (0);\ \Gamma(A, \mathcal{H}_A^q(r)) \cong H^q(k,r)$.

Finally, computing cohomology via the acyclic resolution (2.1) gives

(2.8) $\quad 0 \longrightarrow \Gamma(A, \mathcal{H}_A^q(r)) \longrightarrow H^q(k(t),r) \longrightarrow \coprod_{x \in A^1} H^{q-1}(k(x), r-1) \longrightarrow$

$$\longrightarrow H^1(A, \mathcal{H}_A^q(r)) \longrightarrow 0 .$$

The desired sequence follows from (2.7), (2.8). Q.E.D.

3. The main theorems

Recall that for F a field and n an integer prime to char.F, we have defined

$$R_{n,F} : K_2/nK_2(F) \longrightarrow H^2(F,2) .$$

Theorem (3.1). Let $F = k(t_1,\ldots,t_m)$ be the rational function field in m variables over a ground field k . Then

$$\text{Ker}(R_{n,k}) \cong \text{Ker}(R_{n,F})$$

$$\text{Coker}(R_{n,k}) \cong \text{Coker}(R_{n,F}) .$$

Proof. By induction, it suffices to consider the case $F = k(t)$. Using (1.10), (2.3), and (2.5), there is a commutative diagram

$$\begin{array}{ccccccccc} 0 & \longrightarrow & K_2/nK_2(k) & \longrightarrow & K_2/nK_2(F) & \longrightarrow & \coprod k(x)^*/k(x)^{*n} & \longrightarrow & 0 \\ & & \downarrow R_{n,k} & & \downarrow R_{n,F} & & \| & & \\ 0 & \longrightarrow & H^2(k,2) & \longrightarrow & H^2(F,2) & \longrightarrow & \coprod k(x)^*/k(x)^{*n} & \longrightarrow & 0 \end{array}$$

The theorem follows by the snake lemma. Q.E.D.

Corollary (3.2). Assume k is algebraically closed. Then $R_{n,F}$ is an isomorphism.

Proof. The Galois cohomology of k is zero, and $K_2(k)$ is divisible.

Let X be a smooth algebraic k-scheme with function field F. From (2.4) one gets an exact sequence

(3.3) $$0 \longrightarrow \Lambda_n(F)_X \longrightarrow \underline{\underline{K}}_2/n\underline{\underline{K}}_2 \longrightarrow \mathcal{H}^2_X(2) \longrightarrow \theta_n(F)_X \longrightarrow 0$$

where $\Lambda_n(F) = \mathrm{Ker}(R_{n,F})$, $\theta_n(F) = \mathrm{Coker}(R_{n,F})$. Using that the higher Zariski cohomology of a constant sheaf vanishes one gets from (3.3) an exact sequence

(3.4) $$0 \longrightarrow \Lambda_n(F) \longrightarrow \Gamma(\underline{\underline{K}}_2/n\underline{\underline{K}}_2) \longrightarrow \Gamma(\mathcal{H}^2_X(2)) \longrightarrow \theta_n(F) \longrightarrow$$
$$\longrightarrow H^1(\underline{\underline{K}}_2/n\underline{\underline{K}}_2) \longrightarrow H^1(\mathcal{H}^2_X(2)) \longrightarrow 0 .$$

Proposition (3.5). The spectral sequence

$$E_2^{p,q} = H^p(X, \mathcal{H}^q_X(2)) \Longrightarrow H^{p+q}_{\text{ét}}(X,2)$$

gives rise to a surjection

$$H^2_{\text{ét}}(X,2) \longrightarrow\!\!\!\!\rightarrow \Gamma(X, \mathcal{H}^2_X(2))$$

and an injection

$$H^1(X,\mathcal{H}^2_X(2)) \hookrightarrow H^3_{ét}(X,2) \ .$$

Proof. As a consequence of (2.1), one has

$$H^p(X,\mathcal{H}^q_X(r)) = (0) \qquad p > q \ ,$$

so the above spectral sequence looks like

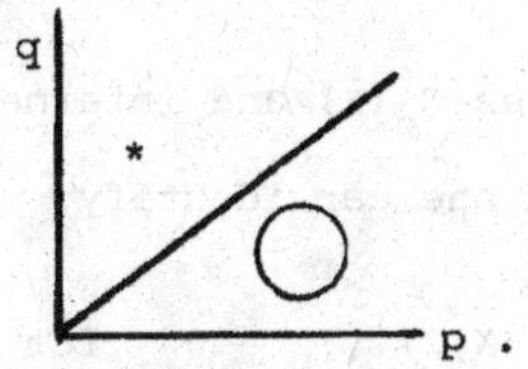

It follows easily that

$$E_\infty^{0,2} \cong \Gamma(X,\mathcal{H}^2_X(2))$$

$$E_\infty^{3,0} = E_\infty^{2,1} = (0) \ ; \ E_\infty^{1,2} = H^1(X,\mathcal{H}^2_X(2)). \qquad \text{Q.E.D.}$$

Corollary (3.6). If the étale cohomology groups $H^2_{ét}(X,2)$ and $H^3_{ét}(X,2)$ are finite (e.g. $k=\bar{k}$ or k finite), then so are the groups $\Gamma(\mathcal{H}^2_X(2))$ and $H^1(\mathcal{H}^2_X(2))$. In this case, one has implications

$$\Lambda_n(F) \text{ finite} \iff \Gamma(X,\underline{\underline{K}}_2/n\underline{\underline{K}}_2) \text{ finite}$$

$$\theta_n(F) \text{ finite} \iff H^1(X,\underline{\underline{K}}_2/n\underline{\underline{K}}_2) \text{ finite} \implies {}_nCH^2(X) \text{ finite.}$$

Proof. Combine (3.4), (3.5), and (1.11).

Let Δ_n denote the composition

$$H^1(X,\underline{\underline{K}}_2) \longrightarrow H^1(X,\underline{\underline{K}}_2/n\underline{\underline{K}}_2) \longrightarrow H^1(X,\mathcal{H}^2_X(2))$$

and write

$$G_n(X) = H^1(X,\mathcal{H}_X^2(2))/\mathrm{Im}\Delta_n .$$

<u>Corolarry</u> (3.7). There is an exact sequence

$$\theta_n(F) \longrightarrow {}_nCH^2(X) \longrightarrow G_n(X) \longrightarrow 0 .$$

$G_n(X)$ is finite if $H^3_{\acute{e}t}(X,2)$ is finite.

The relation between $G_n(X)$ and intermediate jacobians is roughly as follows : using (3.5) one can identify

$$H^1(X,\mathcal{H}_X^2)) = \{\gamma \in H^3_{\acute{e}t}(X,2) \,|\, \gamma|_U = 0 \quad \text{for some open } U \subset X\} =$$

$$= \{\gamma|\gamma \text{ supported on some divisor in } X\} .$$

Given a smooth, proper k-scheme T and a cycle Ψ on $T \times_k X$ of codimension = codimension T + 1, one gets a correspondence

$$H^1_{\acute{e}t}(T,1) \xrightarrow{\Psi_*} H^1(X,\mathcal{H}_X^2(2)) \subset H^3_{\acute{e}t}(X,2) .$$

Assume now $k = \bar{k}$ and that we have resolution of singularities for divisors on X. In this case for suitable (T,Ψ), Ψ_* will be surjective (at least if n is prime to a finite number of bad primes).

$$\Psi_* : H^1_{\acute{e}t}(T,1) = {}_n\mathrm{Pic}(T) \twoheadrightarrow H^1(X,\mathcal{H}_X^2(2)) .$$

Suppose, moreover, X is projective and $k = \mathbb{C}$, so the Griffith's intermediate jacobian $J^2(X)$ is defined (or else assume X projective of dimension 2, so $J^2(X) = \mathrm{A\ell b}(X)$ is always defined). Then $\Psi_*({}_n\mathrm{Pic}^0(T)) \subset H^1(X,\mathcal{H}_X^2))$ is isomorphic to ${}_nJ^2$.

I have no interpretation for the map

$$\Delta_n : H^1(X,\underline{\underline{K}}_2) \longrightarrow H^1(X,\mathcal{H}_X^2(2)) .$$

When $k = \bar{k}$, one can hope that $H^1(X,\underline{\underline{K}}_2)$ is divisible, in which case $\Delta_n = 0$. For example, when is the map

$$H^1(X,\underline{\underline{K}}_1) \otimes k^* \longrightarrow H^1(X,\underline{\underline{K}}_2)_X$$

surjective ? More generally, when is the "universal determinant" map

$$K_0(X) \otimes K_*(k) \longrightarrow K_*(X)$$

surjective ?

4. <u>The case $n = 2^r$</u>

Recall a field F is a C_n-field if a homogeneous polynomial of degree q in $> q^n$ variables with coefficients in F always has a non-trivial zero. For example, if F has transcendence degree n over an algebraically closed field k, then F is C_n.

<u>Theorem</u> (4.1). Let F be a C_2-field of characteristic $\neq 2$. Then $\Lambda_2(F) = (0)$. If, in addition, F contains the 2^r-th roots of 1 for some r, then $\Lambda_{2^r}(F) = (0)$.

<u>Corollary</u> (4.2). Let X be a smooth algebraic surface over an algebraically closed field k. Then the group

$$\Gamma(X,\underline{\underline{K}}_2/2^r\underline{\underline{K}}_2)$$

is finite for any r.

<u>Proof of</u> (4.2). Our assumptions imply the étale cohomology of X is finite, so (4.2) follows from (3.6) and (4.1).

<u>Proof of</u> (4.1). Suppose $\Lambda_n(F) = (0)$ for some fixed n, and that F contains a primitive n^r-th root of 1. Then I claim $\Lambda_{n^r}(F) = (0)$. Indeed, we can identify

$$H^1(F, \mu_{n^r}^{\otimes 2}) \cong F^*/F^{*n^r} .$$

Now consider the diagram with exact rows

$$\begin{array}{ccccccccc}
 & & K_2/n^{r-1}K_2(F) & \longrightarrow & K_2/n^rK_2(F) & \longrightarrow & K_2/nK_2(F) & \longrightarrow & 0 \\
 & & \cap\big\downarrow R_{n^{r-1}} & & \big\downarrow R_{n^r} & & \cap\big\downarrow R_n & & \\
F^*/F^{*n^r} \longrightarrow F^*/F^{*n} & \xrightarrow{\ 0\ } & H^2(F,\mu_{n^{r-1}}^{\otimes 2}) & \hookrightarrow & H^2(F,\mu_{n^r}^{\otimes 2}) & \longrightarrow & H^2(F,\mu_n^{\otimes 2}) & &
\end{array}$$

(exercise : check commutivity.) Assuming $R_{n^{r-1}}$ injective, a diagram chase shows R_{n^r} injective as well.

It remains now to show $\Lambda_2(F) = (0)$.

The proof uses the theory of quadratic forms [7], [11]. A quadratic module (V,q) is a finite dimensional vector space V together with a quadratic function $q : V \longrightarrow F$. The set of isomorphism classes of such objects is a semigroup under $\oplus$, and we write $R(F)$ for the associated Grothendieck group. Tensor product gives $R(F)$ a ring structure, and the map $R(F) \xrightarrow{\dim} \mathbb{Z}$, $(V,q) \longmapsto \dim V$, is a ring homomorphism. Define the ideal $J = \mathrm{Ker}(\dim)$.

Let $\nu_1,\ldots,\nu_n$ be an orthogonal basis for V with respect to q. We write $(V,q) = \langle a_1,\ldots,a_n\rangle$ if $q(\nu_i) = a_i$. With this notation, the ring operations in R(F) are given by

$$\langle a_1,\ldots,a_n\rangle + \langle b_1,\ldots,b_m\rangle = \langle a_1,\ldots,a_n,b_1,\ldots,b_m\rangle$$

$$\langle a_1,\ldots,a_n\rangle \cdot \langle b_1,\ldots,b_m\rangle = \langle a_1b_1,a_1b_2,\ldots,a_nb_m\rangle .$$

Let $H \subset R(F)$ be the cyclic subgroup generated by the <u>hyperbolic plane</u> $\langle 1,-1\rangle$. One can show that a quadratic module $\langle a_1,\ldots,a_n\rangle$ "represents 0" (i.e. there exist $x_1,\ldots,x_n \in F$ not all = 0 such that $a_1x_1^2 + \ldots + a_nx_n^2 = 0$) if and only if

$$\langle a_1,\ldots,a_n\rangle = \langle b_1,\ldots,b_{n-2}\rangle + \langle 1,-1\rangle$$

for suitable $b_1,\ldots,b_{n-2}$. It follows that $H \subset R$ is an ideal, so the Witt ring

$$W(F) = R(F)/H$$

is defined.

We write I for the kernel of the map $\dim : W \longrightarrow \mathbb{Z}/2\mathbb{Z}$. Since $H \cap J = (0)$, the surjection $R \longrightarrow\!\!\!\!\rightarrow W$ induces an isomorphism $J \cong I$. In particular $I^n/I^{n+1} \cong J^n/J^{n+1}$ for all $n \geq 1$. In [7], Milnor shows that the assignment

$$\{a,b\} \longmapsto (\langle a\rangle - \langle 1\rangle)\cdot(\langle b\rangle - \langle 1\rangle) \tag{4.3}$$

defines an isomorphism $\rho : K_2/2K_2(F) \xrightarrow{\cong} I^2/I^3 \cong J^2/J^3$.

Write

$$(a,b) = R_{2,F}\{a,b\} \in H^2(F,\mu_2^{\otimes 2}) \cong {}_2Br(F) \quad .$$

If $\psi = \langle a_1,\ldots,a_n\rangle$ is a quadratic module, the determinant $\det\psi$ is defined by $\det\psi = \overline{a_1a_2 \ldots a_n} \in F^*/F^{*2}$. The Hasse invariant is $S(\psi) = \prod_{i<j} (a_i,a_j) \in {}_2Br(F)$.

These functions are well-defined and satisfy relations

$$\det(\psi \oplus \phi) = \det(\psi)\cdot\det(\phi)$$

$$S(\psi \oplus \phi) = S(\psi)\cdot S(\phi)\cdot(\det\psi\,,\,\det\phi\,)\ .$$

It follows that one has homomorphisms

$$\det : R(F) \longrightarrow F^*/F^{*2}$$

$$S : \mathrm{Ker}(\det) \longrightarrow {}_2Br(F)\ .$$

(The existence of det follows from the universal mapping property of the Grothendieck group. For S, once one grants that it is well-defined on the level of quadratic modules, one can define a homomorphism, via the u.m.p., of R(F) into a twisted product of F^*/F^{*2} and ${}_2Br(F)$ and then restrict to Ker(det).)

Notice

$$\det[(\langle a\rangle - \langle 1\rangle)(\langle b\rangle - \langle 1\rangle)] = \det(\langle ab\rangle - \langle a\rangle - \langle b\rangle + \langle 1\rangle)$$
$$= 1$$

so $J^2 \subset \mathrm{Ker}(\det)$. Also

$$S[(\langle a\rangle - \langle 1\rangle)(\langle b\rangle - \langle 1\rangle)] = S\langle ab,1\rangle\cdot(S\langle a,b\rangle)^{-1}$$
$$= (a,b)^{-1} = (a,b)\quad .$$

It follows from this and the shape of Milnor's isomorphism (4.3) that $S(J^3) = 0$ and there is a commutative diagram

$$\begin{array}{ccc} & K_2/2K_2(F) & \\ \swarrow^{\rho} & & \searrow^{R_{2,F}} \\ J^2/J^3 & \xrightarrow{\ S\ } & {}_2Br(F) \end{array} .$$

Theorem (4.1) now follows from

<u>Lemma</u> (4.4) (Scharlau, [11]). Let F be a C_2-field, char$(F) \neq 2$, and let $\zeta \in R(F)$ be such that $\dim \zeta = \det \zeta = S(\zeta) = 0$. Then $\zeta = 0$.

<u>Proof</u> : By the C_2-property, any quadratic module of dimension > 4 is isotropic (represents 0). If (V,q) is a quadratic module of dimension 4, $(V,q) \oplus \langle -1 \rangle$ represents zero, so there exists a $v \in V$ with $q(v) = 1$. Taking v as the first vector in an orthogonal basis, we can write $(V,q) = \langle 1,a,b,c \rangle$. By induction, any quadratic module of dimension ≥ 4 can be written in the form $\langle 1,1,\ldots,1,a,b,c \rangle$. The crucial point is now

<u>Lemma</u> (4.5) (Scharlau, op. cit.). Let ψ, ϕ be quadratic modules of dimension 3 over any field F. If ϕ and ψ have the same determinant and the same Hasse invariant, they are isomorphic.

<u>Proof</u>. Multiplying by $-\det \psi$, we may assume $\det \psi = -1 = \det \phi$, so

$$\psi = \langle a,b,-ab \rangle \qquad \phi = \langle c,d,-cd \rangle .$$

By assumption

$$S(\psi) = (a,b) = S(\phi) = (c,d) ,$$

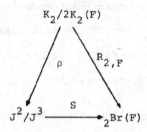

so the quaternion algebras (a,b) and (c,d) are isomorphic.

The algebra (a,b) has generators x, y and relations $x^2 = a$, $y^2 = b$, $xy = -yx$. There is an anti-automorphism $\bar{\ }$ defined by $\bar{x} = -x$, $\bar{y} = -y$, $\overline{xy} = \bar{y}\,\bar{x} = -xy$. There is a quadratic function q on (a,b) defined by $q(z) = z\cdot\bar{z} \in F$, and the associated quadratic module ((a,b),q) is isomorphic to $\langle -a, -b, ab\rangle$.

One checks easily that any isomorphism $(a,b) \cong (c,d)$ is necessarily compatible with the additional quadratic structure, so in our case

$$\langle -a,-b,ab\rangle = \langle -c,-d,cd\rangle \ .$$

The lemma follows by tensoring with $\langle -1\rangle$. Q.E.D.

<u>Question</u> (4.6). Let X be a regular algebraic k-scheme, where k is a algebraically closed field. For what values of p,q,n are the cohomology groups

$$H^o(X,\underline{\underline{K}}_q/n\underline{\underline{K}}_q)$$

finite ? Finiteness for all p,q,n would imply finiteness of ${}_nCH^*(X)$ and $CH^*(X)/nCH^*(X)$.

5. <u>The case $\Lambda_\ell(F) = (0)$</u>

Let F_o be a field containing all ℓ-power roots of 1 for some fixed prime ℓ distinct from char.F_o. Taking the limit over ℓ^ν for finite ν we obtain an exact sequence

$$0 \longrightarrow \Lambda_\infty(F_o) \longrightarrow K_2(F_o) \otimes \mathbb{Q}_\ell/\mathbb{Z}_\ell \xrightarrow{R_\infty} Br(F_o) \otimes \mathbb{Z}_\ell(1) \longrightarrow \theta_\infty(F_o) \longrightarrow 0$$

with

$$\mathbb{Z}_\ell(1) = \varprojlim \mu_{\ell^\nu} \; ; \; \Lambda_\infty = \varinjlim \Lambda_{\ell^\nu} \; ; \; \theta_\infty = \varinjlim \theta_{\ell^\nu} \; .$$

Theorem (5.1). Assume $\Lambda_\ell(F) = (0)$ for all F algebraic over F_o. Then $\theta_\infty(F) = (0)$ for all such F if and only if the Galois cohomology groups

$$H^1(G, K_2'(F')) = (0)$$

where $K_2' = K_2/\text{torsion}$ and F'/F is a Galois extension of fields algebraic over F_o, with

$$G = \mathrm{Gal}(F'/F) \cong \mathbb{Z}/\ell\mathbb{Z} \; .$$

Proof. The critical point is the following lemma.

Lemma (5.2). Let F'/F be cyclic with Galois group $G \cong \mathbb{Z}/\mathbb{Z}$. Then there is an exact sequence of Brauer groups

$$0 \longrightarrow \mathrm{Br}(F'/F) \longrightarrow \mathrm{Br}(F) \longrightarrow \mathrm{Br}(F')^G \longrightarrow 0 \; .$$

Proof. The only point which is not obvious is surjectivity on the right. Let $\bar{F}$ be the algebraic closure of F. There is a spectral sequence

$$E^{p,q} = H^p(F'/F, H^q(\bar{F}/F', \bar{F}^*)) \Longrightarrow H^{p+q}(\bar{F}/F, \bar{F}^*)$$

$$\mathrm{Br}(F) = H^2 \; , \; \mathrm{Br}(F')^G = E_2^{0,2} \; .$$

This is a first-quadrant spectral sequence, so the only possible non-zero differentials with domain $E^{0,2}$ are

$$E_2^{0,2} \xrightarrow{d_2} E_2^{2,1} \ , \ E_3^{0,2} \xrightarrow{d_3} E_3^{3,0} \ .$$

But $E_2^{2,1} = H^2(F'/F, H^1(\bar{F}/F', \bar{F}^*)) = (0)$ by Hilbert's theorem 90, and $E_3^{3,0}$ is a subquotient of

$$E_2^{3,0} = H^3(F'/F, F'^*) \cong H^1(F'/F, F'^*) = (0)$$

because Gal(F'/F) is cyclic.

It follows that $H^2 \longrightarrow E_2^{0,2}$ is surjective. Q.E.D.

Returning to the proof of (5.1), consider the diagram

$$\begin{array}{ccccccc} 0 \longrightarrow & (K_2(F') \otimes \mathbb{Q}_\ell/\mathbb{Z}_\ell)^G & \xrightarrow{R_\infty(F')} & Br(F')(\ell)^G & & & \\ (5.3) & \uparrow \alpha & & \uparrow & (5.2) & & \\ 0 \longrightarrow & K_2(F) \otimes \mathbb{Q}_\ell/\mathbb{Z}_\ell & \xrightarrow{R_\infty(F)} & Br(F)(\ell) & \longrightarrow & \theta_\infty(F) & \longrightarrow 0 \end{array}$$

where Br(F)(ℓ) denotes the subgroup of ℓ-torsion elements (we are ignoring the twisting by roots of 1 since these are in the ground field). Thus $\theta_\infty(F) = (0)$ implies α is surjective. But α factors

$$\begin{array}{ccc} K_2(F) \otimes \mathbb{Q}_\ell/\mathbb{Z}_\ell & \xrightarrow{\alpha} & (K_2(F') \otimes \mathbb{Q}_\ell/\mathbb{Z}_\ell)^G \\ & \beta \searrow \quad \nearrow \gamma & \\ & K_2(F')^G \otimes \mathbb{Q}_\ell/\mathbb{Z}_\ell \ . & \end{array}$$

Some remarks on this diagram :

(i) β is surjective. Indeed there is a transfer

$$\mathrm{tr} : K_2(F') \longrightarrow K_2(F)$$

and $\beta \circ \mathrm{tr}$ is multiplication by ℓ on $K_2(F')^G$. Thus the cokernel of β is ℓ-torsion, and killed by tensoring with $\mathbb{Q}_\ell/\mathbb{Z}_\ell$.

(ii) The cokernel of γ is the cokernel of

$$(K_2(F') \otimes \mathbb{Q}_\ell)^G \xrightarrow{\gamma'} (K_2(F') \otimes \mathbb{Q}_\ell/\mathbb{Z}_\ell)^G .$$

Indeed, if σ is a generator of G, we have by flatness

$$K_2(F')^G \otimes \mathbb{Q}_\ell = \mathrm{Ker}(1-\sigma : K_2(F') \longrightarrow K_2(F')) \otimes \mathbb{Q}_\ell$$

$$\cong \mathrm{Ker}((1-\sigma) \otimes 1 : K_2(F') \otimes \mathbb{Q}_\ell \longrightarrow K_2(F') \otimes \mathbb{Q}_\ell)$$

$$\cong (K_2(F') \otimes \mathbb{Q}_\ell)^G .$$

Using (i), (ii) and the cohomology sequence of

$$0 \longrightarrow K_2'(F') \otimes \mathbb{Z}_\ell \longrightarrow K_2(F') \otimes \mathbb{Q}_\ell \longrightarrow K_2(F') \otimes \mathbb{Q}_\ell/\mathbb{Z}_\ell \longrightarrow 0 .$$

we get

(iii) $\mathrm{Coker}(\alpha) \cong H^1(G, K_2'(F') \otimes \mathbb{Z}_\ell) \cong H^1(G, K_2'(F'))$

In sum

$$\theta_\infty(F) = (0) \Longrightarrow \alpha \text{ surjective} \Longrightarrow H^1(G, K_2'(F')) = (0).$$

It remains to do the converse.

Lemma (5.5). Let F be algebraic over F_o, and let D be a division algebra with center F such that $[D] \varepsilon Br(F)(\ell)$. Assume

(i) There exists a splitting field F''/F for D such that F''/F is Galois of degree ℓ^{ν}, some ν.

(ii) The cohomology vanishing hypotheses of (5.1) hold.

Then $[D] \varepsilon$ Image $R_{\infty}(F)$.

Proof. We proceed by induction on ν. That is, if (F_1, D_1, ν_1) satisfy the above conditions with $\nu_1 < \nu$, we assume $[D_1] \varepsilon \operatorname{Im} R_{\infty}(F_1)$. Given (F,D,ν) with Galois splitting field F'', $[F'':F] = \ell^{\nu}$, let $G = Gal(F''/F)$. G is solvable, so there exists an $F' \subset F''$ with F'/F cyclic of order ℓ. By inductive hypothesis, $[D \otimes F'] \varepsilon \operatorname{Im} R_{\infty}(F')$, say $[D \otimes F'] = R_{\infty}(F')(x')$. Since $R_{\infty}(F')$ is injective, x' is invariant under G. Also our hypotheses imply the map α in (5.3) is surjective, so $x' = \alpha(x)$, for some $x \varepsilon K_2(F) \otimes \mathbb{Q}_{\ell}/\mathbb{Z}_{\ell}$. Thus

$$[D] - R_{\infty}(F)(x) \varepsilon Br(F'/F)(\ell) = \operatorname{Ker}(Br(F)(\ell) \longrightarrow Br(F')(\ell)).$$

The group $Br(F'/F)$ is known classically to lie in the image of $R_{\infty}(F)$ (cf. [12], p.), so $[D] \varepsilon \operatorname{Im} R_{\infty}(F)$ as desired. Q.E.D.

For the proof of (5.1), it remains only to show hypothesis (i) in (5.5) can be suppressed.

Lemma (5.6). Let F_1/F be an algebraic extension of fields. Assume F contains the ℓ-th power roots of 1. Then the diagram

$$\begin{array}{ccc} K_2(F_1) \otimes \mathbb{Q}_\ell/\mathbb{Z}_\ell & \xrightarrow{R_\infty(F_1)} & Br(F_1)(\ell) \otimes \mathbb{Z}_\ell(1) \\ \downarrow{\scriptstyle Tr \otimes 1} & & \downarrow{\scriptstyle Cor \otimes 1} \\ K_2(F) \otimes \mathbb{Q}_\ell/\mathbb{Z}_\ell & \xrightarrow{R_\infty(F)} & Br(F)(\ell) \otimes \mathbb{Z}_\ell(1) \end{array} \tag{5.7}$$

is commutative, where Tr and Cor are respectively the transfer on K_2 and the corestriction on the Brauer group.

Assuming (5.6), the proof of (5.1) is completed as follows : given $\zeta \in Br(F)(\ell) \otimes \mathbb{Z}_\ell(1)$, we want to show $\zeta \in \mathrm{Im}\, R_\infty(F)$. For this it suffices to show $m\zeta \in \mathrm{Im}\, R_\infty(F)$ with m some integer prime to p. Choose a Galois splitting field F"/F for ζ, and let $G = \mathrm{Gal}(F''/F)$. Let $G_1 \subset G$ be an ℓ-Sylow subgroup and let $F_1 = F''^{G_1}$. Then the pair $(F_1, \zeta \otimes F_1)$ satisfy the conditions of (5.5), so $\zeta \otimes F_1 \in \mathrm{Im}\, R_\infty(F_1)$. By (5.6), we know therefore that $\mathrm{Cor}(\zeta \otimes F_1) \in \mathrm{Im}\, R_\infty(F_1)$. Finally, the projection formula gives $\mathrm{Cor}(\zeta \otimes F_1) = m\cdot\zeta$ with $m = |G|/|G_1|$ prime to ℓ. This completes the proof of the theorem.

Proof of (5.6) (sketch). The argument here is due to Tate. The details will appear in [2].

Step 1. Let $G = \mathrm{Gal}(\bar{F}/F)$ be the Galois group of the algebraic closure of F. Let $G_\ell \subset G$ be a (profinite) ℓ-Sylow subgroup, and let $F' = \bar{F}^{G_\ell}$. By a base extension argument, one can replace F by F' and hence assume every extension of F has ℓ-power order.

Step 2. Using the solvability of groups of ℓ-power order, one may assume that $[F_1:F] = \ell$. Thus $F_1 = F(\alpha)$ and any element $\beta = P(\alpha)$ where

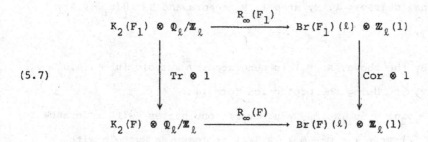

$P \in F[X]$ has degree $< \ell$. By step 1, P factors and $\beta = \prod_i (u_i + v_i\alpha)$; $u_i, v_i \in F_1$.

<u>Step</u> 3. By the above, $K_2(F_1)$ is generated by symbols $\{u_1 + v_1\alpha, u_2 + v_2\alpha\}$ with $u_i, v_i \in F$. Using the projection formula $Tr\{u_1, u_2 + v_2\alpha\} = \{u_1, Norm_{F_1/F}(u_2 + v_2\alpha)\}$ one proves (with reference to (5.7)) $R_\infty(F) \cdot Tr \otimes 1 = Cor \otimes 1 \cdot R_\infty(F_1)$ at least on symbols with $v_1 v_2 = 0$. A general symbol $\{u_1 + v_1\alpha, u_2 + v_2\alpha\}$ can be reduced to a product of such by means of the Steinberg identity $\{x, 1-x\} = 1$. Q.E.D.

<u>Remark</u> (5.8). We have $\Lambda_2(F) = (0)$ when F is a C_2-field of characteristic $\neq 2$. For example :

<u>Corollary</u> (5.9). Let k be an algebraically closed field of characteristic $\neq 2$. Assume $H^1(F'/F, K_2'(F')) = (0)$ whenever $\text{tr deg}_k F = 2$ and F'/F is cyclic of degree 2. Then for any smooth surface X/k, the 2-torsion subgroup of the Chow group, $CH^2(X)(2)$, is cofinite, i.e. $CH^2(X)(2) \hookrightarrow \bigoplus_r \mathbb{Q}_2/\mathbb{Z}_2$ for some finite r.

Added in proof : A sharper version of theorem (4.1) has been proved independently of the author by Lam and Elman [14]. Using their result, one can show that $\Gamma(X, \underline{K}_2/2^r\underline{K}_2)$ is finite when X has dimension ≤ 3 over an algebraically closed field k of characteristic $\neq 2$.

BIBLIOGRAPHY

1. Artin, M., Grothendieck topologies, mimeo. notes published by Harvard University, 1962.

2. Bass, H. and Tate, J., The Milnor ring of a global field, Algebraic K-theory I, Springer lecture notes 341 (1973).

3. Bloch, S., K_2 and algebraic cycles, Annals of Math., Vol. 99, p. 349-379 (1974).

4. Bloch, S., and Ogus, A., Gersten's conjecture and the homology of schemes, Ann. Scient. Ec. Norm. Sup. t.7, p. 181-202 (1974).

5. Brown K., and Gersten S., Algebraic K-theory as generalized sheaf cohomology, Algebraic K-theory I, Springer lecture notes (341) (1973).

6. Gersten, S., Some exact sequences in the higher K-theory of rings, Algebraic K-theory I, Springer lecture notes (341) (1973).

7. Milnor J., Algebraic K-theory and quadratic forms, Inventiones Math. (169).

8. Milnor J., Introduction to Algebraic K-theory, Annals of Math. studies (72).

9. Quillen, D., Higher algebraic K-theory I, Algebraic K-theory I, Springer lecture notes 341 (1973).

10. Samuel, P., Relations d'équivalence en géométrie algébrique, Proceedings of the International Congress, Cambridge, pp. 470-487 (1958).

11. Scharlau, W., Lectures on quadratic forms, Queens University lecture notes in math.

12. Serre, J.P., Corps Locaux, Hermann, (1962).

13. Tate, J., Hour talk, Actes, Congrès intern. Math. 1970.

14. Elman, R., and Lam, T.Y., On the quaternion symbol homomorphism $g_F : k_2F \rightarrow B(F)$, Algebraic K-theory I, Springer lecture notes in math. 341 (1973).

THE BRAUER GROUP AND UNIRATIONALITY : AN EXAMPLE OF ARTIN-MUMFORD

Allen TANNENBAUM

Introduction

This paper is an exposition of a counterexample of Artin-Mumford [4] to the Lüroth problem in dimension 3. We will try to keep our exposition as elementary as possible and thus shall restrict ourselves to working over the complex numbers $\mathbb{C}$. We will thereby be able to avoid using some of the finer points of the étale cohomology.

Finally while making use of some pretty techniques, the Artin-Mumford paper [4] is a bit tersely written. Therefore we will put in some details that are lacking in [4] as well as at certain points attempt to simplify the constructions.

Section 1. The Lüroth Problem.

In this section we give a very brief sketch of the Lüroth problem. We begin with the following standard definitions :

Definition (1.1). Let V be a complex projective variety of dimension n.

(i) V is unirational if there exists a rational dominating map $f : \mathbb{P}^n \to V$ (or equivalently an inclusion of function fields $K(V) \subseteq K(\mathbb{P}^n) \cong \mathbb{C}(X_1,\ldots,X_n)$).

(ii) V is rational if there exists a birational map $f : \mathbb{P}^n \to V$ (or equivalently $K(V) \xrightarrow{\sim} K(\mathbb{P}^n) \cong \mathbb{C}(X_1,\ldots,X_n)$).

The Lüroth problem is when unirationality implies rationality.

For n = 1 or n = 2, this is always the case :

Proposition (1.2). A unirational curve or surface is always rational.

Proof. Let C be a unirational curve. We can clearly assume C is smooth. Then there exists a surjective morphism $\varphi : \mathbb{P}^1 \to C$. If there existed a non-zero holomorphic differential form on C, its pull-back to $\mathbb{P}^1$ would be also non-zero which cannot occur. Therefore the genus of C is zero, i.e. $C \cong \mathbb{P}^1$.

Let S be a unirational surface which we can take to be smooth. Then by blowing up the points of indeterminacy [22], we see that there exists, a surjective morphism $\psi : S' \to S$ where S' is a smooth rational surface. But then $h^1(\mathcal{O}_{S'}) = h^0(\mathcal{O}_{S'}(2K_{S'})) = 0$ and thus by the previous argument the same holds for S. Hence by the Castelnuovo Criterion of rationality [22], S is rational. Q.E.D.

In Section 7 we will construct an example of Artin-Mumford of a smooth unirational 3-fold which is irrational.

Section 2. Birational Invariance of Torsion.

This section contains the key idea behind the Artin-Mumford construction. Namely :

Proposition (2.1). Let V be a smooth complex projective variety. Then the torsion subgroup T(V) of $H^3(V,\mathbb{Z})$ is a birational invariant.

Proof. See Artin-Mumford [4] pages 77-78, or Deligne [9] pages 48-49. Q.E.D.

Remarks (2.2). (i) If V is rational, then from (2.1) we have that T(V) = 0.

(ii) For V a 3-fold, using Poincaré duality and the universal coefficient theorem, it is an easy exercise to show that the torsion part of $H^4(V,\mathbb{Z}) \cong$ torsion part of $H^3(V,\mathbb{Z}) := T(V)$.

Key Idea (2.3). We will construct a smooth, unirational 3-fold V with 2-torsion in $H^4(V,\mathbb{Z})$. By (2.2) then, V cannot be rational.

Remark (2.4). The fact that we are using torsion of $H^3(V,\mathbb{Z})$, motivates the introduction of the Brauer group since the topological Brauer group of V is isomorphic to the torsion part of $H^3(V,\mathbb{Z})$. See Grothendieck [13] page 50.

Section 3. Some Facts from the Classical Theory of Maximal Orders.

In this section we outline the key facts from the classical theory of maximal orders which we will need in the sequel.

Definitions (3.1). Let R be an integrally closed Noetherian domain with quotient field K. Let D be a central simple K-algebra. Then an R-order A in D is a subring which is a finitely generated R-module which generates D over K.

A maximal R-order is an order not properly contained in any other order.

Remark (3.2). The condition of integral closure for R is made so that every order is contained in a maximal order. See Reiner [18] page 127.

Some Facts about Maximal Orders (3.3). Let R be a discrete valuation ring with local uniformizing parameter t. Suppose that R/tR is C_1 (i.e. if f is a homogeneous form with coefficients in R/tR of degree d in n variables, and $n > d$, then f has a non-trivial zero in R/tR). For example R/tR may be a function field of transcendence degree 1, or a finite field. See e.g. Greenberg [12]. Then by Tsen's theorem ([12]), $\mathrm{Br}\, R/tR = 0$.

Let K be the field of quotients of R, D a central simple K-algebra, and A a maximal R-order in D. We want to show how to associate a cyclic extension L of R/tR to D.

We let $\hat{R}$ be the completion of R with respect to tR, $\hat{K}$ the quotient field of $\hat{R}$, and $\hat{D} := \hat{K} \otimes_K D$. Then $\hat{D} = M_n(E)$ where E is a skew field with center $\hat{K}$ and $M_n(E)$ denotes the set of $n \times n$ matrices with entries in E.

Now E has a unique maximal $\hat{R}$-order, namely the integral closure of $\hat{R}$ in E. (See Reiner [18] pages 135-138). Call this maximal order Λ. Then from [18] page 179

$$A/\mathrm{rad}\, A \cong M_n(\Lambda/\mathrm{rad}\, \Lambda)$$

and $\Lambda/\mathrm{rad}\, \Lambda$ is a skew field.

We claim however that $\Lambda/\mathrm{rad}\, \Lambda$ is a field (i.e. is commutative). Indeed let L be the center of $\Lambda/\mathrm{rad}\, \Lambda$. Then L being a finite algebraic extension of R/tR, L is C_1 ([12] page 21), so that $\mathrm{Br}\, L = 0$, and so $\Lambda/\mathrm{rad}\, \Lambda = L$.

Moreover from Schilling [19] page 156, we have that E contains an unramified maximally commutative subfield which is a cyclic extension of $\hat{K}$, say F. Suppose $(E : \hat{K}) = n^2$. Then from Reiner [18] pages 97 and 145 we have that $(F : \hat{K}) = (L : R/tR) = n$. Since F is unramified, if $\bar{R} :=$ integral closure of $\hat{R}$ in F, and p is the maximal ideal of $\bar{R}$, from the above we see that $(\bar{R}/p : R/tR) = n$. Thus $R/p = L$, and thus by [18] page 72, L is a cyclic extension of R/tR.

Section 4. An Exact Sequence.

We will prove in this section the existence of an exact sequence for Br K(S), where K(S) is the function field for a simply connected smooth complex projective surface S . The proof of Artin-Mumford [4] is a complicated exercise in the étale cohomology which we shall discuss in (4.2) below. However the first three terms of the exact sequence hold for an arbitrary smooth irreducible complex projective surface and can be proven using elementary means. This seems worthwhile to do, so we begin with :

Lemma (4.1). Let S be a smooth irreducible complex projective surface. Let Br S be the Brauer group of Azumaya algebras on S . Then there exists a canonical exact sequence of groups

$$0 \to \operatorname{Br} S \xrightarrow{i} \operatorname{Br} K(S) \xrightarrow{a} \bigoplus_{\substack{C \subset S \\ C \text{ irreducible curve}}} H^1(K(C), \mathbb{Q}/\mathbb{Z})$$

where the terms and maps are as follows :

(a) i is the restriction of a global Azumaya algebra on S to the generic point.

(b) $H^1(K(C),\mathbb{Q}/\mathbb{Z})$ is Galois cohomology (Serre [20]) and the direct sum runs over all the irreducible curves of S . We note that $H^1(K(C),\mathbb{Q}/\mathbb{Z})$ is precisely the group of isomorphism classes of cyclic extensions of K(C). Indeed, to see this let G = Galois group of $\overline{K(C)}$ over K(C). Then since G acts trivially on $\mathbb{Q}/\mathbb{Z}$,

$H^1(K(C),\mathbb{Q}/\mathbb{Z}) = H^1(G,\mathbb{Q}/\mathbb{Z}) \cong \mathrm{Hom}(G,\mathbb{Q}/\mathbb{Z})$. Let $\varphi \in \mathrm{Hom}(G,\mathbb{Q}/\mathbb{Z})$. Then by the fundamental theorem of Galois theory (Gruenberg [14] page 120), corresponding to the subgroup $\ker\varphi$, one has a field extension $K(\varphi)$ of $K(C)$. If G' = Galois group of $K(\varphi)$ over $K(C)$, then $G' \cong G/\ker\varphi$ and hence G' is isomorphic to a subgroup of $\mathbb{Q}/\mathbb{Z}$, i.e. $K(\varphi)$ is a cyclic extension of $K(C)$.

(c) The map a is given as follows : Let D be any central simple $K(S)$-algebra. Let $C \subset S$ be an irreducible curve and $c \in C$ the generic point. Then $\mathcal{O}_{S,c}$ is a discrete valuation ring with residue field $K(C)$. Since $K(C)$ is C_1, as in (3.3) we can associate a cyclic extension of $K(C)$ to D. Let d be the class of D in $\mathrm{Br}\,K(S)$. One can easily check that this cyclic extension is independent of the choice of representative for d and thus this defines the map a.

If the extension is non-trivial one says that d (or D) is <u>ramified</u> on C. We note that this ramification has properties analogous to the classical ramification of number theory (with a different and a discriminant), and hence in particular d can ramify at only finitely many irreducible curves. For details see Reiner [18].

<u>Proof</u> of (4.1). We give here a completely elementary proof. A proof using étale cohomology will be indicated in (4.2) below.

The fact that i is injective is proven in Auslander-Goldman [6] page 388 (note the fact that S non-singular is essential here).

We now show that $i(\mathrm{Br}\,S) \subseteq \mathrm{Ker}\,(a)$. Let A be a global Azumaya algebra. Then from Auslander-Goldman [6] page 387, A is a maximal $\mathcal{O}_S$-order in $A \otimes K(S)$. Therefore for any $c \in S$ a point of codimension 1,

A_c is a maximal $\mathcal{O}_{S,c}$-order in $A \otimes K(S)$ (see [18] page 133). But then by definition of Azumaya algebra, $A_c \otimes K(C)$ is a central simple $K(C)$-algebra, and since $\text{Br } K(C) = 0$ we have by definition of the map a, that $a(\text{class of } A \otimes K(S)) = 0$.

Conversely let $d \in \text{Br } K(S)$, $a(d) = 0$. Let D be a representative of d. Choose a maximal order A over S for D. Then as before, for each $c \in S$ a point of codimension 1, A_c will be a maximal $\mathcal{O}_{S,c}$-order in D. But since $a(d) = 0$ and $\text{Br } K(C) = 0$, by the definition of a, $A_c \otimes K(C) \cong M_n(K(C))$ and thus by definition of Azumaya algebra $A_c \in \text{Br}\mathcal{O}_{S,c}$. Identifying $\text{Br} S \hookrightarrow \text{Br}\mathcal{O}_{S,c}$, from [6] page 389 we have $\text{Br } S = \bigcap_{\substack{c \text{ of} \\ \text{codim}=1}} \text{Br}\mathcal{O}_{S,c}$. Thus $A \in \text{Br } S$ i.e. A is a global Azumaya algebra such that $A \otimes K(S) \cong D$. Q.E.D.

In case S is simply connected we can complete the exact sequence of (4.1) :

<u>Theorem</u> (4.2). Let S be a smooth simply connected complex projective surface. Then there exists a canonical exact sequence of groups

$$0 \longrightarrow \text{Br } S \xrightarrow{\ i\ } \text{Br } K(S) \xrightarrow{\ a\ } \bigoplus_{\substack{C \subset S \\ C \text{ irreducible} \\ \text{curve}}} H^1(K(C), \mathbb{Q}/\mathbb{Z}) \xrightarrow{\ r\ } \bigoplus_{\substack{p \in S \\ p \text{ closed}}} \mu^{-1} \xrightarrow{\ s\ } \mu^{-1} \longrightarrow 0$$

where the terms and the maps are as follows :

(a) i, a and the first three terms of the exact sequence are as in (4.1).

(b) $\mu^{-1} := \bigcup_n \text{Hom}(\mu_n, \mathbb{Q}/\mathbb{Z})$ where μ_n are the nth roots of unity which we can canonically identify with $\mathbb{Z}/n\mathbb{Z}$ since we are working over $\mathbb{C}$.

(c) The map r "measures" ramification and is explicitly defined as follows : First recall that $H^1(K(C),\mathbb{Q}/\mathbb{Z})$ is the group of isomorphism classes of cyclic extensions of $K(C)$ or equivalently the group of isomorphism classes of ramified cyclic coverings of the normalization $\bar{C}$ of C. Let $\bar{c} \in \bar{C}$ be a closed point. Then $\mathcal{O}_{\bar{C},\bar{c}}$ is a discrete valuation ring, and consequently determines a valuation v on its field of quotients $K(C)$.

Let $K(C)_v$ = completion of $K(C)$ with respect to v, and let G be the Galois group of $\overline{K(C)_v}$ over $K(C)_v$. Then we have the natural homomorphism of cohomology groups

$$H^1(K(C),\mathbb{Q}/\mathbb{Z}) \xrightarrow{r_{\bar{c}}} H^1(K(C)_v,\mathbb{Q}/\mathbb{Z}) \cong \mathrm{Hom}(G,\mathbb{Q}/\mathbb{Z}) .$$

But $K(C)_v \cong \mathbb{C}((T))$ and from Serre [21] page 76 $\overline{\mathbb{C}((T))} = \bigcup_n \mathbb{C}((T^{1/n}))$, so that $G \cong \hat{\mathbb{Z}} := \varprojlim_n \mathbb{Z}/n\mathbb{Z}$. Consequently

$$\begin{aligned} H^1(K(C)_v,\mathbb{Q}/\mathbb{Z}) &\cong \mathrm{Hom}(\hat{\mathbb{Z}},\mathbb{Q}/\mathbb{Z}) \\ &\cong \bigcup_n \mathrm{Hom}(\mathbb{Z}/n\mathbb{Z},\mathbb{Q}/\mathbb{Z}) \\ &\cong \mu^{-1} . \end{aligned}$$

We will identify $H^1(K(C)_v,\mathbb{Q}/\mathbb{Z})$ with μ^{-1}.

We claim that $r_{\bar{c}}$ "measures" the ramification of $\bar{C}$ at $\bar{c}$ with respect to a cyclic ramified covering of $\bar{C}$. Indeed let $L \in H^1(K(C),\mathbb{Q}/\mathbb{Z})$ and $L_v = r_{\bar{c}}(L)$ be its image in $H^1(K(C)_v,\mathbb{Q}/\mathbb{Z})$. Since L is a Galois extension, $(L : K(C)) = efr$ where e is the ramification degree, and f is the residue degree. But here $f = 1$ since the residue fields are isomorphic to $\mathbb{C}$, and so $(L : K(C)) = er$. Hence $(L_v : K(C)_v) = e$, i.e. L_v is a cyclic extension of degree e = ramification degree, and thus in this sense $r_{\bar{c}}$ measures the ramification at $\bar{c}$ of the extension L.

Then r is defined to be the sum of the ramification of all the closed points of the various $\bar{C}$ lying over the closed point $p \in S$.

(d) The map s is the sum.

<u>Proof</u> of (4.2). We will only sketch the proof as a complete proof is given in Artin-Mumford [4] pages 84-87. However we will include some details which are not in their paper and we want to emphasize exactly where the simple connectivity of S comes into the proof.

First let I denote the filtering system of finite sets π of closed points of S. One defines the cohomology of the "pro-object" $U = \{S-\pi\}_{\pi \in I}$ as

$$H^q(U,F_U) := \varinjlim_{\pi} H^q(S-\pi,F)$$

where F is a sheaf on S. (Cohomology here is always étale cohomology.)

Then one can compute ([4] page 84) that $R^3 i_{\pi *}\mathbb{G}_m = \bigoplus_{p \in \pi} \mu_p^{-1}$, and $R^q i_{\pi *}\mathbb{G}_m = 0$ for $q > 0$, $q \neq 3$, where $i_\pi : S-\pi \rightarrow S$ is the inclusion, and μ_p^{-1} means extension of μ^{-1} by zero outside p.

Applying the Leray spectral sequence to i_π (which by the above degenerates) we get by Godement [11] page 85 exact sequences,

$$H^2(S,i_{\pi *}\mathbb{G}_m) \xrightarrow{\sim} H^2(S-\pi,\mathbb{G}_m)$$

and

$$0 \rightarrow H^3(S,i_{\pi *}\mathbb{G}_m) \rightarrow H^3(S-\pi,\mathbb{G}_m) \rightarrow H^o(S,E^3 i_{\pi *}\mathbb{G}_m) \rightarrow H^4(S,i_{\pi *}\mathbb{G}_m) \rightarrow 0 .$$

Passing to the limit (and using the fact that $\operatorname{Br} S \cong H^2(S,\mathbb{G}_m)$; see Grothendieck [13] page 76) we see that

$$\operatorname{Br} S \xrightarrow{\sim} H^2(U,\mathbb{G}_m)$$

and we have an exact sequence

$$(1) \qquad 0 \to H^3(S,\mathbb{G}_m) \to H^3(U,\mathbb{G}_m) \to \bigoplus_p \mu^{-1} \to H^4(S,\mathbb{G}_m) \to 0.$$

Next one can compute that $R^q\varphi_*\mathbb{G}_m = 0$ for all $q > 0$ where $\varphi : \operatorname{Spec} K(S) \to U$ (see [4] page 86 and Artin [1] page 97) and thus again we are in the degenerate case of the Leray spectral sequence from which we derive

$$H^q(K(S),\mathbb{G}_m) \cong H^q(U,\varphi_*\mathbb{G}_m) .$$

Now $\varphi_*\mathbb{G}_m$ can be regarded as the sheaf of nowhere non-vanishing meromorphic functions on S and $\bigoplus_C \mathbb{Z}_{K(C)}$ the sheaf of divisors. Hence we have an exact sequence

$$(2) \qquad 0 \to \mathbb{G}_m \to \varphi_*\mathbb{G}_m \to \bigoplus_C \mathbb{Z}_{K(C)} \to 0 .$$

We claim $H^1(K(C),\mathbb{Z}) = 0$. Indeed setting $G :=$ the Galois group of $\overline{K(C)}$ over $K(C)$, we have that

$$\begin{aligned} H^1(K(C),\mathbb{Z}) &= H^1(G,\mathbb{Z}) \\ &= \varinjlim_V H^1(G/V,\mathbb{Z}) \\ &= \varinjlim_V \operatorname{Hom}(G/V,\mathbb{Z}) \end{aligned}$$

where the direct limit runs through all open, normal subgroups V of G. But G/V is finite and $\mathbb{Z}$ is torsion free, and hence $\operatorname{Hom}(G/V,\mathbb{Z}) = 0$.

Moreover since $\mathbb{Q}$ is a uniquely divisible G-module, $H^r(K(C),\mathbb{Q}) = 0$ for all $r > 0$. So from the exact sequence

$$0 \to \mathbb{Z} \to \mathbb{Q} \to \mathbb{Q}/\mathbb{Z} \to 0$$

we have that $H^1(K(C),\mathbb{Q}/\mathbb{Z}) \cong H^2(K(C),\mathbb{Z})$.

Next we claim that $H^3(K(S),\mathbb{G}_m) = 0$. Indeed consider the Kummer sequence (Artin [1] page 102)

$$0 \to \mu_n \to \mathbb{G}_m \xrightarrow{n} \mathbb{G}_m \to 0$$

on Spec K(S). Then we have that

(3) $\quad H^3(K(S),\mu_n) \to H^3(K(S),\mathbb{G}_m) \xrightarrow{n} H^3(K(S),\mathbb{G}_m) \to H^4(K(S),\mu_n)$

is exact. But since the cohomological dimension of K(S) is 2 we have that $H^3(K(S),\mu_n) = H^4(K(S),\mu_n) = 0$. (See Artin [3] and [1] pages 95-100.) Now $H^3(K(S),\mathbb{G}_m)$ is torsion (see Shatz [23] page 35; we are using the fact that over a field Galois cohomology and étale cohomology are the same; for this see Deligne [10] pages 24-26). We conclude then from (3) that $H^3(K(S),\mathbb{G}_m) = 0$ as claimed.

Finally it is easy to see that $H^q(U,\mathbb{Z}_{K(C)}) = H^q(K(C),\mathbb{Z})$. Applying our remarks and computations above to the long exact cohomology sequence associated to (2) we get an exact sequence

(4) $\quad 0 \to \text{Br } S \to \text{Br } K(S) \to \bigoplus_C H^1(K(C),\mathbb{Q}/\mathbb{Z}) \to H^3(U,\mathbb{G}_m) \to 0.$

Up to now we have not used the fact that S is simply connected. This comes in as follows : Using the fact that $H^q(S,\mathbb{G}_m)$ is torsion for $q \geq 2$ ([13] page 71) and the Kummer sequence applied to S , one sees immediately that $H^q(S,\mathbb{G}_m) \cong H^q(S,\mu)$ (where μ = set of all roots of unity) for $q \geq 3$. But by Poincaré duality (Deligne [10]) one has that $H^4(S,\mu) \cong \mu^{-1}$ and $H^3(S,\mu)$ is dual to $H^1(S,\mu)$. But $H^1(S,\mu) = 0$ since S is simply connected. We may therefore amalgamate the exact sequences (1) and (4) to derive the exact sequence of (4.2).

One must still show the maps have the required interpretations. This is a pretty argument which is given on pages 86-87 of [4]. Q.E.D.

Remark (4.3). The existence of the exact sequence of (4.2) also follows formally from the paper of Bloch-Ogus [7]. However one must still prove that the maps of this exact sequence have the classical interpretations given in (4.1) and (4.2) and it is this fact that gives the exact sequence much of its power.

Section 5. Severi-Brauer Schemes.

One of the prettiest ideas of Artin-Mumford [4] is to generalize the notion of Severi-Brauer schemes corresponding to quaternion Azumaya algebras to certain schemes ("Brauer-Severi schemes") corresponding to maximal orders in quaternion algebras. We will discuss this idea in Section 6 below. However to motivate this, we will now briefly look at the construction of Severi-Brauer schemes.

Definition (5.1). Let k be any field, and V a k-variety. Then V is called a Severi-Brauer variety if $V \otimes_k \bar{k} \cong \mathbb{P}^{n-1}_{\bar{k}}$.

Theorem (5.2). There exists a canonical isomorphism between isomorphism classes of Severi-Brauer varieties over k of dimension n-1 and isomorphism classes of central simple k-algebras of rank n^2. In particular there exists a canonical isomorphism between isomorphism classes of Severi-Brauer curves over k and quaternion k-algebras.

Moreover each such curve admits an embedding in $\mathbb{P}_k^2$ as a conic and this conic is unique up to projective transformation.

Proof. We will sketch two arguments proving the existence of the general 1-1 correspondence both of which will be useful later on. For details see Serre [21] pages 160-168.

The first argument is by descent. Indeed it is well-known that every automorphism of the ring $M_n(\bar{k})$ of $n \times n$ matrices over $\bar{k}$ is interior. Thus one may identify the group of automorphisms of $M_n(\bar{k})$ with $PGL(n,\bar{k}) \cong GL(n,\bar{k})/\bar{k}^*$. But by descent [21] pages 160-162, the pointed set of isomorphism classes of central simple k-algebras of rank n^2 is isomorphic to the non-abelian cohomology $H^1(G,PGL(n,\bar{k}))$ where G = Galois group of $\bar{k}$ over k. But $H^1(G,PGL(n,\bar{k}))$ clearly also classifies Severi-Brauer varieties of dimension n-1. (See [21] page 168.)

The second proof goes as follows : Let D be a central simple k-algebra of rank n^2. Let Gr(D) be the Grassmannian of n-dimensional linear subspaces of D. Define V = {left ideals $L \subset D$ of rank n over k}. Then V is a Severi-Brauer variety corresponding to D and one can check this defines a 1-1 correspondence ([21] page 168).

Finally for quaternion k-algebras and Severi-Brauer curves we can argue as follows : It is well-known that a quaternion k-algebra is determined up to isomorphism by its norm (see e.g. O'Meara [17] pages 145-146).

Let $D = (\frac{\alpha,\beta}{k})$ (notation as in [17] page 142). Then given $x \in D$

$$x = x_0 1 + x_1 i + x_2 j + x_3 ij$$

(where $i^2 = \alpha$, $j^2 = \beta$, $ij = -ji$, $x_\ell \in k$ for $0 \leq \ell \leq 3$), if $Nx :=$ norm of x, we have

$$Nx = x_o^2 - x_1^2\alpha - x_2^2\beta + x_3^2\alpha\beta$$

and thus the norm determines the conic in $\mathbb{P}_k^2$ defined by

$$x_o^2 - x_1^2\alpha - x_2^2\beta = 0 .$$

Since any non-trivial conic over an algebraically closed field is isomorphic to $\mathbb{P}^1$, we have the required facts about Severi-Brauer curves and quaternion k-algebras. Q.E.D.

Now the notion of Severi-Brauer variety has been relativized (of course by Grothendieck [13] pages 63-65). Indeed we have the following definition :

Definition (5.3). Let $k = \bar{k}$ be an arbitrary algebraically closed field and let X be a k-scheme. Then a Severi-Brauer scheme $V \xrightarrow{\pi} X$, is an X-scheme, which is locally isomorphic to $\mathbb{P}_X^{n-1}$ in the étale topology.

Theorem (5.4). There exists a canonical 1-1 correspondence between isomorphism classes of Severi-Brauer X-schemes of relative dimension n-1 over X and isomorphism classes of Azumaya $\mathcal{O}_X$-algebras of rank n^2. In particular there exists a canonical 1-1 correspondence between isomorphism classes of Severi-Brauer X-schemes of relative dimension 1 and quaternion $\mathcal{O}_X$-algebras. Each such Severi-Brauer scheme of relative dimension 1 admits an essentially unique embedding as a bundle of conics in a $\mathbb{P}^2$-bundle over X .

Proof. We have essentially the same arguments as in (5.2).

Indeed again by descent both sets described in the first part of the theorem are classified by $H^1(X, PGL(n, \mathcal{O}_X))$. See [13] pages 57-65.

To mimic the second argument of (5.2) we need the following result from [13] page 64 : $\pi : V \to X$ is a Severi-Brauer scheme if and only if π is a proper, flat morphism whose geometric fibers are isomorphic to projective spaces. So let A be an Azumaya algebra of rank n^2 defined over X . To A we want to associate a Severi-Brauer scheme V. Then the idea is to define V as a functor on X-schemes by :

$$V(X') = \{\text{left ideals } L \subset A \otimes \mathcal{O}_{X'} \text{ which are locally free of rank } n\}.$$

As for the Grassmannian functor ([15] pages 282-284), it is easy to show that V is representable by an X-scheme V which is a closed subscheme of the Grassmannian of rank n submodules of A . Then it is clear that the geometric fibers of $V \to X$ are $\mathbb{P}^{n-1}$'s, so by the above V is a Severi-Brauer scheme.

Concerning the conic bundles and quaternion Azumaya algebras we have the following : Let A be a quaternion algebra defined over X and let V be constructed as in the previous paragraph. For simplicity we let X = Spec R where R is a local ring. Then we want to write an equation for V embedded in $\mathbb{P}^2_R$. It is an easy exercise to conclude from the argument below that in the general case V admits an embedding as a conic bundle in some $\mathbb{P}^2$-bundle over X .

Now A being a quaternion algebra over R a local ring, A may be presented as the R-algebra generated by elements i,j with relations

$$i^2 = a$$

$$j^2 = b$$

$$ij = -ji$$

where a,b are units in R.

At any point $p \in X$, an element of A/mpA generates a left ideal of dimension ≥ 2. Thus any $L \subset A \otimes \mathcal{O}_{X'}$ of rank 2 will be principal for any X'. Since L is closed under left multiplication, L cannot lie in the subspace spanned by $\{1,i,j\}$. Consequently there exists a non-zero element $w = x + iy + jz \in L$ unique up to scalar multiplication. Clearly w generates a left ideal of rank 2 if and only if w, iw, jw are linearly dependent and computation (using the norm!) shows this is true if and only if

$$f(x,y,z) = x^2 - ay^2 - bz^2 = 0.$$

But $f(x,y,z)$ is homogeneous in x,y,z and hence defines V as a closed subscheme of $\mathbb{P}^2_R$. Moreover over each point $p \in X$ we get a conic $x^2 - a(p)y^2 - b(p)z^2 = 0$ contained in $\mathbb{P}^2_{k(p)}$. Q.E.D.

<u>Section 6</u>. <u>Brauer-Severi Schemes</u>.

Throughout this section S will denote a smooth irreducible complex projective surface with function field $K(S)$.

<u>Definition</u> (6.1). Let $d \in \mathrm{Br}\, K(S)$, and let the map a be as in (4.1). Let $C_1,\ldots,C_n$ be the irreducible curves on S at which $a(d)$ is not zero. Then $C = C_1 \cup \ldots \cup C_n$ is called the <u>ramification curve</u> of d.

We now study the behavior of maximal orders in central simple

K(S)-algebras over the complement of C :

Proposition (6.2). Let U = S-C, and let D be any representative of $d \in \operatorname{Br} K(S)$ (notation as in (6.1)). Then U is the maximal Zariski open subset of S over which D extends to an Azumaya algebra. Moreover the maximal orders A for D over U are precisely the Azumaya algebras extending D .

Proof. The first part of the proposition is immediate from the exact sequence of (4.1). As for the part about maximal orders, first note that if A is an Azumaya algebra extending D over U , since U is smooth, from Auslander-Goldman [6] page 387 A is automatically a maximal order in D over U .

Conversely, if A is a maximal order in D over U , for each point $c \in U$ of codimension 1, A_c is a maximal $\mathcal{O}_{U,c}$-order ([18] page 133). But since c is in the complement of the ramification curve, by definition of the map a , $A_c \otimes (\mathcal{O}_{U,c}/\mathcal{M}_{U,c})$ is similar (in the Brauer group sense) to $\mathcal{O}_{U,c}/\mathcal{M}_{U,c}$ (where $\mathcal{M}_{U,c}$ is the maximal ideal of $\mathcal{O}_{U,c}$), and this implies that A_c is an Azumaya $\mathcal{O}_{U,c}$-algebra i.e. (class of A_c) $\in \operatorname{Br}\mathcal{O}_{U,c}$. But by [6] page 389 $\operatorname{Br} U = \bigcap_{\substack{c \in U \\ c \text{ of} \\ \text{codim}=1}} \operatorname{Br}\mathcal{O}_{U,c}$. Thus A is an Azumaya algebra over U . Q.E.D.

Having described maximal orders A in D over U , we now want to understand what happens over the ramification curve C . This we will see is the key step in generalizing the results of Section 5. First we have the following :

Proposition (6.3). Let D , S be as in (6.2). Let A be a maximal order for D over all of S . Then A is a locally free $\mathcal{O}_S$-module.

Proof. Let $p \in S$ be any point.

Then we must show that A_p is a free $\mathcal{O}_{S,p}$-module. First note that $\mathcal{O}_{S,p}$ is a local regular ring of dimension ≤ 2. Moreover by Reiner [18] (page 132 Cor. (11.2) and page 133 Theorem (11.4)), A_p is a maximal $\mathcal{O}_{S,p}$-order. Moreover again by [18] page 133, A_p is a reflexive $\mathcal{O}_{S,p}$-module i.e. A_p is isomorphic to its double dual. Therefore by Auslander-Goldman [5] (Corollary to (4.7) pages 17-18), A_p is a free $\mathcal{O}_{S,p}$-module. Q.E.D.

Remarks-Notation (6.4). (i) We now restrict ourselves to the case where D is a quaternion algebra over $K(S)$ with non-singular ramification curve C. A will denote a maximal order for D over all of S.

(ii) In the situation of (i) above Artin-Mumford [4] pages 88-90 prove that A may be presented locally at a point $p \in C$ as the $\mathcal{O}_{S,p}$-algebra generated by elements i, j with relations $i^2 = a$, $j^2 = bt$, $ij = -ji$, where $t = 0$ is a local equation for C and a, b are units in $\mathcal{O}_{S,p}$. They also show that a is not congruent to a square mod t (a very important fact for (6.5) below!). Conversely, if a is not congruent to a square mod t, an algebra with such a presentation will be a maximal order in some non-trivial quaternion algebra.

We now come to a very nice result of [4] :

Theorem (6.5). Notation as in (6.4). Then there exists a canonical 1-1 correspondence between :

(i) maximal orders A in quaternion algebras D whose ramification curves are non-singular;

(ii) smooth S-schemes $\pi : V \to S$ proper and flat over S all of whose geometric fibers are isomorphic to $\mathbb{P}^1$ or $\mathbb{P}^1 \vee \mathbb{P}^1$ (two P^1's meeting transversally at 1 point) such that for every irreducible curve C' along which the fibers of π are reducible, the irreducible components of $\pi^{-1}(c')$ (c' the generic point of C') define a quadratic extension of K(C'). These quadratic extensions are those given by a(d) as in (4.1).

<u>Definition</u> (6.6). V is called the <u>Brauer-Severi</u> scheme associated to A.

<u>Proof</u> of (6.5). We just give the main idea of the proof. For details see [4] pages 90-92.

The main point is to essentially copy the last part of the proof of (5.4). Indeed given A as above, we define V as the functor of left ideals locally free of rank 2, and one can show that V is representable by an S-scheme V. From the presentation of A of (6.4) (ii) and again from the proof of (5.4), one can show that V has a local equation of the form $x^2 - ay^2 - btz^2 = 0$ from which the properties stated in (ii) immediately follow.

This construction gives a functor from algebras A with the presentation of (6.4) (ii) to S-schemes V as described in the first part of (6.5) (ii). By (6.4) (ii) A is a maximal order if and only if a is not congruent to a square mod t which by the above local equation for V is equivalent to the condition of the irreducible components of $\pi^{-1}(c')$ defining the quadratic extension.

From deformation theory (using the fact that $X_1X_2 - tX_o^2 = 0$ is a versal deformation of $\mathbb{P}^1 \vee \mathbb{P}^1$; see Artin [2]), given any $V \to S$ as described in the first part of (6.5) (ii), it is easy to show that V can be put locally (in the étale topology) in the form $X_1X_2 - tX_0^2 = 0$. But this is isomorphic to the scheme V_s associated to the standard algebra A_s having the presentation of (6.4) (ii) with a = b = 1. Moreover any algebra with a presentation as in (6.4) (ii) is locally isomorphic in the étale topology to A_s.

One then can show that the defined map of étale sheaves $\underline{\text{Aut}}\ A_s \to \underline{\text{Aut}}\ V_s$ is an isomorphism. But then by descent (see Knus-Ojanguren [16]) we have that $H^1(S, \underline{\text{Aut}}\ V_s)$ classifies isomorphism classes of S-schemes which are locally isomorphic to V_s in étale topology, and $H^1(S, \underline{\text{Aut}}\ A_s)$ classifies isomorphism classes of $\mathcal{O}_S$-algebras which are locally isomorphic to A_s. Consequently these two pointed sets are isomorphic. Moreover by our above remarks, under this isomorphism maximal orders correspond to those S-schemes V which are described in (6.5) (ii). Q.E.D.

Finally concerning Brauer-Severi schemes we have the following topological fact :

<u>Proposition</u> (6.7). Let A,D,C,S be as in (6.4) (i). Suppose that C is reducible (and therefore disconnected). Let V be the Brauer-Severi scheme associated to A . Then $H^4(V,\mathbb{Z})$ has 2-torsion.

<u>Proof</u>. Let $C = C_1 \cup C_2 \cup \ldots$ where C_1 and C_2 are disjoint non-singular curves. Let $p_i \in C_i$ be closed points (i = 1,2) and if $\pi : V \to S$, let $\pi^{-1}(p_i) = \ell_i + \ell_i'$ where ℓ_i and ℓ_i' are the two lines comprising $\pi^{-1}(p_i)$. We claim that $2(\ell_1 - \ell_2) \sim 0$ ("~" denotes "is homologous to"), i.e. that the class of $\ell_1 - \ell_2$ in $H^4(V,\mathbb{Z})$ (Poincaré duality) is 2-torsion. Indeed since by moving around a closed loop in C_i we can always interchange ℓ_i and ℓ_i' (see the local equation for V in the

proof of (6.5)), we have that $\ell_i \sim \ell_i'$. Moreover $\pi^{-1}(p_1) \sim \pi^{-1}(p_2)$. Therefore

$$2(\ell_1 - \ell_2) \sim (\ell_1 + \ell_1') - (\ell_2 + \ell_2')$$
$$\sim 0 \ .$$

Of course, one must check now that $\ell_1 - \ell_2$ is not homologous to 0. This is an exercise in the Leray spectral sequence for the map $\pi : V \to S$ which is carried out in [4] pages 92-93. Q.E.D.

Section 7. Construction of a Unirational Irrational 3-fold.

We are almost ready to construct the counter-example of Artin-Mumford. We first however need to make the following remark :

Remark (7.1). We will be using the fact that if S is a smooth rational irreducible surface, then Br $S = 0$. We sketch a proof of this fact.

First to avoid confusion the symbol H^i_{et} will be used for étale cohomology, while H^i_{cl} will be used for classical cohomology. Then from the Kummer sequence on S and the fact that $H^2_{et}(S,\mathbb{G}_m)$ is torsion, it is easy to deduce the existence of an exact sequence

$$(*) \qquad 0 \to H^1_{et}(S,\mathbb{G}_m) \otimes \mathbb{Q}/\mathbb{Z} \to H^2_{et}(S,\mu) \to H^2_{et}(S,\mathbb{G}_m) \to 0$$

where μ = roots of unity. See Grothendieck [13] pages 71 and 80.

Next since S is rational, $H^1_{cl}(S,\mathcal{O}) = H^2_{cl}(S,\mathcal{O}) = 0$. Thus from the exponential sequence (we are working over $\mathbb{C}$), $H^1_{cl}(S,\mathcal{O}^*) \cong H^2_{cl}(S,\mathbb{Z})$. But

$$H^1_{et}(S,\mathbb{G}_m) \cong H^1_{cl}(S,\mathcal{O}^*) \cong \text{Pic } S \qquad \text{(Deligne [10] page 20)}$$

and by the universal coefficient theorem and [10] page 51, we have

$$H^2_{cl}(S,\mathbb{Z}) \otimes \mu \cong H^2_{cl}(S,\mu)$$
$$\cong H^2_{et}(S,\mu) \; .$$

Since $\mu \cong \mathbb{Q}/\mathbb{Z}$, from the above computations we have $H^1_{et}(S,\mathbb{G}_m) \otimes \mathbb{Q}/\mathbb{Z} \cong H^2_{et}(S,\mu)$, which implies from the exact sequence (*) that $H^2_{et}(S,\mathbb{G}_m) = 0$. But from [13] page 76, Br S $\cong H^2_{et}(S,\mathbb{G}_m)$.

At long last, here is the example of Artin-Mumford :

Example (7.2). In $\mathbb{P}^2$, let C_1 and C_2 be two smooth cubics meeting transversally in nine distinct points. Let Q be a smooth conic which meets each C_i tangentially in three distinct points, and let q be a rational function on $\mathbb{P}^2$ with divisor (q) = Q-2L where L is the line at infinity. Then one can show (see the lemma on page 93 of [4]) that the restriction of q to C_i is not a square in $K(C_i)$ i = 1,2.

Set $L_i := K(C_i)(\sqrt{q})$. It is easy to check from the definition of q that L_i is an unramified extension of $K(C_i)$ for each i = 1,2.

Next blow up $\mathbb{P}^2$ at the points of intersection of the C_i to separate the C_i, and call the resulting surface S and the proper transforms of the C_i again C_i. Now S being rational, from (7.1) Br S = 0. Thus since L_i is an unramified extension of $K(C_i)$ for i = 1,2, by (4.2) there exists a unique class $d \in$ Br K(S) such that a(d) has class L_i in $H^1(K(C_i),\mathbb{Q}/\mathbb{Z})$ and is trivial elsewhere.

Now we claim that d is the class of a quaternion algebra D . To see this, we let $K(S') := K(S)(\sqrt{q})$, and we let S' be a non-singular projective model for K(S'). Then it is easy to see that by construction $d \otimes K(S') \in \mathrm{Br}\, K(S')$ is unramified. Moreover S' is rational; indeed, it is birational to a double covering of $\mathbb{P}^2$ with branch curve Q and such a double covering will be a quadric surface in $\mathbb{P}^3$, which is always rational. Thus Br S' = 0, and so applying (4.2) to S', we see that $d \otimes K(S')$ is trivial, i.e. d splits over K(S'). But by a classical theorem of Albert (Cohn [8] page 371), the index of d must then divide $(K(S') : K(S)) = 2$ which means that d is the class of a quaternion algebra as claimed.

Let V be the Brauer-Severi scheme associated to a maximal order in D. The generic fiber of V over S, say V_g, is clearly the Severi-Brauer variety corresponding to D. Under the correspondence of (5.2), the fact that $D \otimes K(S')$ is trivial, implies that $V_g \underset{K(S)}{\otimes} K(S') \cong \mathbb{P}^1_{K(S')}$, and this implies that $K(V)(\sqrt{q})$ is rational and so K(V) is unirational. But by (2.1) and (6.7), V is irrational.

<u>Remark</u> (7.3). Spencer Bloch has pointed out that the quaternion algebra D described above in (7.2) can be written down explicitly as follows : Let f_1, f_2 be equations for the cubic curves C_1, C_2 in $\mathbb{P}^2$ respectively. Then $f_1/f_2 \in K(\mathbb{P}^2) \cong K(S)$. One can then easily check that $D \cong \left(\frac{f_1/f_2, q}{K(S)}\right)$, i.e. D is the quaternion K(S)-algebra generated by i,j with relations $i^2 = f_1/f_2$, $j^2 = q$, $ij = -ji$.

R E F E R E N C E S

[1] M. Artin. Grothendieck topologies. Harvard Math. Dept. Lecture Notes (1962).

[2] M. Artin. Deformations of singularities. Lecture Notes of the Tata Institute of Fundamental Research, Vol. 54, Bombay (1976).

[3] M. Artin, A. Grothendieck, and J.L. Verdier. Théorie des topos et cohomologie étale des schémas. Lecture Notes in Math. 305, Springer-Verlag, Heidelberg (1973).

[4] M. Artin and D. Mumford. Some elementary examples of unirational varieties which are not rational. Proc. London Math. Soc. 25, 3rd series, pp. 75-95 (1972).

[5] M. Auslander and O. Goldman. Maximal orders. Trans. Amer. Math. Soc. 97, pp. 1-24 (1960).

[6] M. Auslander and O. Goldman. The Brauer group of a commutative ring. Trans. Amer. Math. Soc. 97, pp. 367-409 (1960).

[7] S. Bloch and A. Ogus. Gersten's conjecture and the homology of schemes. Ann. Scient. Ec. Norm. Sup., 4e série, t. 7, pp. 181-202 (1974).

[8] P.M. Cohn. Algebra, Vol. 2. John Wiley and Sons. London and New York (1977).

[9] P. Deligne. Variétés unirationnelles non rationnelles (d'après M. Artin et D. Mumford). Sém. Bourbaki, exposé 402, pp. 45-57. In Lecture Notes in Math. 317, Springer-Verlag, Heidelberg (1973).

[10] P. Deligne et al. Cohomologie étale (SGA 4 $\frac{1}{2}$). Lecture Notes in Math. 569, Springer-Verlag, Heidelberg (1977).

[11] R. Godement. Topologie algébrique et théorie des faisceaux. Hermann, Paris (1958).

[12] M.J. Greenberg. Lectures on forms in many variables. W.A. Benjamin, Inc. New York and Amsterdam (1969).

[13] A. Grothendieck. Le groupe de Brauer I, II, III. In Dix exposés sur la cohomologie des schémas, pp. 46-188. North-Holland, Amsterdam (1968).

[14] K. Gruenberg. Profinite groups. In Algebraic number theory, edited by J.W. Cassels and A. Fröhlich, pp. 116-127. Academic Press, London, and Thompson Publ. Co., Washington, D.C. (1967).

[15] S.L. Kleiman. Geometry on grassmannians and applications to splitting budles. I.H.E.S. Publ. Math. 36, pp. 281-298 (1969)

[16] M.-A. Knus and M. Ojanguren. Théorie de la descente et algèbres d'Azumaya. Lecture Notes in Math. 389, Springer Verlag, Heidelberg (1974).

[17] O.T. O'Meara. Introduction to quadratic forms. Springer, Berlin and New York (1963).

[18] I. Reiner. Maximal orders. Academic Press, London, New York, and San Francisco (1975).

[19] O.F. Schilling. The theory of valuations. Math. Surveys IV, Amer. Math. Soc., New York (1950).

[20] J.-P. Serre. Cohomologie galoisienne. Lecture Notes in Math. 5, Springer-Verlag, Heidelberg (1965).

[21] J.-P. Serre. Corps locaux. Hermann, Paris (1968).

[22] I.R. Shaferevich. Algebraic surfaces. Proc. Steklov Inst. Math. 75 (1965) (trans. by A.M.S. 1967).

[23] S. Shatz. Profinite groups, arithmetic, and geometry. Annals of Math. Studies 67, Princeton Univ. Press (1972).

SOME THEOREMS ON AZUMAYA ALGEBRAS

Ofer GABBER

CHAPTER I : A PURITY THEOREM FOR VECTOR BUNDLES AND AZUMAYA ALGEBRAS IN DIMENSION 3

§ 0. Notations and Statements of Results

Let R be a regular local ring of dimension 3 and m its maximal ideal, $u \in m - m^2$, $W = \mathrm{Spec}(R)$, $X = \mathrm{Spec}(R) - \{m\}$, $V = \mathrm{Spec}(R/uR)$, $Y = \mathrm{Spec}(R[u^{-1}])$, $Z_n = \mathrm{Spec}(R/u^nR) - \{m\}$ for $n \geq 1$, $Z = Z_1$, $\hat{X}$ = the formal scheme obtained by completing X along Z, $\hat{R}$ = the completion of R in the (u)-adic topology. If M is an R-module of finite length we denote length(M) also by lg(M). If F is a sheaf of $\mathcal{O}_X$-modules let $\hat{F} = (F|_Z) \underset{(\mathcal{O}_X|_Z)}{\otimes} \mathcal{O}_{\hat{X}}$. By a vector bundle on a locally and commutatively ringed space we mean a locally free sheaf of finite rank. $\mathbb{Z}_0$ denotes the set of non-negative integers. Those objects and notations will be fixed in this paper unless otherwise specified.

THEOREM 1 : Every vector bundle on Y is free.

THEOREM 2 : If R is henselian [5] with respect to the ideal (u) and A is an Azumaya algebra [2, I 2][1] on X and $A \underset{\mathcal{O}_X}{\otimes} \mathcal{O}_Z$ is trivial

---

(1) In this paper we follow the definition (which differs from that of [2] and [3,V4]) that an Azumaya algebra on a commutatively ringed topos $(T,\mathcal{O}_T)$ is an $\mathcal{O}_T$-algebra B (associative, with 1) such that there is a covering family U_α of the final object of T and Azumaya algebras B_α over

$$\Gamma(U_\alpha,\mathcal{O}_T) \overset{\mathrm{def}}{=} R_\alpha \text{ such that } B_\alpha \underset{R_\alpha}{\otimes} (\mathcal{O}_T|U_\alpha) \simeq B|U_\alpha .$$

(i.e., an endomorphism algebra of a vector bundle) on Z, then $A|_Y$ is isomorphic to a matrix algebra. It follows that A is trivial. (Use that $H^2(X_{ét}, \mathbb{G}_m) \to H^2(Y_{ét}, \mathbb{G}_m)$ is injective.)

Remark (not needed in this paper) : Theorem 1 can be deduced by applying Theorem 2 to endomorphism algebras. However, we will prove each theorem directly. Also, Theorem 1 and Theorem 2' (below) imply Theorem 2.

SKETCH OF THE PROOFS OF THEOREMS 1 AND 2 : (I) If G is a vector bundle on Y we can extend it to a vector bundle E on X . We associate to E the integer ℓ_E = length $[\mathrm{Cok}(H^0(X,\pi))]$ where $\pi : E \to E/uE$ is the natural projection. If $\Phi \subset E/uE$ is a direct summand then $E_1 = \pi^{-1}(\Phi)$ is a vector bundle extending G. We show (in § 3) that : (*) $\ell_{E_1} = \ell_E$ - length $[\mathrm{Cok}(H^0(Z,q))]$ where $q : E_1 \to \Phi$ is defined by π. If ℓ_E is minimal (among all extensions of G) it follows that $H^0(Z,q)$ is surjective and that similar surjectivities occur after we apply to E the process of passing from E to E_1 any number of times. We will show (in § 6) that those surjectivities imply that $\hat{E}$ is a power L^n of a line bundle L on $\hat{X}$. Using that L is generated by its sections (if $E \neq 0$), one shows L can be extended to an open set in Spec$(\hat{R})$ and hence $L \simeq \mathcal{O}_{\hat{X}}$. Hence $\Gamma(X,E) \otimes_R \hat{R} \simeq \Gamma(\hat{X},\hat{E}) \simeq \Gamma(\hat{X},\mathcal{O}_{\hat{X}}^n) = \hat{R}^n$. Hence $\Gamma(X,E) \simeq R^n$. Hence $E \simeq \mathcal{O}_X^n$, so $G \simeq \mathcal{O}_Y^n$. (II) If A is an Azumaya algebra on X one shows that $\hat{A} \simeq \underline{\mathrm{End}}(E)$ for some vector bundle E on $\hat{X}$. To E we apply the process of (I). We show (in § 5) (**) : $\ell_{A_1} = \ell_A - 2n \cdot$ length$[\mathrm{Cok}(H^0(Z,q))]$ where n = degA and A_1 is the Azumaya algebra on X with compatible isomorphisms $A_1|_Y \xrightarrow{\sim} A|_Y$ and $\hat{A}_1 \xrightarrow{\sim} \underline{\mathrm{End}}(E_1)$. Hence as in (I) if ℓ_A is minimal among all A with isomorphic $A|_Y$ we get $E \simeq L^n$ so $\hat{A} \simeq M_n(\mathcal{O}_{\hat{X}})$, so $\Gamma(A) \otimes_R \hat{R} \simeq \Gamma(\hat{A}) \simeq M_n(\hat{R}) \simeq M_n(R) \otimes_R \hat{R}$. One shows (using the henselianity of (R,uR)) that $\Gamma(A) \simeq M_n(R)$. Hence $A \simeq M_n(\mathcal{O}_X)$. We remark that § 4 is not vital for proving the theorems. Also much of § 5 can be avoided at the cost of having instead of (**) only the inequality $\ell_{A_1} \leq \ell_A - n\cdot\text{length}[\text{Cok}(H^0(Z,q))]$.

<u>THEOREM 2'</u> : If X is a regular scheme (or algebraic space) of dimension ≤ 3 and $Z \subset |X|$ is a closed subset of codimension ≥ 2, then $H^i_Z(X_{\text{ét}},G_m) = 0$ if $0 \leq i \leq 3$. Hence (by the long exact derived functor sequence (see [1] V 6.5.3) obtained from the functor sequence $H^0_Z(X,F) \to H^0(X,F) \to H^0(X-Z,F)$), $H^2_{\text{ét}}(X,G_m) \xrightarrow{\sim} H^2_{\text{ét}}(X-Z,G_m)$.

<u>Proof that Theorem 2 implies Theorem 2'</u> : We refer to ([1] Exposé V 6) for basic facts on local étale cohomology (local cohomology in the case of topological spaces is discussed in [4]). We have (by [1] V 6.4.3) a local to global spectral sequence for étale cohomology $H^p(Z,\underline{H}^q{}_Z(X,G_m)) \to H^{p+q}_Z(X,G_m)$; using it we reduce to show $\underline{H}^i{}_Z(X,G_m) = 0$ for $0 \leq i \leq 3$. The case $i < 3$ is easy (and does not use the dimension restriction on X) ([2], III.6, formulas 6.3, 6.4, 6.5). The case $i = 3$ is known if $\dim X \leq 2$ ([2], III, Theorem 6.1(b)). Suppose ξ is a geometric point of X, $\mathcal{O}_\xi$ the stalk of the étale structure sheaf of X at ξ, $X' = \text{Spec}(\mathcal{O}_\xi)$, $j : X' \to X$ the natural morphism, $Z' = j^{-1}(Z)$, $U = X' - \{\text{the closed point } Q\}$. By ([1] VII 5.2, V 6.5) the stalk of $\underline{H}^i{}_Z(X,G_m)$ at ξ is isomorphic to $H^i_{Z'}(X',G_m)$. We have to show those are zero. We can assume $i = 3$ and $\dim\mathcal{O}_\xi = 3$ and $Z' \neq \emptyset$. Now $H^3_{Z'}(X',G_m) \xleftarrow{\sim} H^2(X'-Z',G_m)$. As $X'-Q$ is regular of dimension 2, Theorem 2' for $\dim X \leq 2$ gives $H^2(X'-Q,G_m) \xrightarrow{\sim} H^2(X'-Z',G_m)$. If $c \in H^2(X'-Q,G_m)$ then by ([2] II, Cor. 2.2[1]) c comes by the

---

(1) The proof given in [2] (page 77) is incomplete in the "one dimensional" step.

coboundary of ([3], IV 4.2, V 4) from an Azumaya algebra A on $X' - Q$. Now choose $u \in m - m^2$ where m is the maximal ideal of $\mathcal{O}_\xi$ and let $V = \mathrm{Spec}(\mathcal{O}_\xi / u\mathcal{O}_\xi)$. $A \otimes_{\mathcal{O}_U} \mathcal{O}_{V-Q}$ is trivial on $V-\Omega$ ([2] III, proof of Theorem 6.1 (b)). Using Theorem 2 $A_{|U-V}$ is trivial. Hence $c_{|U-V} = 0$. But the map $H^2(U,G_m) \to H^2(U-V,G_m)$ is injective because by ([2] II Cor. 1.8) both sides inject into $H^2(\mathrm{fract}(\mathcal{O}_\xi), G_m)$. So $c = 0$. Q.E.D.

§ 1. Some Lemmas on Vector Bundles

We collect here some facts which are used in this paper. We first recall the following :

Lemma 1 : Every vector bundle E on an open set Ω_0 in a regular scheme X can be extended to an open set Ω whose complement is of codimension ≥ 3. In particular every vector bundle on Z is free. (This last fact is used to show the fact used in § 0 that R/uR strictly henselian $\Rightarrow$ Br(Z) = 0.)

Proof : By ([9] IV 1.7.7, I 9.4.8) E has a coherent extension F to X. Let $H \overset{\mathrm{def}}{=} F^{**} = \underline{\mathrm{Hom}}_{\mathcal{O}_X}(\underline{\mathrm{Hom}}_{\mathcal{O}_X}(F,\mathcal{O}_X),\mathcal{O}_X)$. We use the general fact that for a coherent sheaf F (on a space with a coherent structure sheaf) $\underline{\mathrm{Hom}}(F,\mathcal{O}_X)$ is coherent and its stalks are $\mathrm{Hom}(F_x,\mathcal{O}_{X,x})$. Define $\Omega = \{x \in X | H_x$ is a free $\mathcal{O}_{X,x}$-module$\}$. Ω is open. $H_{|\Omega_0} \simeq E$ implies that $\Omega \supset \Omega_0$. If $x \in X$ and $R \overset{\mathrm{def}}{=} \mathcal{O}_{X,x}$ we can choose an exact sequence of R-modules $R^p \overset{\alpha}{\to} R^q \overset{\beta}{\to} (F^*)_x \to 0$. By dualizing we get the exact sequence $0 \to H_x \xrightarrow{\beta^t} R^q \xrightarrow{\alpha^t} R^p \to \mathrm{Cok}(\alpha^t) \to 0$. If $\dim R \leq 2$ then by ([8] III 5.1 or [9] 0_{IV} 17.3.1) $\mathrm{proj.\,dim}_R(\mathrm{Cok}(\alpha^t)) \leq 2$. Hence H_x is projective, hence is free, hence $x \in \Omega$. $H_{|\Omega}$ is the required extension.

More general results than the following two lemmas are proved in SGA2 VIII, IX.

Lemma 2 : If E is a vector bundle on X (or on Z), then $\Gamma(X,E)$ is finitely generated over R. The map $\hat{R} \otimes_R \Gamma(X,E) \to \Gamma(\hat{X},\hat{E})$ is an isomorphism.

Proof : One can choose a finitely generated R-module M such that if $\tilde{M} = M \otimes_R \mathcal{O}_W$ where $W = \mathrm{Spec}(R)$ then $\tilde{M}_{|X} \simeq E^*$. Then if $R^q \xrightarrow{g} R^p \xrightarrow{\varepsilon} M \to 0$ is a resolution of M we get an exact sequence $0 \to E \xrightarrow{\tilde{\varepsilon}^t} \mathcal{O}_X^p \xrightarrow{\tilde{g}^t} \mathcal{O}_X^q$. Hence $\Gamma(X,E) = \mathrm{Ker}(\Gamma(X,\mathcal{O}_X)^p \to \Gamma(X,\mathcal{O}_X)^q)$ and $\Gamma(\hat{X},\hat{E}) = \mathrm{Ker}(\Gamma(\hat{X},\mathcal{O}_{\hat{X}})^p \to \Gamma(\hat{X},\mathcal{O}_{\hat{X}})^q)$. As depth $(R) = 3 \geq 2$ we get[(1)] $\Gamma(X,\mathcal{O}_X) = R$. Hence $\Gamma(X,E)$ is finitely generated and a similar argument works on Z. We have $\mathcal{O}_{\hat{X}} = \varprojlim_n \mathcal{O}_X/u^n\mathcal{O}_X$. As depth $(R/u^nR) = 2$ we have[(2)] $\Gamma(X,\mathcal{O}_X/u^n\mathcal{O}_X) \xrightarrow{\sim} R/u^nR$. Hence $\Gamma(\hat{X},\mathcal{O}_{\hat{X}}) \xrightarrow{\sim} \varprojlim_n R/u^nR = \hat{R}$. The flatness of $\hat{R}$ over R proves then the second statement, q.e.d.

Lemma 3 : If R is complete with respect to (u) then the functor $E \to \hat{E}$ from the category of vector bundles on X to the category of vector bundles on $\hat{X}$ is fully faithful and its essential image consists of those vector bundles on $\hat{X}$ which are generated by their sections.

Proof : If E and F are vector bundles on X we apply Lemma 2 to the homomorphism bundle $\underline{\mathrm{Hom}}_{\mathcal{O}_X}(E,F)$ to get $\mathrm{Hom}_{\mathcal{O}_X}(E,F) \xrightarrow{\sim} \mathrm{Hom}_{\mathcal{O}_{\hat{X}}}(\hat{E},\hat{F})$. Now suppose E is a vector bundle on $\hat{X}$ generated by $M = \Gamma(\hat{X},E)$. M is separated in the (u)-adic topology and $M/uM \subset \Gamma(Z,E/uE)$ which is finitely generated. Hence[(3)] M is finitely generated over $R = \hat{R}$. Let

---

(1) E.g., by the local cohomology interpretation of depth ([4] V 3.3, especially V 3.4 (1) $\Leftrightarrow$ (2)).

(2) See the previous footnote.

(3) Using Prop. 10.24 in M.F. Atiyah and I.G. Macdonald, Introduction to Commutative Algebra, Addison-Wesley Pub. Co., 1969.

$F = M \otimes_R \mathcal{O}_{\hat{X}}$. F is coherent and $F \xrightarrow{h} E$ is an epimorphism. $h \otimes_{\mathcal{O}_{\hat{X}}} \mathcal{O}_Z$ is the map $(M/uM) \otimes_{R/uR} \mathcal{O}_Z \to E \otimes_{\mathcal{O}_{\hat{X}}} \mathcal{O}_Z = E/uE$. As $M/uM \subset \Gamma(Z, E/uE)$, $h \otimes_{\mathcal{O}_{\hat{X}}} \mathcal{O}_Z$ is a monomorphism hence an isomorphism. Hence $\mathrm{Ker}(h) \subset uF \cap \mathrm{Ker}(h) = u\,\mathrm{Ker}(h)$, as E is u-torsion free. By Nakayama $\mathrm{Ker}(h) = 0$ and so h is an isomorphism. Let $E_0 = M \otimes_R \mathcal{O}_X$. Then $\hat{E}_0 \simeq E$. In particular E_0 is locally free in a neighborhood Ω of Z in X. We restrict E_0 to Ω and then extend it by Lemma 1 to a vector bundle on X with the required property. (A posteriori, by Lemma 2, E_0 is locally free on X.)

Lemma 4 : A line bundle L on $\hat{X}$ which is generated by its sections is trivial.

Proof : By Lemma 3 L comes (after replacing R by $\hat{R}$) from a line bundle L_0 on X. But $\mathrm{Pic}\, X = 0$ by the proof of [8] VII 3.14 or [9] IV 21.11.1.

Lemma 5 : Let S be a formal scheme with subschemes of definition (we follow the definitions of [9] I 10.4, 10.5) $S_n \subset S$, $n \geq 1$, s.t. $S_n \subset S_{n+1}$ and $\mathcal{O}_S \xrightarrow{\sim} \varprojlim_{n \geq 1} \mathcal{O}_{S_n}$ topologically. Suppose E_n is a vector bundle on S_n for every n and we are given an isomorphism $t_n : E_{n+1} \otimes_{\mathcal{O}_{S_{n+1}}} \mathcal{O}_{S_n} \xrightarrow{\sim} E_n$ for every n. Then $E = \varprojlim_{\text{via } t's} E_n$ is a vector bundle on S and $\forall_{n \geq 1}$ the map $E \otimes_{\mathcal{O}_S} \mathcal{O}_{S_n} \to E_n$ is an isomorphism.

Proof : This is a local problem on $|S|$. Thus we can assume that we have a trivialization $E_1 \xrightarrow{\phi_1} \mathcal{O}^m_{S_1}$ and S_1 is affine. As $\Gamma(E_{i+1}) \to \Gamma(E_i)$ we can lift $\psi_1 \overset{\text{def}}{=} \phi_1^{-1}$ successively to maps $\mathcal{O}^m_{S_n} \xrightarrow[\psi_n]{} E_n$ which are necessarily isomorphism. Thus the system $(E_n, t_n)_{n \geq 1}$ is isomorphic to the system $(\mathcal{O}^m_{S_n}, \text{can.})_{n \geq 1}$ for which the assertions are clear.

Lemma 6 : The map $\mathrm{Pic}(\hat{X}) \to \mathrm{Pic}(Z_n)$ is surjective for every $n \geq 1$.

Proof : We use the long exact cohomology sequence of the sheaf sequence $0 \to \mathcal{O}_Z \overset{\eta}{\to} \mathcal{O}^*_{Z_{m+1}} \to \mathcal{O}^*_{Z_m} \to 0$ (where $\eta(f \bmod (u)) = 1 + u^m f$ for a local section f of $\mathcal{O}_{Z_{m+1}}$) and that $H^2(Z,\mathcal{O}_Z) = 0$ because dim.Krull$(Z) < 2$ to obtain that $\mathrm{Pic}(Z_{m+1}) \to \mathrm{Pic}(Z_m)$ is onto. Now given a line bundle L_n on Z_n we find by induction on $m \geq n$ line bundles L_m on Z_m and isomorphisms $t_m : L_{m+1} \otimes_{\mathcal{O}_{Z_{m+1}}} \mathcal{O}_{Z_m} \overset{\sim}{\to} L_m$. By Lemma 5 $L = \varprojlim_{\text{via } t\text{'s}} L_m$ is a line bundle on $\hat{X}$ with $L \otimes_{\mathcal{O}_{\hat{X}}} \mathcal{O}_{Z_n} \simeq L_n$.

Lemma 7 : Suppose that A is an Azumaya algebra on $\hat{X}$, E_1 is a vector bundle on Z, and $A \otimes_{\mathcal{O}_{\hat{X}}} \mathcal{O}_Z \xrightarrow[\psi_1]{\sim} \underline{\mathrm{End}}_{\mathcal{O}_Z}(E_1)$ is an isomorphism of algebras. Then there is a vector bundle E on $\hat{X}$ and isomorphisms $A \underset{\psi}{\overset{\sim}{\to}} \underline{\mathrm{End}}(E)$ and $E \otimes_{\mathcal{O}_{\hat{X}}} \mathcal{O}_{Z_1} \underset{t}{\overset{\sim}{\to}} E_1$ such that $\psi_1 = \underline{\mathrm{End}}(t) \circ (\psi \otimes_{\mathcal{O}_{\hat{X}}} \mathcal{O}_Z)$.

Proof : Define $A_n = A \otimes_{\mathcal{O}_{\hat{X}}} \mathcal{O}_{Z_n}$. We first want to find vector bundles E_n on Z_n and isomorphisms $A_n \xrightarrow[\psi_n]{\sim} \underline{\mathrm{End}}(E_n)$ and $E_{n+1} \otimes_{\mathcal{O}_{Z_{n+1}}} \mathcal{O}_{Z_n} \xrightarrow[t_n]{\sim} E_n$ for all $n \geq 1$ such that they agree with the E_1 and ψ_1 given above and such that $\psi_n = \underline{\mathrm{End}}(t_n) \circ (\psi_{n+1} \otimes_{\mathcal{O}_{Z_{n+1}}} \mathcal{O}_{Z_n})$ for all n. Suppose E_i and ψ_i are already constructed for $i \leq n$ and t_i for $i < n, n \geq 1$. We recall (see [3] V 4.2) that one has a G_m-gerbe Λ_{A_n}[(1)] on the étale site of Z_n whose objects are triples (h,E,ψ) where $h : \Omega \to Z_n$ is étale, E is a vector bundle on Ω, and $h^*A_n \xrightarrow[\psi]{\sim} \underline{\mathrm{End}}_{\mathcal{O}_\Omega}(E)$. Let $i : Z_n \to Z_{n+1}$ be the natural morphism. Then one has a morphism $T : \Lambda_{A_{n+1}} \longrightarrow i_{\text{ét}*}\Lambda_{A_n}$ of gerbes such that $T(h,E,\psi) = (h \times_{Z_{n+1}} Z_n, E \otimes_{\mathcal{O}_\Omega} \mathcal{O}_{i^{-1}(\Omega)}, \psi \otimes_{\mathcal{O}_\Omega} \mathcal{O}_{i^{-1}(\Omega)})$. By ([1] VIII 1.1) the morphism of étale sites $i_{\text{ét}}$ (= $i^*_{\text{ét}}$ in [1] VII 1.4) defined by i is an equivalence of sites. Hence $i^{\mathrm{ch}}_{\text{ét}*}$ (defined in [3] II 3.1.5) carries gerbes (defined in [3] III 2.1.1) to gerbes. T induces the map represented by $G_{m_{Z_{n+1}}} \to i_* G_{m_{Z_n}}$ on liens, which is surjective.

---

(1) Λ_{A_n} is $d(A_n)$ in [3] (approximately).

By ([3] IV 2.5) we have a gerbe L of liftings of the given $\sigma = (id_{Z_n}, E_n, \psi_n) \in \Lambda_{A_n}(Z_n) = (i_{\text{ét}*}\Lambda_{A_n})(Z_{n+1})$ with respect to T. The lien of L is the lien defined by the sheaf of groups $\underline{\text{Ker}}(\mathbb{G}m_{Z_{n+1}} \to \mathbb{G}m_{Z_n}) \underset{\eta}{\simeq} \mathcal{O}_Z$. As $H^2_{\text{ét}}(Z_{n+1}, \mathcal{O}_Z) = 0$ and using [3] IV 3.4, L has a cartesian section τ. τ gives a "lifting" of σ to $(id_{Z_{n+1}}, E_{n+1}, \psi_{n+1}) \in \Lambda_{A_{n+1}}(Z_{n+1})$ up to an isomorphism given by the required t_n. Thus all the E_i, ψ_i, t_i can be constructed by induction. Define $E = \varprojlim E_n$, $\psi = \varprojlim \psi_n$, t the natural projection $E \otimes_{\mathcal{O}_X} \mathcal{O}_{Z_1} \to E_1$. By Lemma 5, E is a vector bundle and t is an isomorphism. Note also that $\varprojlim \underline{\text{End}}(E_n) \xrightarrow{\sim} \underline{\text{End}}(E)$ as $E_n \xrightarrow{\sim} E \otimes_{\mathcal{O}_{\hat{X}}} \mathcal{O}_{Z_n}$.

§ 2. A Diagram Lemma

Lemma 8 : Suppose

$$\begin{array}{ccccccc} 0 & \to & A & \xrightarrow{f} & B & \xrightarrow{g} & C \\ & & {\scriptstyle h}\uparrow & & {\scriptstyle k}\uparrow & \nearrow{\scriptstyle n} & \\ & & A' & \xrightarrow[m]{} & B' & & \end{array}$$

is a commutative diagram with exact rows in an abelian category. Then if the maps in the first row of the following diagram(*) are defined by the requirement that (*) will commute, then the first row of (*) is exact.

$$\begin{array}{lccccccccc} (*) : 0 \to & \text{Cok}(h) & \xrightarrow{\alpha} & \text{Cok}(k) & \to & \text{Cok}(n) & \to & \text{Cok}(g) & \to & 0 \\ & {\scriptstyle p}\uparrow & & {\scriptstyle q}\uparrow & & \uparrow & & \uparrow & & \\ & A & \xrightarrow{f} & B & \xrightarrow{g} & C & \xrightarrow{id} & C & & \\ & {\scriptstyle h}\uparrow & & \uparrow{\scriptstyle k} & & \uparrow{\scriptstyle n} & & \uparrow{\scriptstyle g} & & \\ & A' & \xrightarrow{m} & B' & \xrightarrow{id} & B' & \xrightarrow{k} & B & & \end{array}$$

Proof : Straightforward diagram chasing after reducing to the case of abelian groups. Let us check for example that α is injective. Suppose $\alpha x = 0$. Choose y s.t. $x = py$. Then $0 = \alpha x = \alpha py = qfy$. Hence $fy = kz$ for some $z \in B'$. $nz = gkz = gfy = 0$. Hence $z = mw$ for some $w \in A'$. So $fy = kz = kmw = fhw$. As $\mathrm{Ker}(f) = 0$ we get $y = hw$. Hence $x = py = phw = 0$.

§ 3. The ℓ_E-Index

In this section we obtain parallel results for vector bundles on X and on $\hat{X}$ respectively.

Definition 1 : If E is a vector bundle on X (respectively on $\hat{X}$) define the R-module $L_E = \mathrm{Cok}(H^0(X,E) \to H^0(X,E/uE))$ (respectively replace X by $\hat{X}$). Define $\ell_E = \lg(L_E)$ if $\lg(L_E) < \infty$, $\ell_E = +\infty$ if L_E is not of finite length.

Lemma 9 :

(a) L_E is finitely generated.

(b) In the case of vector bundles on X, L_E is of finite length.

(c) If E is a vector bundle on X, then $L_E \tilde{\to} L_{\hat{E}}$ and $\ell_E = \ell_{\hat{E}}$.

(d) If $R = \hat{R}$ and E is a vector bundle on $\hat{X}$, then $\ell_E < \infty$ iff $E \simeq \hat{E}_0$ for some vector bundle E_0 on X. A posteriori, using Theorem 1 and Lemma 11, the assumption $R = \hat{R}$ can be dropped.

Proof :

(a) $H^0(X,E/uE)$ is finitely generated by Lemma 2.

(b) Use (a) and that the localization of L_E at primes $P \neq m$ is 0 by a flat base change property by $\mathrm{Spec}(R_p) \to \mathrm{Spec}(R)$. See ([9] III 1.4.15 and 1.4.10). The statement there is corrected in [9] (ERR_{III}, 25) and is strengthened in [9] IV 1.7.21.

(c) Use Lemma 2.

(d) If $E = \hat{E}_0$ for E_0 a v.b. on X then by (c) and (b) $\ell_E = \ell_{E_0} < \infty$. Conversely if L_E is of finite length then $L_E \underset{R}{\otimes} \mathcal{O}_Z = 0$, so considering the maps $H^0(E) \underset{R}{\otimes} \mathcal{O}_Z \underset{\alpha}{\to} H^0(E/uE) \underset{R}{\otimes} \mathcal{O}_Z \tilde{\to} E/uE$, α and hence $\beta\alpha$ are epimorphisms. By Nakayama E is generated by its sections. By Lemma 3 E is "algebraizable".

Lemma 10 :

(a) E on X (respectively on $\hat{X}$) is free iff $L_E = 0$.

(b) A vector bundle E on X is free iff $\hat{E}$ is free.

Proof : (b) follows from (a) and 9(c).

(a) Necessity : If E is free then $E \simeq \mathcal{O}_X^n$, $n = \mathrm{rk}E$. Then $L_E \simeq \mathrm{Cok}(\Gamma(X,\mathcal{O}_X)^n \to \Gamma(Z,\mathcal{O}_Z)^n) \simeq \mathrm{Cok}(R^n \to (R/uR)^n) = 0$. A similar proof works when X is replaced by $\hat{X}$ and R by $\hat{R}$.

(a) Sufficiency : Suppose $L_E = 0$. As $E \underset{\mathcal{O}_X}{\otimes} \mathcal{O}_Z$ is free we can choose a basis $(e_i)_{1 \le i \le n}$ of $E \underset{\mathcal{O}_X}{\otimes} \mathcal{O}_Z$. Lift e_i to $\bar{e}_i \in \Gamma(X,E)$ (use $L_E = 0$). The collection $(\bar{e}_i)_{1 \le i \le n}$ defines a morphism $\phi : \mathcal{O}_X^n \to E$. The stalk ϕ_x of ϕ at each $x \in Z$ is an isomorphism. (When E is on $\hat{X}$ an analogous statement suffices to complete the proof.) If U is the maximal open set on which ϕ is an isomorphism then $U \supset Z$. It is enough to prove that $X = U$. Suppose on the contrary that $X \ne U$. Let $S = X - U$ and $\bar{S}$ = the closure of S in Spec(R). Then the locus of u on S is {m}, hence $\dim \bar{S} \le 1$. On the other hand S is defined locally on X by one equation, hence $\mathrm{codim}\, \bar{S} = 1$. This is a contradiction.

<u>Definition 2</u> : A modification of vector bundles on X (respectively on $\hat{X}$) is a triple (E,F,ϕ) where E and F are vector bundles on X (respectively on $\hat{X}$) and ϕ is an $\mathcal{O}_X$- (respectively $\mathcal{O}_{\hat{X}}$-) module isomorphism $F[u^{-1}] \overset{\phi}{\underset{\sim}{\to}} E[u^{-1}]$. The pair (F,ϕ) is called then a modification of E. The modification (F,ϕ) is called simple iff $uE \subset \phi(F) \subset E$.

<u>Lemma 11</u> : If E is a vector bundle on X then the map $(F,\phi) \mapsto (\hat{F}, \phi \otimes_{\mathcal{O}_X} \mathcal{O}_{\hat{X}})$ induces a bijection between the isomorphism classes of modifications of E and the isomorphism classes of modifications of $\hat{E}$. A similar statement holds for simple modifications.

<u>Proof</u> : We exhibit an inverse map as follows : Let (H,ψ) be a modification of $\hat{E}$. Take $N \geq 0$ s.t. $u^N\hat{E} \subset \psi(H) \subset u^{-N}\hat{E}$. $\psi(H)/u^N\hat{E}$ is a coherent $\mathcal{O}_{Z_{2N}}$-submodule of $u^{-N}\hat{E}/u^N\hat{E}$. As $u^{-N}\hat{E}/u^N\hat{E} \underset{\lambda}{\overset{\sim}{\leftarrow}} u^{-N}E/u^NE$ we can define F to be the subsheaf of $u^{-N}E$ such that $F \supset u^NE$ and $\lambda(F/u^NE) = \psi(H)/u^N\hat{E}$. F is a coherent $\mathcal{O}_X$-module. It is a vector bundle because $F_x \tilde{\to} E_x$ for $x \in X - Z$ and $F_x \otimes_{\mathcal{O}_{X,x}} \mathcal{O}_{\hat{X},x} \tilde{\to} H_x$ [while $\mathcal{O}_{X,x} \to \mathcal{O}_{\hat{X},x}$ is faithfully flat] for $x \in Z$ (we apply [7] VIII 1.11). We define $F[u^{-1}] \underset{\phi}{\tilde{\to}} E[u^{-1}]$ using that $u^NE \subset F \subset u^{-N}E$. $(H,\psi) \mapsto (F,\phi)$ is an inverse map.

<u>Remark</u> : If (F,ψ) is a modification of E then F^* forms a modification of E^*, $F \otimes H$ forms a modification of $E \otimes H$ for any vector bundle H, etc. We omit the definitions of such modifications in the future.

<u>Lemma 12</u> : If E is a vector bundle on X (respectively on $\hat{X}$) and F is a coherent $\mathcal{O}_X$-(respectively $\mathcal{O}_{\hat{X}}$-) module such that $uE \subset F \subset E$, then :
(a) F/uE is a locally free $\mathcal{O}_Z$-module. (b) E/F is a locally free $\mathcal{O}_Z$-module iff F is a locally free $\mathcal{O}_X$- (respectively $\mathcal{O}_{\hat{X}}$-) module.

<u>Proof</u> : (a) Z is regular of dimension 1 and $F/uE \subset E/uE$ is a torsion free coherent sheaf on Z.

(b) Necessity : Suppose E/F is locally free over $\mathcal{O}_Z$. In a neighborhood of each point $x \in Z$ we can choose a base $e_1,\dots,e_n$ of E. After renumbering, $e_1,\dots,e_p$ project to a basis of E/F over $\mathcal{O}_Z$ for some p. After adding to each e_j for $j > p$ a local section of $\sum_{i=1}^{p} \mathcal{O}_X e_i$ we can assume $e_j \in F$ $\forall j > p$. Then $ue_1,\dots,ue_p,e_{p+1},\dots,e_n$ form a basis of F near x.

(b) Sufficiency : Suppose that F is locally free, then at each $x \in Z$ we have $\text{proj.dim}_{\mathcal{O}_{X,x}}(E_x/F_x) \le 1$. For a regular local ring S we have ([8] III 5.19) the formula depth (M) + $\text{p.d.}_S(M)$ = dim S for any non-zero finitely generated S-module M. If $E_x/F_x \neq 0$ we apply it to E_x/F_x over the two rings $\mathcal{O}_{X,x}$ and $\mathcal{O}_{Z,x}$ and obtain that $\text{p.d.}_{\mathcal{O}_{Z,x}}(E_x/F_x) = \text{p.d.}_{\mathcal{O}_{X,x}}(E_x/F_x) - 1 \le 0$. Hence E_x/F_x is free over $\mathcal{O}_{Z,x}$. A reversal of the order in this proof proves the necessity again.

Definition 3 : If (E,F,ϕ) is a simple modification on X or on $\hat{X}$ and $j : F \rightarrowtail E$ is induced from ϕ we define

$$L_1(j) = \text{Cok}(H^0(E/uE) \to H^0(E/F))$$

$$L_2(j) = \text{Cok}(H^0(F/uF) \to H^0(F/uE)).$$

Remark : By Lemma 12 E/F and F/uE are locally free $\mathcal{O}_Z$-modules. Then as in Lemma 9(b) one deduces that $L_1(j)$ and $L_2(j)$ are R-modules of finite length. We also note that $L_\alpha(j) \xrightarrow{\sim} L_\alpha(j \otimes_{\mathcal{O}_{\hat{X}}} \mathcal{O}_X)$ $(\alpha = 1,2)$ if j is on X.

Lemma 13 : Under the notations of Definition 3, $\ell_E < \infty$ if and only if $\ell_F < \infty$ and if $\ell_E < \infty$ then

$$\ell_F = \ell_E + \lg(L_1(j)) - \lg(L_2(j)) .$$

<u>Proof</u> : We consider the diagrams [of $\mathcal{O}_X$-(or $\mathcal{O}_{\hat{X}}$-) modules] with exact rows :

(a)
$$\begin{array}{ccccccc} 0 & \rightarrow & F/uE & \rightarrow & E/uE & \rightarrow & E/F \\ & & \uparrow & & \uparrow & \nearrow & \\ 0 & \rightarrow & F & \longrightarrow & E & & \end{array}$$

(b)
$$\begin{array}{ccccccc} 0 & \rightarrow & uE/uF & \rightarrow & F/uF & \rightarrow & F/uE \\ & & \uparrow & & \uparrow & \nearrow & \\ 0 & \rightarrow & uE & \longrightarrow & F & & \end{array}$$

We apply the left exact functor H^0 into the category of R-modules. The resulting diagrams of R-modules satisfy the hypothesis of Lemma 8. Applying Lemma 8 we obtain the following exact sequences respectively :

(1) $0 \rightarrow K_1 \rightarrow L_E \rightarrow K_2 \rightarrow L_1(j) \rightarrow 0$

(2) $0 \rightarrow K_2 \rightarrow L_F \rightarrow K_1 \rightarrow L_2(j) \rightarrow 0$

where $K_1 = \mathrm{Cok}(H^0(F) \rightarrow H^0(F/uE))$

$K_2 = \mathrm{Cok}(H^0(E) \rightarrow H^0(E/F))$.

As $L_2(j)$, $L_1(j)$ are of finite length (1) implies $\lg(L_E) < \infty \Leftrightarrow \lg(K_2)$ and $\lg(K_1)$ are finite, and (2) implies $\lg(L_F) < \infty \Leftrightarrow \lg(K_1)$ and $\lg(K_2)$ are finite. This proves the first statement. Now suppose $\ell_E < \infty$. Then all the modules in the exact sequences (1) and (2) are of finite length. The alternating sums $\lg(L_E) + \lg(L_1(j)) - \lg(K_1) - \lg(K_2)$ and $\lg(L_F) + \lg(L_2(j)) - \lg(K_2) - \lg(K_1)$ of lengths in the exact sequences (1) and (2) are thus zero. Thus $\ell_F + \lg(L_2(j)) = \lg(K_1) + \lg(K_2) = \ell_E + \lg(L_1(j))$.

§ 4. The ℓ_E^F-Index

In this section we construct for modifications (E,F,ϕ) on $\hat{X}$ an integer which generalizes $\ell_F - \ell_E$ if both L_E and L_F are of finite length.

Lemma 14 : There is a unique way to assign an integer ℓ_E^F to every isomorphism class of modifications (E,F,ϕ) on $\hat{X}$ such that the following properties hold : (a) If (E,F,ϕ) is a simple modification then [in the notation of Definition 3] $\ell_E^F = \lg(L_1(j)) - \lg(L_2(j))$. (b) If (E,F,ϕ) and (F,G,ψ) are modifications, then $\ell_E^G = \ell_F^G + \ell_E^F$. Moreover, those integers satisfy the following properties : (c) $\ell_F < \infty$ iff $\ell_E < \infty$ and in this case we have $\ell_E^F = \ell_F - \ell_E$. In particular (by Lemma 9(c) if (E,F,ϕ) is a modification on X then $\ell_{\hat{E}_0}^{\hat{F}_0} = \ell_{F_0} - \ell_{E_0}$. (d) $\ell_E^{u^{N}E} = 0$ for every integer N.

Proof : We first make the following definitions :

Definition 4 : If $F \subset E$ "is" a modification on $\hat{X}$ (or on X) then by an admissible filtration from E to F we mean a chain $E = \Phi^0 \supset \Phi^1 \supset \dots \supset \Phi^p = F$, $p \geq 0$, of locally free coherent $\mathcal{O}_{\hat{X}}$-(or $\mathcal{O}_X$-) modules such that $u\Phi^i \subset \Phi^{i+1}$ for $0 \leq i < p$.

Example : Suppose $F \subset E$ are vector bundles on $\hat{X}$ such that $u^pE \subset F$ for some $p \geq 0$. For $0 \leq i \leq p$ define $\Phi^i = E \cap u^{i-p}F$. Clearly Φ^i is coherent, $\Phi^0 = E$, $\Phi^p = F$, and $u\Phi^i \subset \Phi^{i+1}$ for $0 \leq i < p$. We show that Φ^i is locally free by induction on i. This is true for $i = 0$. Suppose $0 < i \leq p$ and that Φ^{i-1} is locally free. Now $\Phi^{i-1}/\Phi^i = (E \cap u^{i-1-p}F)/(u^{i-p}F \cap E) \subset u^{i-1-p}F/u^{i-p}F \simeq F/uF$. Thus Φ^{i-1}/Φ^i is a torsion free and hence locally free $\mathcal{O}_Z$-module. By Lemma 12(b) Φ^i is a locally free $\mathcal{O}_{\hat{X}}$-module. This example shows that for any modification (E,F,ϕ) such that $\phi(F) \subset E$

there is an admissible filtration from E to $\phi(F)$.

Definition 5 : If $\Phi^0 \supset \Phi^1 \supset \dots \supset \Phi^p$ is an admissible filtration and if $i_j : \Phi^j \rightarrowtail \Phi^{j-1}$ is the natural inclusion for $0 < j \leq p$, we define

$$\ell(\Phi^0,\Phi^1,\dots,\Phi^p) = \sum_{j=1}^{p} [\lg(L_1(i_j)) - \lg(L_2(i_j))] \ .$$

Proof of the Uniqueness of ℓ_E^F and Properties (c),(d) :

Suppose (E,F,ϕ) is a modification on $\hat{X}$ such that $\phi(F) \subset E$. We can choose an admissible filtration $(\Phi^0,\dots,\Phi^p)$ from E to $\phi(F)$. By property (b) $\ell_E^F = \sum_{i=1}^{p} \ell_{\Phi^{i-1}}^{\Phi^i} \overset{(a)}{=} \ell(\Phi^0,\dots,\Phi^p)$. This proves the uniqueness in this case. Property (d) for $N \geq 0$ follows by looking at the admissible filtration $E \supset uE \supset \dots \supset u^N E$. For $N < 0$ one can use the rule $\ell_E^F = -\ell_F^E$ which follows from (b) to reduce to $N > 0$. Once we have (d) and an arbitrary modification (E, F, ϕ), then picking $N \geq 0$ s.t. $\phi(u^N F) \subset E$ we get $\ell_E^F \overset{(b)}{=} \ell_{u^N F}^{F} + \ell_E^{u^N F} \overset{(d)}{=} \ell_E^{u^N F}$. This proves the uniqueness. To show (c) we can assume that $F \subset E$. Then we choose an admissible filtration from E to F and apply Lemma 13 to each $\Phi^j \subset \Phi^{j-1}$ modification. Then add the resulting equations and use (a) and (b).

Proof of the Existence : We first give the following :

Sublemma 15 : $\ell(\Phi^0,\dots,\Phi^p)$ depends only on the inclusion $\Phi^p \rightarrowtail \Phi^0$.

Lemma 15 $\Rightarrow$ the Existence : If $\phi(F) \subset E$ we can define, unambiguously by Lemma 15, $\ell_E^F = \ell(\Phi^0,\dots,\Phi^p)$ for all admissible filtrations from E to $\phi(F)$. Property (b) (transitivity) is obvious when $\psi(G) \subset F$ and $\phi(F) \subset E$. (a) holds by definition. We extend the definition to an arbitrary modification (E,F,ϕ) by putting $\ell_E^F = \ell_E^{u^N F}$, if $N \geq 0$ is such that $\phi(u^N F) \subset E$. We have to show independence from N. This is done by first proving as before (d) for $N \geq 0$, and then if $N' \geq N$ and $u^N F \subset E$ we get $\ell_E^{u^{N'}F} \overset{(\underline{b})}{=} \ell_{u^N F}^{u^{N'}F} + \ell_E^{u^N F} \overset{(\underline{d})}{=} \ell_E^{u^N F}$. (b) in general follows from the definitions and the case before.

<u>Proof of Lemma 15</u> : We prove it by induction on $\lg_{\mathcal{O}_{\hat{X},\xi}}(\Phi_\xi^0/\Phi_\xi^p) = m$. It is obvious when $m = 0$ (by Lemma 12(b) $\Phi_\xi^0 = \Phi_\xi^p \Rightarrow \Phi^0 = \Phi^p$). Assume that $m > 0$, and that $\Phi^0 \supset \Phi^1 \supset \ldots \supset \Phi^p$ and $\Psi^0 \supset \Psi^1 \supset \ldots \supset \Psi^q$ are both admissible filtrations for $F \subset E$. It can be assumed that $\Phi^0 \neq \Phi^1$ and $\Psi^0 \neq \Psi^1$. Define $\Phi'^1 = \Phi^1 \cap \Psi^1$. Φ'^1 is locally free by Lemma 12(b) [$uE \subset \Phi'^1$ and $E/\Phi'^1 \subset E/\Phi^1 \oplus E/\Psi^1$ so E/Φ'^1 is a torsion free $\mathcal{O}_Z$-module]. Find an admissible filtration $\Phi'^1 \supset \Phi'^2 \supset \ldots \supset \Phi'^{p'} = F$. By the induction hypothesis $\ell(\Phi^1,\ldots,\Phi^p) = \ell(\Phi^1,\Phi'^1,\ldots,\Phi'^{p'}) = \ell(\Phi^1,\Phi'^1) + \ell(\Phi'^1,\ldots,\Phi'^{p'})$ and $\ell(\Psi^1,\ldots,\Psi^q) = \ell(\Psi^1,\Phi'^1,\ldots,\Phi'^{p'}) = \ell(\Psi^1,\Phi'^1) + \ell(\Phi'^1,\ldots,\Phi'^{p'})$. Define $M = \ell(\Phi'^1,\ldots,\Phi'^{p'})$. Then $\ell(\Phi^0,\Phi^1,\ldots,\Phi^p) = \ell(\Phi^0,\Phi^1) + \ell(\Phi^1,\Phi'^1) + M$ and $\ell(\Psi^0,\Psi^1,\ldots,\Psi^q) = \ell(\Psi^0,\Psi^1) + \ell(\Psi^1,\Phi'^1) + M$. Thus to prove the lemma it is enough to show the equalities : $\ell(\Phi^0,\Phi^1) + \ell(\Phi^1,\Phi'^1) = \ell(\Phi^0,\Phi'^1) = \ell(\Phi^0,\Psi^1) + \ell(\Psi^1,\Phi'^1)$. Up to changes of notation, we have to show the following :

<u>Sublemma 16</u> : Let $E \overset{i}{\supset} F \overset{j}{\supset} G \supset uE$ be an admissible filtration. Then $\ell(E,G) = \ell(E,F) + \ell(F,G)$.

<u>Proof</u> : We consider the following diagrams [of $\mathcal{O}_{\hat{X}}$-modules] with exact rows :

(a)
$$\begin{array}{ccccccc} 0 & \to & F/G & \to & E/G & \to & E/F \\ & & \uparrow & & \uparrow & \nearrow & \\ 0 & \to & F/uE & \to & E/uE & & \end{array}$$

(b)
$$\begin{array}{ccccccc} 0 & \to & uE/uF & \to & G/uF & \to & G/uE \\ & & \uparrow & & \uparrow & \nearrow & \\ 0 & \to & uE/uG & \to & G/uG & & \end{array}$$

(c) $$\begin{array}{ccccccc} 0 & \to & G/uE & \to & F/uE & \to & F/G \\ & & \uparrow & & \uparrow & \nearrow & \\ 0 & \to & G/uF & \to & F/uF & & \end{array}$$

We then apply the left exact functor H^0 into the category of $\hat{R}$-modules. We then apply Lemma 8 to the resulting diagrams. We obtain the following exact sequences of $\hat{R}$-modules respectively :

(1) $0 \to N_1 \to L_1(ij) \to L_1(i) \to N_2 \to 0$

(2) $0 \to N_2 \to L_2(j) \to L_2(ij) \to N_3 \to 0$

(3) $0 \to N_3 \to L_2(i) \to L_1(j) \to N_1 \to 0$

where $N_1 = \mathrm{Cok}(H^0(F/uE) \to H^0(F/G))$

$N_2 = \mathrm{Cok}(H^0(E/G) \to H^0(E/F))$

$N_3 = \mathrm{Cok}(H^0(G/uF) \to H^0(G/uE))$.

All modules in the exact sequences (1), (2), (3) are of finite length. The alternating sum of the lengths is zero in each of (1), (2), (3). If we add those equations the length (N_i)'s cancel and we get the required conclusion.

§ 5. Behaviour under ⊗ and *

In this section we prove the following :

Lemma 17 : Let (F,ϕ) be a modification of E on $\hat{X}$ and G any vector bundle on $\hat{X}$. Then

(1) $\ell_{E\otimes G}^{F\otimes G} = \mathrm{rk}G \cdot \ell_E^F$

(2) $\ell_{E^*}^{F^*} = \ell_E^F$.

Proof : Following the proof of Lemma 14 uniqueness we can, using property (b) only, reduce to the case $uE \subset F \overset{j}{\subset} E$. We assume from now on we are in this case. We note that $F \otimes G$ and E^* "are" simple modifications of $E \otimes G$ and F^* respectively.

Proof of (1) : We recall the definitions :

$$\ell_{E\otimes G}^{F\otimes G} = lg(L_1(j \otimes id_G)) - lg(L_2(j \otimes id_G))$$

$$\ell_E^F = lg(L_1(j)) - lg(L_2(j)) .$$

As $G/uG \simeq \mathcal{O}_Z^{rkG}$ we get $L_1(j \otimes id_G) \simeq Cok(\Gamma(Z,(E/uE) \underset{\mathcal{O}_{\hat{X}}}{\otimes} G) \to \Gamma(Z,(E/F) \underset{\mathcal{O}_{\hat{X}}}{\otimes} G)) \simeq Cok(\Gamma(Z,(E/uE)^{rkG}) \to \Gamma(Z,(E/F)^{rkG})) \simeq L_1(j)^{rkG}$. Similarly for L_2. Taking lengths and subtracting we get the result.

Proof of (2) : Using Lemma 14(b) $\ell_{E^*}^{F^*} = -\ell_{F^*}^{E^*} = -lg(L_1(j^*)) + lg(L_2(j^*))$. Hence in order to get the required equation it is enough to prove the following :

(+) : $lg(L_\alpha(j^*)) = lg(L_{3-\alpha}(j))$ for $\alpha = 1,2$. To prove (+) we recall (see [4] IV 3.1 (iv), Defn. 4.1, 4.7(c), 5.4) that if $W = Spec(R/uR)$ then $J \overset{defn.}{=} H^2_{\{m\}}(W,\mathcal{O}_W) \simeq H^1(Z,\mathcal{O}_Z)$ is an injective hull of the residue field of R/uR, and that if M is an R/uR-module of finite length then $Hom(M,J)$ is of finite length equal to length(M). Thus to prove (+) it is enough to prove the following :

($\neq$) : $L_1(j^*) \simeq Hom_{R/uR}(L_2(j),J)$, $L_1(j) \simeq Hom_{R/uR}(L_2(j^*),J)$.

Proof of ($\neq$) : The second statement in ($\neq$) follows from the first by applying the latter to j^* instead of j. We now prove the first statement. Using the short exact sequence of $\mathcal{O}_Z$-vector bundles

(I) : $0 \to uE/uF \to F/uF \to F/uE \to 0$, we obtain that

$$(*) : L_2(j) = \mathrm{Cok}(\Gamma(Z,F/uF) \to \Gamma(Z,F/uE)) \simeq \mathrm{Im}(\Gamma(Z,F/uE) \underset{\delta_I}{\to} H^1(Z,E/F)).$$

Similarly, using the exact sequence

(II): $0 \to E^*/uF^* \to F^*/uF^* \to F^*/E^* \to 0$, we obtain

$$(**): L_1(j^*) = \mathrm{Cok}(\Gamma(Z,F^*/uF^*) \to \Gamma(Z,F^*/E^*)) \to \mathrm{Im}(H^0(Z,F^*/E^*) \underset{\delta_{II}}{\to} H^1(Z,E^*/uF^*).$$

We now prove

Sublemma 18 : The sequence (II) is isomorphic to the sequence obtained from (I) by applying the functor $F \mapsto \underline{\mathrm{Hom}}_{\mathcal{O}_Z}(F,\mathcal{O}_Z)$.

Proof of Lemma 18 : We define the following pairings (on local sections) :

λ_1 : $F^*/uF^* \otimes F/uF \to \mathcal{O}_Z$ by $(f^* \bmod uF^*) \otimes (f \bmod uF) \mapsto f^*(f) \bmod (u)$

λ_2 : $F^*/E^* \otimes E/F \to \mathcal{O}_Z$ by $(f^* \bmod E^*) \otimes (e \bmod F) \mapsto f^*(ue) \bmod (u)$

λ_3 : $E^*/uF^* \otimes F/uE \to \mathcal{O}_Z$ by $(e^* \bmod uF^*) \otimes (f \bmod uE) \mapsto e^*(f) \bmod (u)$.

One checks that those are well defined and that the termwise map(II) $\to \underline{\mathrm{Hom}}_{\mathcal{O}_Z}((I),\mathcal{O}_Z)$ they induce commutes with the arrows of those sequences. One has to check that for each $1 \le i \le 3$ the map μ_i, defined using λ_i, from a term in (II) to the dual of a term in (I), is an isomorphism. This is easy for μ_1. The assertion for μ_2 can be obtained from the assertion for μ_3 by applying the latter to (F,uE) instead of (E,F). We now consider the stalk $\mu_{3,x} : E^*_x/uF^*_x \to \underline{\mathrm{Hom}}_{\mathcal{O}_{Z,x}}(F_x/uE_x,\mathcal{O}_{Z,x})$ of μ_3 at $x \in Z$.

Injectivity : If $e^* \in E^*_x$ is such that $\mu_{3_x}(e^* \bmod uF^*_x) = 0$, then

$e^*(F_x) \subset u\mathcal{O}_{\hat{X},x}$ so $u^{-1}e^*$ gives a homomorphism $f^* : F_x \to \mathcal{O}_{\hat{X},x}$. Thus, $e^*|_{F_x} = uf^* \in uF_x^*$.

Surjectivity : Given $\phi : F_x/uE_x \to \mathcal{O}_{Z,x}$ consider the diagram (of $\mathcal{O}_{\hat{X},x}$-modules) :

$$\begin{array}{ccc} F_x/uE_x & \to & \mathcal{O}_{Z,x} \\ \uparrow & & \uparrow \\ F_x & \overset{\bar{\phi}}{\cdots\to} & \mathcal{O}_{\hat{X},x} \end{array}$$

The dotted arrow $\bar{\phi}$ can be fitted [such that the square commutes] because F_x is projective. $\bar{\phi}(uE_x) \subset u\mathcal{O}_{\hat{X},x}$ by looking at the induced map of the kernels of the vertical arrows. So we can define $\bar{\bar{\phi}} \in E_x^*$ by $\bar{\bar{\phi}}(e) = u^{-1}\bar{\phi}(ue)$. $\bar{\phi} = \bar{\bar{\phi}}|_{F_x}$. So $\mu_{3,x}(\bar{\bar{\phi}} \bmod uF_x^*) = \phi$. Hence $\mu_{3,x}$ is surjective as well.

Now using Lemma 18 and (*) and (**) and a change of notation in sequence (I) we see that in order to complete the proof of ($\neq$) it is enough to prove the following :

Sublemma 19 : Let S be a two-dimensional regular local ring. Let $W = \mathrm{Spec}(S)$, m the maximal ideal of S, $Z = \mathrm{spec}(S) - \{m\}$, $J = H^1(Z,\mathcal{O}_Z)$, $(\Delta) : 0 \to E' \underset{j}{\to} E \underset{\pi}{\to} E'' \to 0$ an exact sequence of locally free coherent $\mathcal{O}_Z$-modules, $(\Delta^*) : 0 \to E''^* \underset{\pi^*}{\to} E^* \underset{j^*}{\to} E'^* \to 0$ the dual sequence, $\delta : H^0(E'') \to H^1(E')$ and $\delta^* : H^0(E'^*) \to H^1(E''^*)$ be the connecting homomorphisms obtained from (Δ) and (Δ^*) respectively. Then, $\mathrm{Hom}_S(\mathrm{Im}(\delta),J) \simeq \mathrm{Im}(\delta^*)$ as S-modules.

Proof : We first show that it is enough to prove Lemma 19 for the pull back of the sequence Δ to $\mathrm{Spec}(\hat{S}) - \{mS\} \overset{\mathrm{defn}}{=\!=} Z'$. We recall (see note to (b) in the proof of Lemma 9 in § 3) that [as $S \to \hat{S}$ is flat] if $p : Z' \to Z$ is the natural morphism and F is a quasi coherent sheaf on Z then $H^i(Z,F) \underset{S}{\otimes} \hat{S} \xrightarrow{\sim} H^i(Z'p^*F)$. If δ' and δ'^* are the connecting homomorphisms from H^0 to H^1 for $p^*(\Delta)$ and $p^*(\Delta^*)$, we get S-module isomorphisms

(1) $J' \overset{\text{defn}}{=\!=} H^1(Z', \mathcal{O}_{Z'}) \tilde{\leftarrow} J \underset{S}{\otimes} \hat{S} \tilde{\leftarrow} J$;

(2) $\mathrm{Im}(\delta') \underset{\alpha}{\tilde{\leftarrow}} \mathrm{Im}(\delta) \underset{S}{\otimes} \hat{S} \tilde{\leftarrow} \mathrm{Im}(\delta)$;

(3) $\mathrm{Im}(\delta'^*) \tilde{\leftarrow} \mathrm{Im}(\delta^*) \underset{S}{\otimes} \hat{S} \tilde{\leftarrow} \mathrm{Im}(\delta^*)$.

Here we use that $\underset{S}{\otimes}\hat{S}$ commutes with images, and that if M is an m-torsion S-module then the map $M \to M \underset{S}{\otimes} \hat{S}$ is an isomorphism. We note also that

(4) $\mathrm{Hom}_{\hat{S}}(\mathrm{Im}(\delta'), J') \overset{(2)\alpha}{\tilde{\to}} \mathrm{Hom}_S(\mathrm{Im}(\delta), J') \overset{(1)}{\tilde{\leftarrow}} \mathrm{Hom}_S(\mathrm{Im}(\delta), J)$ [as S-modules].

(4) and (3) achieve the desired reduction to the case that S is complete.

Proof of Lemma 19 when $S = \hat{S}$: We consider the pairings by cup-product

$B_1 : H^0(E'') \otimes H^1(E''^*) \to H^1(\mathcal{O}_Z) = J$;

$B_2 : H^1(E') \otimes H^0(E'^*) \underset{g}{\to} H^1(\mathcal{O}_Z) = J$.

B_1 gives a map $H^1(E''^*) \underset{g}{\to} \mathrm{Hom}_S(H^0(E''), J))$;

B_2 gives a map $H^0(E'^*) \underset{h}{\to} \mathrm{Hom}_S(H^1(E'), J))$.

As E" is free (or by local duality [4] V 2.1) one checks that g is an isomorphism. We recall that for a finitely generated module M over a complete local noetherian ring the map $M \overset{\kappa_M}{\to} \mathrm{Hom}\,(\mathrm{Hom}(M,J),J)$ (where J is an injective hull of the residue field of the ring) is an isomorphism. (See [4] Expose IV Prop. 5.1 (more or less).) Thus the isomorphism Hom(g,J) takes the form $\mathrm{Hom}(g,J)\circ\kappa : H^0(E'') \to \mathrm{Hom}_S(H^1(E''^*), J)$. One checks that it is the one induced by B_1. Replacing E" by E' we obtain that h is an isomorphism. Now we consider the following diagram :

$$(*)\qquad \begin{array}{ccc} \mathrm{Hom}_S(H^1(E'),J) & \xrightarrow[\mathrm{Hom}(\delta,J)]{} & \mathrm{Hom}_S(H^0(E''),J) \\ \wr\Big\uparrow h & & g\Big\uparrow\wr \\ H^0(E'^*) & \xrightarrow[\delta^*]{} & H^1(E''^*) \end{array} .$$

Applying the exact functor $\mathrm{Hom}(?,J)$ to the exact sequences $H^0(E'') \to \mathrm{Im}(\delta) \to 0$ and $0 \to \mathrm{Im}(\delta) \to H^1(E'')$ we obtain that $\mathrm{Im}(\mathrm{Hom}_S(\delta,J)) \simeq \mathrm{Hom}_S(\mathrm{Im}(\delta),J)$. Hence (by comparing the images of the horizontal arrows of (*)) Lemma 19 will be proven if we show the following :

<u>Sublemma 20</u> : Under the assumptions of Lemma 19, the diagram (*) is anti-commutative.

<u>Proof</u> : We have to show the following identity : If $x \in H^0(E'^*)$ and $y \in H^0(E'')$ then $B_1(y \otimes \delta^* x) = -B_2(\delta y \otimes x)$. This can be derived from general principles on derived categories. We give instead a direct proof. Choose an open cover $\{U_\alpha\}$ of Z where x lifts to $\bar{x}_\alpha \in \Gamma(U_\alpha, E^*)$ and y lifts to $\bar{y}_\alpha \in H^0(U_\alpha, E)$. $\delta^*(x)$ is given by the 1-cocycle $\pi^{*-1}(\bar{x}_\beta - \bar{x}_\alpha)$. δy is given by the 1-cocycle $j^{-1}(\bar{y}_\beta - \bar{y}_\alpha)$. $B_1(y \otimes \delta^* x)$ is given by the 1-cocycle of $\mathcal{O}_Z$ $(\pi^{*-1}[\bar{x}_\beta - \bar{x}_\alpha])(y) = (\bar{x}_\beta - \bar{x}_\alpha)(\bar{y}_\alpha)$. $B_2(\delta y \otimes x)$ is given by the 1-cocycle of $\mathcal{O}_Z$ $x(j^{-1}(\bar{y}_\beta - \bar{y}_\alpha)) = \bar{x}_\beta(\bar{y}_\beta - \bar{y}_\alpha)$. Hence the sum is given by the 1-cocycle $\bar{x}_\beta(\bar{y}_\beta) - \bar{x}_\alpha(\bar{y}_\alpha)$ which is a coboundary and thus gives the 0 cohomology class.

<u>Corollary of Lemma 17</u> : If (E,F,ϕ) is a modification on $\hat{X}$ then $\ell_{\underline{\mathrm{End}}(F)/\underline{\mathrm{End}}(E)} = 2 \cdot \mathrm{rk}E \cdot \ell_E^F$.

Proof : Use the isomorphism $\underline{\mathrm{End}}(E) \simeq E \otimes E^*$ and Lemma 17 repeatedly :

$$\ell_{E\otimes E^*}^{F\otimes F^*} \overset{14(b)}{=\!=} \ell_{E\otimes F^*}^{F\otimes F^*} + \ell_{E\otimes E^*}^{E\otimes F^*} \overset{17(1)}{=\!=}$$

$$= (\mathrm{rk}F) \cdot \ell_E^F + (\mathrm{rk}E) \cdot \ell_{E^*}^{F^*} \overset{17(2)}{=\!=} (\mathrm{rk}E) \cdot \ell_E^F + (\mathrm{rk}E) \cdot \ell_E^F .$$

§ 6. Minimal Bundles

In this section which continues § 4 we study the vector bundles on $\hat{X}$ which are "minimal" in the sense of the following definition :

Definition 6 : A vector bundle E on $\hat{X}$ is called minimal if and only if for every modification (F,ϕ) of it we have $\ell_E^F \geq 0$.

Lemma 21 : E is minimal iff $E \simeq L^n$ for some line bundle L on $\hat{X}$, where $n = \mathrm{rk}E$.

Proof : We first give the following :

Claim : If E is free then E is minimal.

Proof of the Claim : By Lemma 10 $L_E = 0$. Hence $\ell_E = 0$. By Lemma 14(c) if (F,ϕ) is a modification of E then $\ell_F < \infty$ and $\ell_E^F = \ell_F - \ell_E = \ell_F \geq 0$. So E is minimal.

Sublemma 22 : If E is minimal and L is a line bundle then $E\otimes L$ is minimal.

Proof : Suppose (F,ψ) is a modification of $E\otimes L$. Then $F\otimes L^{-1}$ "is" a modification of E. By definition of minimality $\ell_E^{F\otimes L^{-1}} \geq 0$. By Lemma 17(1) $\ell_{E\otimes L}^F = \ell_E^{F\otimes L^{-1}} \geq 0$. Hence $E\otimes L$ is minimal.

<u>Proof of the Sufficiency in Lemma 21</u> : Combine the claim and Lemma 22.

<u>Proof of the Necessity</u> : We first give the following definition :

<u>Definition 7</u> : A diagonalization of a vector bundle E of rank $n \in H^0(W, Z_0)$ on a locally ringed space W is a pair (L, ψ) where L is a line bundle on W and ψ is an $\mathcal{O}_W$-isomorphism $E \underset{\sim}{\overset{\psi}{\to}} L^n$. If E is a vector bundle on $\hat{X}$ we define $E_m = E \otimes_{\mathcal{O}_{\hat{X}}} \mathcal{O}_{Z_m}$ [regarded as a vector bundle over Z_m]. We now state for each $m \geq 1$:

<u>Lemma $21'_m$</u> : If E is a minimal bundle on $\hat{X}$ then every diagonalization of E_m can be lifted to a diagonalization of E_{m+1}.

<u>Proof that Lemma $21'_m$ for all $m \geq 1 \Rightarrow$ the Necessity</u> : We are given a minimal bundle E on $\hat{X}$ of rank n. E_1 is free over $\mathcal{O}_Z$. In particular it is diagonalizable. Choose a diagonalization (L_1, ψ_1) of E_1. Applying Lemma $21'_m$ in succession we can find a system $\{(L_k, \psi_k)\}_{k \geq 1}$ of diagonalizations of E_k, such that for every $k \geq 1$ we are given an isomorphism $L_{k+1} \otimes_{\mathcal{O}_{Z_{k+1}}} \mathcal{O}_{Z_k} \underset{t_k}{\tilde{\to}} L_k$ such that the diagram

$$\begin{array}{ccc}
E_{k+1} & \xrightarrow{\psi_{k+1}} & L^n_{k+1} \\
\Big\downarrow \text{can.} & & \Big\downarrow t^n_k \circ (\mathrm{id} \otimes 1)^n \\
E_k & \xrightarrow{\psi_k} & L^n_k
\end{array}$$

commutes. Note that $E \tilde{\to} \varprojlim E_k$. Define $L = \varprojlim L_k$ and $\psi = \varprojlim \psi_k$: $E \tilde{\to} L^n$. By Lemma 5 L is a line bundle on $\hat{X}$. This proves the necessity.

<u>Proof of Lemma $21'_m$</u> : We consider the following statement : <u>$21''_m$</u> : Lemma $21'_m$ is true when the given diagonalization (L_m, ψ_m) of E_m is such that $L_m = \mathcal{O}_{Z_m}$.

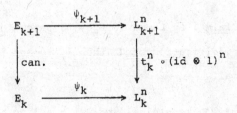

<u>Proof that $21'_m \Leftrightarrow 21''_m$</u> : $\Rightarrow$ is clear. $\Leftarrow$: Suppose we are given a minimal bundle E with a diagonalization $(L^`,\psi_m)$ of $E_m$. By Lemma 6 there exists a line bundle L on $\hat{X}$ together with an isomorphism $L \otimes_{\mathcal{O}_{\hat{X}}} \mathcal{O}_{Z_m} \overset{\sim}{\underset{\lambda}{\leftarrow}} L^`$. By Lemma 22 $E\otimes L^{-1}$ is minimal. $c = (\mathcal{O}_{Z_m}, \lambda^n \psi_m \otimes id_{L_m^{-1}})$ is a diagonalization of $E_m \otimes L_m^{-1}$. By $21''_m$ we can lift c to a diagonalization c' of $E_{m+1} \otimes L_{m+1}^{-1}$. "$c' \otimes_{\mathcal{O}_{Z_{m+1}}} id_{L_{m+1}}$" is a lifting of $(L^`,\psi_m)$ to E_{m+1}.

<u>Proof of $21''_m$</u> : We begin with some preliminaries. If $(S,\mathcal{O}_S)$ is a ringed site and $n \in H^0(S,\mathbb{Z}_0)$ let ${}_nV_{\mathcal{O}_S}$ be "the" gerbe on S whose objects are the pairs (Ω,E) where $\Omega \in S$ and E is a locally free sheaf of rank n on S/Ω. Let σ_0 be "the" cartesian section of ${}_nV_{\mathcal{O}}$ such that $\sigma_0(\Omega) = (\Omega,\mathcal{O}^n|_{S/\Omega})$ on objects. $\underline{\text{Aut}}(\sigma_0) \simeq GL_n(\mathcal{O}_S^{op})$. (<u>Aut</u> is defined in ([3] I 3.5.2) and denoted there by Aut in bold face type). Suppose $\mathcal{O}' \overset{p}{\to} \mathcal{O}$ is a morphism of sheaves of rings s.t. $GL_n(p)$ is onto (e.g., when p is onto and either <u>Ker</u>(p) is nilpotent or $\mathcal{O}'$ is local in the sense of [1] IV 13.9. In the "correct" definition of "locally ringed topos" one has to add to the definition of ([1] IV 13.9) that the locus of 0 = 1 is empty). We get a cartesian morphism $T_p : {}_nV_{\mathcal{O}'} \to {}_nV_{\mathcal{O}}$ such that $T(\Omega,E) = (\Omega, E \otimes_{\mathcal{O}'|S/\Omega} \mathcal{O}|_{S/\Omega})$. We have an obvious isomorphism $\zeta_p : T\circ\sigma'_0 \to \sigma_0$. The map on automorphism sheaves $\underline{\text{Aut}}(\sigma'_0) \to \underline{\text{Aut}}(T\circ\sigma'_0)$ induced by T is essentially $GL_n(p^{op}) : GL_n(\mathcal{O}'^{op}) \twoheadrightarrow GL_n(\mathcal{O}^{op})$. Let ${}_nG_p = \underline{\text{Ker}}(GL_n(p^{op}))$. Let ${}_nR_p$ be the <u>gerbe</u> of liftings ([3] IV 2.5) of σ_0 with respect to T. Thus the objects of ${}_nR_p$ are pairs (x,ϕ) where $x \in \mathcal{O}b({}_nV_{\mathcal{O}'})$ and $\phi : T(x) \overset{\sim}{\to} \sigma_0(\Omega)$ where $x = (\Omega,?)$. The functor $\Omega \mapsto (\sigma'_0(\Omega),\zeta_\Omega)$ gives a cartesian section, which we call ${}_n\tau_p$, of ${}_nR_p$. $\underline{\text{Aut}}({}_n\tau_p) \simeq {}_nG_p$ canonically. We now recall for reference the following (cf. [3, III 2.5.3, 3.6.5(5)]) :

Lemma 23 : If Λ is a gerbe on S and $\sigma_0 \in \mathrm{CART}_S(S,\Lambda)$ (= the category of cartesian sections of Λ, which we will also denote by $\mathrm{CART}(\Lambda)$), then we have a bijection {Isom. classes in $\mathrm{CART}(S,\Lambda)$} $\to \check{H}^1(S,\underline{\mathrm{Aut}}(\sigma_0))$. We denote the image of a section σ by $\mathrm{cl}_{\sigma_0}(\sigma)$. If $T : \Lambda' \longrightarrow \Lambda$ is a cartesian functor and $\zeta : T \circ \sigma_0' \stackrel{\sim}{\to} \sigma_0$ is given where $\sigma_0 \in \mathrm{CART}(S,\Lambda)$, $\sigma_0' \in \mathrm{CART}(S,\Lambda')$ then for every $\sigma \in \mathrm{CART}(S,\Lambda')$ $\check{H}^1(S,h)\mathrm{cl}_{\sigma_0'}(\sigma) = \mathrm{cl}_{\sigma_0}(T\circ\sigma)$ where h is the composed homomorphism $\underline{\mathrm{Aut}}(\sigma_0') \xrightarrow[\text{by } T]{} \underline{\mathrm{Aut}}(T\circ\sigma_0') \xrightarrow[\alpha\to\zeta\alpha\zeta^{-1}]{} \underline{\mathrm{Aut}}(\sigma_0)$.

We now use Lemma 23 to reformulate $21_m''$. S will always be the Zariski site of Z. Consider $\mathcal{O}_{\hat{X}|Z} \xrightarrow{q} \mathcal{O}_{Z_{m+1}} \xrightarrow{p} \mathcal{O}_{Z_m}$. Before that note that [in the general setting before Lemma 23] $\mathrm{CART}({}_nR_p)$ is equivalent to the category of pairs (E,ϕ) where E is vector bundle of rank n over $\mathcal{O}'$ and $\phi : E \otimes_{\mathcal{O}'} \mathcal{O} \longrightarrow \mathcal{O}^n$ (by the functor which sends such pair to its restrictions to each S/Ω). (We agree to omit this functor in what follows.) In our case q induces the morphisms ${}_nR_{pq} \xrightarrow{\alpha} {}_nR_p$ and ${}_nG_{pq} \xrightarrow{\beta_m} {}_nG_p$.[1] For (E,ϕ)" $\in$ "$\mathrm{CART}({}_nR_p)$ $\check{H}^1(\beta)(\mathrm{cl}_\tau(E,\phi)) =$ $\mathrm{cl}_\tau(E \otimes_{\mathcal{O}_{\hat{X}}} \mathcal{O}_{Z_{m+1}},\phi)$ by Lemma 23. We also have the morphism ${}_1R_p \xrightarrow{\gamma} {}_nR_p$ sending $((\Omega,L),\phi)$ to $((\Omega,L^n),\phi^n)$. γ induces the diagonal map ${}_1G_p \xrightarrow{\Delta} {}_nG_p$ on $\underline{\mathrm{Aut}}(\tau)$'s. Hence for $(L,\phi) \in \mathrm{CART}({}_1R_p)$ $\mathrm{cl}(L^n,\phi^n) = \check{H}^1(\Delta)\mathrm{cl}(L,\phi)$. Now, given the diagonalization $(\mathcal{O}_{Z_m},\phi)$ of E_m as in $21_m''$, one checks that it can be lifted to a diagonalization of E_{m+1} iff $(E_{m+1},\phi) \in \mathrm{CART}({}_nR_p)$ is isomorphic to the image by γ of some $(L_{m+1},\psi) \in \mathrm{CART}({}_1R_p)$. Passing to the isomorphism classes, we see that $21_m''$ is equivalent to :

$\overline{21}_m$: If $(E,\phi) \in \mathrm{CART}({}_nR_{pq})$ with E minimal, then $\mathrm{cl}(E_{m+1},\phi) =$ $\check{H}^1(\beta)(\mathrm{cl}(E,\phi)) \in \mathrm{Im}(\check{H}^1(\Delta))$.

---

(1) $\alpha(x,\phi) = (T_q(x),\phi\circ\xi_x)$ where ξ is the canonical isomorphism $T_p\circ T_q \stackrel{\sim}{\to}_{\xi} T_{pq}$. α and $\alpha\circ{}_n\tau_{pq} \stackrel{\sim}{\to}_{\zeta_q} {}_n\tau_p$ induce $\beta_m = GL_n(q)\big|_{{}_nG_{pq}}$ on $\underline{\mathrm{Aut}}(\tau)$'s.

Proof of $\overline{21}_m$: We observe that ${}_nG_p \overset{\text{def}}{=\!=} \underline{\text{Ker}}(GL_n(\mathcal{O}_{Z_{m+1}}) \to GL_n(\mathcal{O}_{Z_m}))$ is isomorphic to $M_n(\mathcal{O}_Z)$ by the map $\eta_{n,m} = \eta_n : M_n(\mathcal{O}_Z) \to {}_nG_p$ defined by $\eta(A) = I_n + u^m\tilde{A}$ where $\tilde{A}$ is a lifting of A to $M_n(\mathcal{O}_{Z_{m+1}})$. The map $\eta_n^{-1}\Delta\eta_1 : \mathcal{O}_Z \to M_n(\mathcal{O}_Z)$ is the diagonal map Δ'. Let $p_{i,j}$ be the projection of $M_n(\mathcal{O}_Z)$ to the (i,j) entry if $1 \le i,\ j \le n$. Via the isomorphism $H^1(Z,M_n(\mathcal{O}_Z)) \tilde{\to} M_n(H^1(Z,\mathcal{O}_Z))$, $\text{Im}(H^1(\Delta'))$ consists of the scalar matrices. Hence $\overline{21}_m$ is equivalent to the following statement :

$\overline{\overline{21}}_m$: If $(E,\phi) \in \text{CART}({}_nR_{pq})$ with E minimal, then :

(a) $\check{H}^1(p_{i,j} \circ \eta_n^{-1} \circ \beta)(\text{cl}(E,\phi)) = 0$ if $i \ne j$, $1 \le i,\ j \le n$.

(b) $\check{H}^1(p_{i,i} \circ \eta_n^{-1} \circ \beta)(\text{cl}(E,\phi)) = \check{H}^1(p_{j,j} \circ \eta_n^{-1} \circ \beta)(\text{cl}(E,\phi))$, if $1 \le i,\ j \le n$.

Proof that $\overline{\overline{21}}_m$(a) [for all (E,ϕ)...] implies $\overline{\overline{21}}_m$: To prove $\overline{\overline{21}}_m$(b) for a given (E,ϕ), we apply $\overline{\overline{21}}_m$(a) (which we call below $(a)_m$) to $(E,M\phi)$ where M runs through $GL_n(R/u^mR)$. Let $T_M : {}_nR_{pq} \to {}_nR_{pq}$ be the morphism such that $T_M((\Omega,F),\phi) = ((\Omega,F),M\phi)$. If $\bar{M} \in GL_n(\hat{R})$ lifts M we can define $\zeta_{\bar{M}} : T_M \circ {}_n\tau_{pq} \tilde{\to} {}_n\tau_{pq}$ by letting $\bar{M}$ act naturally on the vector bundles $\mathcal{O}^n_{\hat{X}|\Omega}$ involved in ${}_n\tau_{pq}$. $(T_M, \zeta_{\bar{M}})$ induce a homomorphism $\underline{\text{Aut}}({}_n\tau_{pq}) \to \underline{\text{Aut}}({}_n\tau_{pq})$, i.e., ${}_nG_{pq} \to {}_nG_{pq}$. This is just conjugation by $\bar{M}$, denoted below by $\underline{\text{Int}}(\bar{M})$. By Lemma 23 we get the identity $\text{cl}_\tau(E,M\phi) = \check{H}^1(\underline{\text{Int}}(\bar{M}))\text{cl}_\tau(E,\phi)$. One verifies that $\eta_n^{-1}\beta\underline{\text{Int}}(\bar{M}) = \underline{\text{Ad}}(M) \circ \eta_n^{-1} \circ \beta$. Thus (a) applied to $(E,M\phi)$ says $0 = H^1(p_{i,j})\ H^1(\underline{\text{Ad}}(M))X = (M X M^{-1})_{i,j}$ where $X = H^1(\eta_n^{-1}\beta)\text{cl}(E,\phi)$ [regarded as an $n \times n$ matrix with values in the R/uR-module $H^1(Z,\mathcal{O}_Z)$] and $i \ne j$ (fixed below). By $(a)_m$ we get $X_{ji} = X_{ij} = 0$, so taking $M = I + e_{ij}$ we get $0 = X_{ij} + X_{jj} - X_{ii} - X_{ji} = X_{jj} - X_{ii}$. This proves (b).

Proof of (a)$_m$: (i,j) such that $i \neq j$ are fixed. We also fix two disjoint subsets I, J of $\{1,2,\dots,n\}$ such that $i \in J$, $j \in I$, and $I \cup J = \{1,2,\dots,n\}$. Let e_k, $1 \leq k \leq n$, be the standard basis of $\mathcal{O}_Z^n$. Given $(E,\varphi) \in \mathrm{CART}({}_n R_{pq})$ with E minimal, let $F_{(E,\varphi)}$ be the locally free sheaf defined by $uE \subset F \overset{\leq}{\subset} E$ and $F/uE = \bigoplus_{k\in I} \mathcal{O}_Z\, \varphi_1^{-1}(e_k)$ where $\varphi_1 : E/uE \tilde{\to} \mathcal{O}_Z^n$ is defined by $\varphi \otimes_{\mathcal{O}_{Z_m}} \mathcal{O}_{Z_1}$. As the exact sequence $0 \to F/uE \to E/uE \to E/F \to 0$ splits, we obtain $L_1(s) = 0$ (see Definition 3). By Lemma 14(a) $\ell_E^F = -\lg(L_2(s)) \leq 0$. By minimality of E $\ell_E^F \geq 0$. So $\ell_E^F = 0$. Using Lemma 14(b) F is minimal too. We will prove statement (a)$_m$ for (i,j) by induction on m.

Proof of (a)$_1$: We use the fact that $L_2(s) = 0$, i.e., that $H^0(Z,F/uF) \to H^0(Z,F/uE)$ is onto. We now need

Lemma 24 : Let $\mathcal{O}' \underset{p}{\to} \mathcal{O}$ be a surjective homomorphism of sheaves of rings on a site S and $n \in H^0(S,\mathbb{Z}_0)$. Let $N = \underline{\mathrm{Ker}}(p)$. Assume $N^2 = 0$. Define $\eta : N \otimes_{\mathcal{O}} M_n\mathcal{O} \to {}_nG_p$ by $\eta(u \otimes A) = I_n + u\tilde{A}$, where $\tilde{A}$ is the lifting of A to $M_n(\mathcal{O}')$ and where in what follows, the letter u will stand for local sections of N. Let $ev_j : M_n\mathcal{O} \to \mathcal{O}^n$ be the homomorphism defined by $pr_i \circ ev_j = pr_{i,j}$. Suppose $(E,\varphi) \in \mathrm{CART}({}_nR_p)$. Define $\varphi' : N{\cdot}E \tilde{\to} N \otimes_{\mathcal{O}} \mathcal{O}^n$ by $\varphi'(ue) = u \otimes \varphi(e\otimes 1)$. Let $\delta : H^0(E/NE) \to H^1(NE)$ be the connecting homomorphism for the exact sequence $0 \to NE \to E \overset{t}{\to} E/NE \to 0$. Then for each $1 \leq j \leq n$

(*) $H^1(\varphi')\delta[\varphi^{-1}(e_j)] = \check{H}^1((id_N \otimes ev_j)\eta^{-1})cl(E,\varphi)$ where $e_j \in H^0(S,\mathcal{O}^n)$ [respectively $e_j' \in H^0(S,\mathcal{O}'^n)$] is the section the by j-th basis element.

Proof of Lemma 24 : Let Λ be the gerbe of (right) torseurs over $N \otimes \mathcal{O}^n$, and $\sigma_0 \in \mathrm{CART}(\Lambda)$ be obtained from the trivial torseur $N \otimes \mathcal{O}^n$. $\underline{\mathrm{Aut}}(\sigma_0) \simeq N \otimes \mathcal{O}^n$ (left translations). We "define" a cartesian functor

${}_n R_p \xrightarrow[Q]{} \Lambda$ by $Q((\Omega,E),\varphi) = ((\Omega,(\varphi t)^{-1}(e_{j|\Omega}))$, action) where we note that $(\varphi t)^{-1}(e_j) \subset E_{|\Omega}$ is an NE-coset and letting $x \in N \otimes \mathcal{O}^n$ act on it by adding $\varphi'^{-1}(x)$. We define $Q \circ {}_n\tau_p \underset{\xi}{\tilde{\to}} \sigma_0$ using in the case of ${}_n\tau_p$ the Λ-isomorphism $N \otimes \mathcal{O}^n \to (\varphi t)^{-1}(e_j)$ sending the zero section to $e_j' \in \Gamma(S,E)$. (Q,ξ) induces the map $(id_N \otimes ev_j) \circ \eta^{-1} : {}_nG_p \to N \otimes \mathcal{O}^n$ on the automorphism sheaves. (Indeed if $x = I + \Sigma u_i A_i \in {}_nG_p$, it acts via Q on $(\varphi t)^{-1}(e_j) = e_j' + N\mathcal{O}'^n$ by the restriction of its action on $\mathcal{O}'^n$. This action sends $e_j' + \Sigma u_\alpha v_\alpha$ $(v_\alpha \in \mathcal{O}'^n)$ to $(I + \Sigma u_i A_i)(e_j' + \Sigma u_\alpha v_\alpha) = e_j' + \Sigma u_\alpha v_\alpha + \Sigma u_i A_i(e_j')$. Hence after applying ξ it acts on $N \otimes \mathcal{O}^n$ by translating by $\varphi'^{-1}[\Sigma u_i A_i(e_j')] = \Sigma u_i \otimes A_i(e_j') = (id_N \otimes ev_j)(\Sigma u_i \otimes A_i) = (id_N \otimes ev_j)(\eta^{-1}(x))$.) We now apply Lemma 23. We obtain that the RHS of (*) is equal to the class of the $N \otimes \mathcal{O}^n$-torseur $t^{-1}[\varphi^{-1}(e_j)]$. By [3, III 3.5.4, 3.6.8] we know that $\delta[\varphi^{-1}(e_j)]$ = the class of the NE-torseur $t^{-1}[\varphi^{-1}(e_j)]$. Q.E.D.

We now apply Lemma 24. Let $\mathcal{O}' = \mathcal{O}_{Z_2}$, $\mathcal{O} = \mathcal{O}_{Z_1}$, $S = Z_{Zar}$. We use that $N = u\mathcal{O}_{Z_2} \simeq \mathcal{O}$ and eliminate $N \underset{\mathcal{O}}{\otimes}$. Recall that $\check{H}^1(\beta)cl(E,\varphi) = cl(E_2,\varphi)$. Thus, if $c \overset{def}{=} \check{H}^1(pr_{i,j}\eta_n^{-1}\beta)cl(E,\varphi)$ then $c = H^1(pr_i)H^1(ev_j\eta_n^{-1})cl(E_2,\varphi) \overset{\text{Lemma 24}}{=\!=} H^1(pr_i \circ \varphi')(\delta[\varphi^{-1}(e_j)])$.
Consider now the commutative diagrams

(a)
$$\begin{array}{ccccccccc}
0 & \longrightarrow & uE/uF & \longrightarrow & F/uF & \xrightarrow{w} & F/uE & \longrightarrow & 0 \\
 & & \uparrow{\scriptstyle \nu} & & \uparrow & & \uparrow & & \\
0 & \longrightarrow & uE/u^2E & \longrightarrow & F/uE & \longrightarrow & F/uE & \longrightarrow & 0 \\
 & & \downarrow & & \downarrow & & \downarrow{\scriptstyle \xi} & & \\
0 & \longrightarrow & uE/u^2E & \longrightarrow & E/u^2E & \xrightarrow{t} & E/uE & \longrightarrow & 0
\end{array}$$

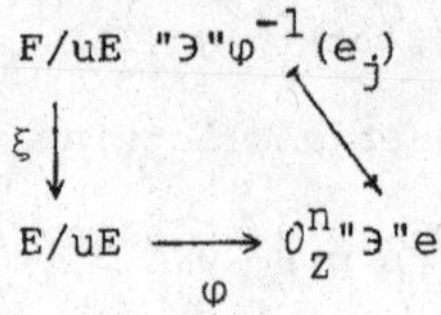

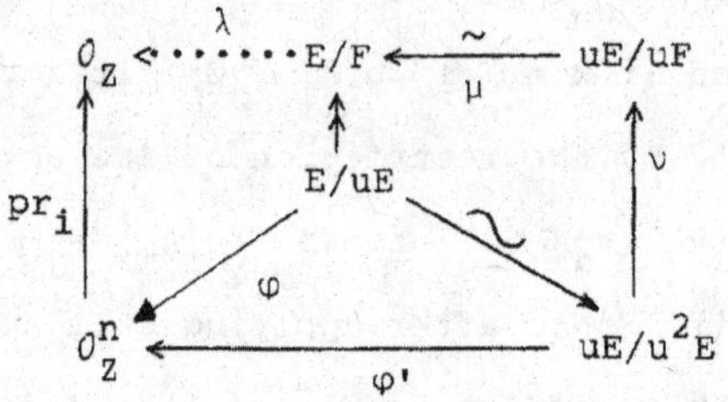

where the dotted arrow λ exists because $pr_i\varphi\xi(F/uE) = pr_i(\bigoplus_{k\in I} O_Z\cdot e_k) = 0$. Thus $c = H^1(pr_i\varphi')\delta(\varphi^{-1}(e_j)) = H^1(\lambda\mu\nu)\delta(\varphi^{-1}(e_j))$. If δ' is the connecting homomorphism from H^0 to H^1 for the exact sequence (a) then $\delta' = 0$ because $\mathrm{Cok}(H^0(w)) = L_2(s) = 0$. $\delta' = H^1(\nu)\delta H^0(\xi)$ by functoriality. As $\varphi^{-1}(e_j) \in \mathrm{Im}\, H^0(\xi)$ by the definition of F, we see that $c = 0$ as desired.

__Proof of (a)$_m$ for $m > 1$__ : We can assume that $(a)_{m-1}$ is true. We define a cartesian functor $T : {}_nR_{q_m} \to {}_nR_{q_{m-1}}$ (where q_ℓ is the projection $O_{\hat{X}}|_Z \xrightarrow{q_\ell} O_{Z_\ell}$) as follows : $T((\Omega,E),\varphi) = ((\Omega,F_{(E,\varphi)}),\psi)$ where $F_{(E,\varphi)}$ is defined on page 156 and $\psi : F_{m-1} \xrightarrow{\sim} O_{Z_{m-1}|\Omega}$ is defined as follows :

Consider the maps $F_{m-1} \xleftarrow{\pi} F/u^mE \xrightarrow{\theta} E/u^mE$. Let $\Delta_k = 1$ for $k \in I$, $\Delta_k = u$ for $k \in J$. $\Delta_k\varphi^{-1}(e_k) \in \mathrm{Im}\,\theta$. The sections $\pi\theta^{-1}(\Delta_k\varphi^{-1}(e_k)) \overset{\text{def}}{=} s_k$ form a basis for F_{m-1}. We define ψ such that $\psi s_k = e_k$. $T \circ {}_n\tau_{q_m}$ is canonically isomorphic (say by ζ) to ${}_n\tau_{q_{m-1}}$ since $F_{{}_n\tau_{q_m}}$ is free with basis $\Delta_k e_k$ lifting $\psi^{-1}(e_k)$. The resulting map ${}_nG_{q_m} \xrightarrow{h} {}_nG_{q_{m-1}}$ is $h(A) = \Delta^{-1}\cdot A\cdot\Delta$, where Δ denotes the diagonal $n\times n$ matrix s.t. $\Delta_{ii} = \Delta_i$, since the map $O_{\hat{X}}^n \xrightarrow[\zeta^{-1}]{\sim} F_{{}_n\tau_{q_m}} \xrightarrow{s} O_{\hat{X}}^n$ is given by the matrix Δ and the action of T on automorphisms is by restriction from E to $F_{(E,\varphi)}$. We saw at the beginning of the proof that if $(E,\varphi) \in \mathrm{CART}({}_nR_{q_m})$ is such that E is minimal then $F_{(E,\varphi)}$ is minimal too. Thus we can apply the induction hypothesis to $T(E,\varphi) \in \mathrm{CART}({}_nR_{q_{m-1}})$. Using Lemma 23 we obtain that $cl(T(E,\varphi)) = \check{H}^1(h)cl(E,\varphi)$. The induction hypothesis gives then

$\check{H}^1(p_{i,j}\eta^{-1}_{n,m-1} \circ \beta_{m-1} \circ h)cl(E,\varphi) = 0$. Now

$$(p_{i,j}\eta^{-1}_{n,m-1} \circ \beta_{m-1} \circ h)(I+u^m A) = pr_{i,j}\eta^{-1}_{n,m-1}\beta_{m-1}(I+u^{m-1}u\Delta^{-1}A\Delta) =$$

$$= pr_{i,j}\eta^{-1}_{n,m-1}[(I+u^{m-1}\; u\Delta^{-1}\cdot A\cdot\Delta) \bmod u^m] = pr_{(i,j)}[\tfrac{u}{\Delta}\cdot A\cdot\Delta \bmod u] =$$

$$= \frac{u}{\Delta_i} A_{ij}\Delta_j \bmod\cdot u = A_{ij} \bmod u = pr_{(i,j)}[A \bmod u] =$$

$$= pr_{(i,j)}[\eta^{-1}_{n,m}[(I+u^m\; A) \bmod u^{m+1}]] = (pr_{i,j} \circ \eta^{-1}_{n,m} \circ \beta_m)(I+u^m A).$$

Substituting this equality of homomorphisms in the previous result we get $(a)_m$.

§ 7. Proof of Theorem 1

We are given a vector bundle G on Y. We can assume that $G \neq 0$. By Lemma 1 we can extend G to a vector bundle E on X. Choose E such that ℓ_E is minimal. Thus if F is a modification of E then $\ell_{\hat{E}}^{\hat{F}} \xlongequal{\text{Lemma 14(c)}} \ell_{\hat{F}} - \ell_{\hat{E}} \xlongequal{9(c)} \ell_F - \ell_E \geq 0$ because $F_{|Y} \simeq G$. Since (Lemma 11) those $\hat{F}$ form (up to isomorphism) all the modifications of $\hat{E}$ we obtain (Definition 6) that $\hat{E}$ is minimal. Hence by Lemma 21 there is a formal line bundle L s.t. $\hat{E} \simeq L^n$ where n = rank E. E, hence $\hat{E}$, hence L, are generated by their sections. By Lemma 4 L is trivial. Hence $\hat{E}$ is free. By Lemma 10(b) E is free. Restricting to Y we obtain that G is free. Q.E.D.

Corollary : If E,F are vector bundles on X then

(i) $\ell_{E\otimes F} = (\mathrm{rk}\ F)\ell_E + (\mathrm{rk}\ E)\ell_F$

(ii) $\ell_{E^*} = \ell_E$.

Proof : Both (i) and (ii) are trivial when E is free. In the general case Theorem 1 implies that E has a modification (E',φ) such that E' is free. Now use Lemma 17 :

$$\ell_{E\otimes F} \underset{\overline{\overline{}}}{\overset{14(c)}{}} \ell^{\hat{E}\otimes\hat{F}}_{\hat{E}'\otimes\hat{F}} + \ell_{E'\otimes F} \overset{17(1)}{=\!=} (\mathrm{rk}\ F)\ \ell^{\hat{E}}_{\hat{E}'} + (\mathrm{rk}\ E)\cdot\ell_F$$

$$\overset{14(c)}{=\!=} (\mathrm{rk}\ F)\ \ell_E - (\mathrm{rk}\ F)\cdot\ell_{E'} + (\mathrm{rk}\ E)\cdot\ell_F$$

$$\overset{10(a)}{=\!=} (\mathrm{rk}\ F)\ \ell_E + (\mathrm{rk}\ E)\cdot\ell_F.$$

$$\ell_{E^*} \overset{14(c)}{=\!=} \ell^{\hat{E}^*}_{\hat{E}'^*} + \ell_{E'^*} \overset{10(a)}{=\!=} \ell^{\hat{E}^*}_{\hat{E}'^*} \overset{17(2)}{=\!=} \ell^{\hat{E}}_{\hat{E}'}$$

$$\overset{14(c)}{=\!=} \ell_E - \ell_{E'} \overset{10(a)}{=\!=} \ell_E .$$

§ 8. Proof of Theorem 2

Let A be an Azumaya algebra of rank n^2 on X such that $A \otimes_{\mathcal{O}_X} \mathcal{O}_Z$ is trivial. Applying Lemma 7 to $\hat{A}$ we obtain that there is a vector bundle E on $\hat{X}$ and an isomorphism of $\mathcal{O}_{\hat{X}}$-algebras $\hat{A} \xrightarrow{\sim} \underline{\mathrm{End}}(E)$. Suppose (F,φ) is a modification of E. Then

$$\ell^{\underline{\mathrm{End}}(F)}_{\underline{\mathrm{End}}(E)} \overset{\text{Cor. to 17}}{=\!=} 2\cdot n\cdot\ell^F_E . \quad \ell_{\hat{A}} \overset{9(c)}{=\!=} \ell_A < \infty.$$ By Lemma 14 (c)

$$\ell_{\underline{\mathrm{End}}(F)} < \infty \text{ and } \ell^{\underline{\mathrm{End}}(F)}_{\underline{\mathrm{End}}(E)} = \ell_{\underline{\mathrm{End}}(F)} - \ell_{\underline{\mathrm{End}}(E)} \geq -\ell_{\hat{A}} .$$

Hence $2n\ell^F_E \geq -\ell_A$. Hence ℓ^F_E is bounded below.

Choose a modification (F,φ) of E such that ℓ_E^F is minimal. If (G,ψ) is a modification of F, then by 14(b) $\ell_F^G = \ell_E^G - \ell_E^F \geq \ell_E^F - \ell_E^F = 0$. By Definition 6, F is minimal. By Lemma 21, $F \simeq L^n$ for some line bundle L on $\hat{X}$. Hence $\underline{\mathrm{End}}(F) \simeq \underline{\mathrm{End}}(L^n) = \underline{\mathrm{End}}(L\otimes\mathcal{O}^n) \simeq \underline{\mathrm{End}}(\mathcal{O}^n) = M_n(\mathcal{O}_{\hat{X}})$. $(\underline{\mathrm{End}}(F), \underline{\mathrm{End}}(\varphi))$ is a modification of $\underline{\mathrm{End}}(E)$. By Lemma 11 there is a locally free sub $\mathcal{O}_X$-module B of $A[u^{-1}]$ such that

$$(B|_Z)\underset{\mathcal{O}_X|_Z}{\otimes}\mathcal{O}_{\hat{X}} = \text{Image}\ (\underline{\mathrm{End}}(F) \xrightarrow[\psi^{-1}[u^{-1}]\underline{\mathrm{End}}(\varphi)]{} \hat{A}[u^{-1}]) \text{ and } B|_Y = A|_Y .$$

We claim that B is closed under multiplication and contains 1. This is clear on Y and follows for the stalk at $z \in Z$ by checking it after applying $? \underset{\mathcal{O}_{X,z}}{\otimes} \mathcal{O}_{\hat{X},z}$ and using that $B_z = B[u^{-1}]_z \cap \hat{B}_z$ inside $B_z[u^{-1}]$. As $\hat{B} \simeq \underline{\mathrm{End}}(F) \simeq M_n(\mathcal{O}_{\hat{X}})$, $\hat{B}$ is free as a vector bundle. By Lemma 10(b) B is free as a vector bundle. Hence $\Gamma(X,B)$ is free of finite type over R. Let $B' = \Gamma(X,B) \underset{R}{\otimes} \mathcal{O}_W$. B' is a sheaf of algebras with 1 on W which restricts to B on X, and is a vector bundle. We claim that B' is an Azumaya algebra. Indeed by [2, I, Thm. 5.1 (ii)] we have to check that the map $B' \underset{\mathcal{O}_W}{\otimes} B' \xrightarrow{h} \underline{\mathrm{Hom}}_{\mathcal{O}_W}(B',B')[h(x \otimes y)(z) \overset{\mathrm{def}}{=} xzy]$ is an isomorphism. But h is an isomorphism on X and looking at $\det(h)$ we see that h is an isomorphism. We consider the contravariant functor $\underline{F} : (\mathrm{Sch})/W \longrightarrow ((\mathrm{sets}))$ which associates to $T \xrightarrow[g]{} W$ the set of isomorphisms of $\mathcal{O}_T$-algebras $g^*(B') \xrightarrow[\psi]{\sim} M_n\mathcal{O}_T$. $\underline{F}$ is trivially represented by an affine scheme $S_{B'} \xrightarrow[\pi]{} W$ of finite type (write the matrix entries of ψ in terms of bases of B' and $M_n\mathcal{O}$ and impose the condition $\psi(xy) = \psi(x)\cdot\psi(y)$ and invert $\det(\psi)$). The formation of $S_{B'} \xrightarrow[\pi]{} W$ commutes with base changes $W' \xrightarrow[p]{} W$. If we choose by [2, I 5.1 (iii)] an étale morphism p onto W such that p^*B' is isomorphic to a matrix algebra, then $S_{p^*B'}$ represents the automorphisms of M_n on schemes over W', i.e., it is $\simeq PGL(n)_{W'}$ which is smooth over W'. By descent [7, II 4.13] $S_{B'}$ is smooth over W. Now

$(B'/uB')\big|_Z \simeq (B/uB)\big|_Z \simeq \underline{\mathrm{End}}(F)/u\ \underline{\mathrm{End}}(F) \xrightarrow[F/uF \text{ free}]{\sim} M_n(\mathcal{O}_Z)$. By taking direct images to V we get $(B'|_V) \otimes_{\mathcal{O}_{W/V}} \mathcal{O}_V = (B'/uB')|_V \simeq M_n(\mathcal{O}_V)$. Hence $\underline{F}(V) \neq \emptyset$.

So there exists a W-morphism from V to $S_{B'}$. We now recall (see [6, page 568]) the following :

Lemma 25 : If $X \xrightarrow[\pi]{} Y$ is a smooth morphism of affine schemes and $Y_0 \subset Y$ is a closed subscheme such that (Y, Y_0) is a henselian couple (i.e., $(\Gamma(Y, \mathcal{O}_Y), \Gamma(Y, I_{Y,Y_0}))$ is a henselian couple in the terminology of [5, XI 2]) then every Y-morphism $Y_0 \xrightarrow[\sigma_0]{} X$ can be extended to a section of π.

Remark. If Y is local (it is in our application) the proof of Lemma 25 is simpler and can be done by extending first Y_0, regarded as a subscheme of $X \times_Y Y_0$ via $\sigma_0 \times \mathrm{id}_{Y_0}$, to a subscheme, étale over Y, of X.

Applying Lemma 25 to our case we find that $S_{B'} \to W$ has a section. So $B' \simeq M_n\mathcal{O}_W$. Now $B'|_Y \simeq B|_Y \simeq A|_Y$ so $A|_Y$ is isomorphic to a matrix algebra. Q.E.D.

CHAPTER II : THE EXISTENCE OF AZUMAYA ALGEBRAS WITH A GIVEN COHOMOLOGY CLASS IN $H^2_{ét}(X,G_m)_{torsion}$ AND AN APPLICATION

Our main theorem is the following

THEOREM 1. Let X be an affine scheme or a union $U \cup V$ where $U,V,U \cap V$ are affine schemes. Let $c \in H^2_{ét}(X,G_m)$ be a class of finite order n. Then there is an Azumaya algebra A of rank d^2 over X s.t. d is divisible by exactly those primes which divide n and $c = \delta(cl(A))$ where $cl(A) \in H^1_{ét}(X,PGL(d))$ classifies the isomorphism class of A and δ is the non-abelian coboundary in the sequence

$$H^1_{ét}(X,GL(d)) \longrightarrow H^1_{ét}(X,PGL(d)) \xrightarrow{\delta} H^2_{ét}(X,G_m)$$

deduced from the central extension of étale group sheaves

$$0 \longrightarrow G_m \longrightarrow GL(d) \longrightarrow PGL(d) \longrightarrow 1$$

Remark. The Brauer group of a scheme X was defined by Grothendieck in [2, I.2] to be $Br(X) \stackrel{def}{=} Az(X)/\sim$ where $Az(X) \stackrel{def}{=}$ {the set of isomorphism classes of Azumaya algebras over X} and $[A] \sim [B]$ iff $\exists$ E,F vector bundles over X of positive rank s.t. $A \otimes \underline{End}(E) \simeq B \otimes \underline{End}(F)$. The rank of the vector bundles and Azumaya algebras is not required to be constant on X. However if X is quasi-compact one sees easily that we can restrict ourselves to things of constant rank without changing Br(X). Grothendieck shows that

$$[A] \sim [B] \Longleftrightarrow \delta(cl(A)) = \delta(cl(B)),\ rk\ A = d^2 \Rightarrow d \cdot \delta(cl(A)) = 0\ ,$$

and that Br(X) is a group using $\otimes$ product of algebras, and that

$$\mathrm{Br}(X) \xhookrightarrow[\alpha_X=\delta\ \mathrm{cl}]{} H^2_{\text{ét}}(X,G_m)_{\text{torsion}} \overset{\text{def}}{=\!=\!=} \{\alpha\in H^2(X,G_m) \mid \exists\ n\in H^0(X,\mathbb{Z}_+)\ \text{s.t.}\ n\alpha = 0\}$$

is an (injective) group homomorphism. Thus for X as in Theorem 1 α_X is an isomorphism. We also mention that if X is regular irreducible with generic point η, then $H^2_{\text{ét}}(X,G_m) \to H^2(\eta,G_m)$ is a monomorphism by [2, II 1.8], hence $H^2(X,G_m)$ is then a torsion group. R. Hoobler proved in [18] that if in addition X is affine and of finite type over a field, then (a) : α_X is an isomorphism and (b) : $\mathrm{Br}(X) = \bigcap_{\substack{x\in X \\ \text{of height 1}}} \mathrm{Br}(\mathrm{Spec}\ \mathcal{O}_{X,x})$ inside $\mathrm{Br}(k(X))$. My proof of the "gluing" Lemma 2 is a refinement of his argument for (a).

Let $P = \{p\in\mathbb{Z} \mid p \text{ prime}\}$. If $\Sigma\subset P$ and $n\in\mathbb{Z}$, let $\Sigma \nmid n$ mean that "if $p\in\Sigma$ then $p\nmid n$". If $n\in\mathbb{Z}$ define $\Sigma(n) = \{p\in P \mid p\nmid n\}$. For every scheme X and $c\in H^2(X,G_m)$ and $\Sigma\subset P$ let $(1\Sigma)c$ be the statement "$\exists [A]\in \mathrm{Az}(X)$ s.t. $\delta(\mathrm{cl}(A)) = c$ and $\Sigma\nmid \mathrm{rk}\ A$", so then Theorem 1 is $(1\Sigma(\mathrm{ord}(c)))c$, noting that $\mathrm{ord}(\alpha_X([A])) \mid \sqrt{\mathrm{rk}\ A}$.

<u>Structure of the Proof</u>. We will show the following facts :

<u>LEMMA 1</u>. If $X = \mathrm{Spec}\ \mathcal{O}$ where $\mathcal{O}$ is local and $c\in \mathrm{Im}(\alpha_X)$, then $(1\Sigma(\mathrm{ord}(c)))c$.

<u>LEMMA 2</u>. If $X = U\cup V$ where $U, V, U\cap V$ are affine, and $c\in H^2(X,G_m)_{\text{torsion}}$ and $\Sigma\subset\Sigma(\mathrm{ord}(c))$, then $(1\Sigma)c|_U$ and $(1\Sigma)c|_V$ imply $(1\Sigma)c$.

<u>COROLLARY</u>. For such X,Σ,c if $\forall_{x\in X}\ (1\Sigma)(c|_{\mathrm{Spec}(\mathcal{O}_{X,x})})$ then $(1\Sigma)c$.

<u>LEMMA 3</u>. If X is as in Lemma 1, then $(1\emptyset)c$ for every $c\in H^2(X,G_m)_{\text{torsion}}$.

Remarks. In all statements involving rk A we assume it is constant if $X \neq \emptyset$ and define those statements to be true if $X = \emptyset$. Clearly the above three lemmas and Corollary to Lemma 2 imply Theorem 1. All sheaves (except possibly quasi-coherent ones) on our schemes are for the étale topology.

LEMMA 4. If $X \xrightarrow{\varphi} Y$ is a surjective finite locally free morphism and $c \in H^2(Y, G_m)$ and $\varphi^* c \in \alpha_X(\mathrm{Br}(X))$ then $c \in \alpha_Y(\mathrm{Br}(Y))$. Moreover, if $\deg \varphi = r$ and $\varphi^* c = \alpha_X([A])$ and $\mathrm{rk}\, A = d^2$, then $\exists\, B \in \mathrm{Az}(Y)$ with $\mathrm{rk}\, B = (rd)^2$ s.t. $c = \delta(\mathrm{cl}(B))$.

Proof of Lemma 4.

Remark. This lemma holds also for strictly ringed topoi.

Let $\varphi^* c = \delta(\mathrm{cl}(A))$. For every $d \in \mathbb{Z}_+ = \{n \in \mathbb{Z} \mid n > 0\}$ let

$X_d = \{x \in X \mid \mathrm{rk}\, A_x = d^2\}$. For every $r \in \mathbb{Z}_0 = \mathbb{Z}_+ \cup \{0\}$ let

$Y_{r,d} = \{y \in Y \mid \mathrm{rk}(\varphi_* \mathcal{O}_{X_d})_y = r\}$ and

$Y'_{r,d} = Y_{r,d} - \varphi(\bigcup_{i=1}^{d-1} X_i)$, $X'_{r,d} = X_d \cap \varphi^{-1}(Y'_{r,d})$.

$X'_{r,d} \xrightarrow[\varphi'_{r,d}]{} Y'_{r,d}$ is of constant degree r, $\mathrm{rk}(A|_{X'_{r,d}}) = d^2$.

$X = \coprod_{r,d \in \mathbb{Z}_+} Y'_{r,d}$. Clearly whenever we have a disjoint union decomposition $Y = \coprod_\alpha U_\alpha$, U_α open, then $H^i_{\text{ét}}(Y, F) \xrightarrow{\sim} \prod_\alpha H^i_{\text{ét}}(U_\alpha, F|_{U_\alpha})$ for every abelian sheaf F. (One uses injective resolutions and the result for $i = 0$.) Also $\mathrm{Az}(Y) \xrightarrow{\sim} \prod_\alpha \mathrm{Az}(U_\alpha)$. Hence it suffices to consider $\varphi'_{r,d}$, i.e., to prove the second part of Lemma 4. So we assume $\varphi = \varphi'_{r,d}$ for some $(r,d) \in \mathbb{Z}^2_+$. We recall [1, VIII 5.6, 5.8] that for an integral morphism $\varphi : X \to Y$, $\varphi_{\text{ét}_*}$ preserves epimorphisms of sheaves (of sets) and that for an abelian sheaf G on $X_{\text{ét}}$ (resp. sheaf of groups) $R^q \varphi_* G = 0$ for every $q > 0$ (resp. for $q = 1$). From this one deduces that the map $H^q(Y,\varphi_*G) \xrightarrow[\varphi^*]{} H^q(\varphi^*\varphi_*G) \longrightarrow H^q(X,G)$ is an isomorphism. For a central extension of group sheaves $0 \to C \to G \overset{\pi}{\to} G' \to 0$ on a site T one can define a map of pointed sets $\delta : H^1(T,G') \to H^2(T,C)$, which is functorial for maps of exact sequences and pullbacks, and satisfies $\mathrm{Ker}(\delta) = \mathrm{Im}(H^1(T,\pi))$. Giraud defined δ (see [3, IV 4.2]) by considering "gerbes of liftings". δ can also be defined using hypercover cocycles. We consider the exact sequence

(I) $\quad 0 \longrightarrow G_m \longrightarrow GL(d)_X \longrightarrow PGL(d)_X \longrightarrow 1 \quad$ on $X_{\acute{e}t}$.

Let $\beta : G_{m_Y} \longrightarrow \varphi_*G_{m_X}$ and $\gamma : \varphi^*G_{m_Y} \longrightarrow G_{m_X}$ be the natural maps. From the morphism of central exact sequences

$$\begin{array}{ccccccccc} \varphi^*\varphi_*(I) = (II)' & 0 \longrightarrow & \varphi^*\varphi_*G_{m_X} & \longrightarrow & \varphi^*\varphi_*GL(d)_X & \longrightarrow & \varphi^*\varphi_*PGL(d)_X & \longrightarrow 1 \\ \downarrow & & \downarrow & & \downarrow & & \downarrow & \\ (I) & 0 \longrightarrow & G_{m_X} & \longrightarrow & GL(d)_X & \longrightarrow & PGL(d)_X & \longrightarrow 1, \end{array}$$

we obtain a commutative diagram (in which δ_k denotes a connecting homomorphism deduced from the exact sequence (k))

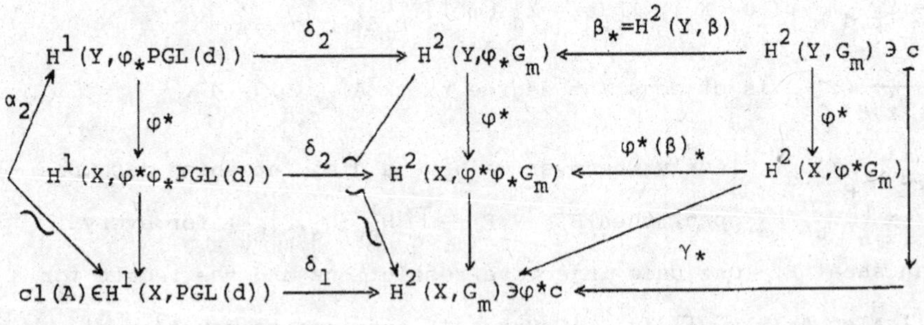

Hence $\beta_*c = \delta_2\alpha_2^{-1}\mathrm{cl}(A) \in \mathrm{Im}\ \delta_2$. (By abuse of notation $H^2(\gamma)\varphi^*c$ was still denoted by φ^*c.)

We now consider the morphism of central exact sequences

$$\begin{array}{ccccccccc}
(III) & 0 \longrightarrow & G_{m_Y} & \xrightarrow{\lambda_1} & \varphi_*GL(d)_X & \longrightarrow & (\varphi_*GL(d))/G_{m_Y} & \longrightarrow & 1 \\
\downarrow & & \downarrow \beta & & \downarrow id & & \downarrow \beta_2 & & \\
\varphi_*(I)=(II) & 0 \longrightarrow & \varphi_*G_m & \xrightarrow{\lambda_2} & \varphi_*GL(d) & \longrightarrow & \varphi_*PGL(d) & \longrightarrow & 1
\end{array}$$

Assertion. $c \in \mathrm{Im}[\delta_3 : H^1(Y,(\varphi_*GL(d))/G_m) \longrightarrow H^2(Y,G_m)]$.

1st proof. Giraud proved [3, Thm. IV 3.4.2] that for an abelian sheaf A on a site E

$$H^2(E,A) \xrightarrow[\zeta]{\sim} \{A\text{-equivalence classes of A gerbes on E}\}.$$

Let $cl(\mathcal{G}) \overset{def}{=\!=} \zeta^{-1}([\mathcal{G}])$. For

$A \xrightarrow{u} B$ and an A-gerbe $\mathcal{G}$ and a B-gerbe $\mathcal{G}'$ we have

$H^2(E,u)(cl(\mathcal{G})) = cl(\mathcal{G}')$ iff there exists a u-equivariant cartesian morphism $\mathcal{G} \to \mathcal{G}'$. Moreover by [3, IV 4.2.8] if we have a central extension $0 \to C \to H \to G \to 1$, then $\mathrm{Im}[\delta : H^1(E,G) \to H^2(E,C)]$ consists of the classes of those C-gerbes $\mathcal{G}$ for which there exists a C-morphism $\mathcal{G} \to TORS(E,H)$. Now to prove the assertion suppose that $c = cl(\mathcal{G})$ and $\beta_*c = cl(\mathcal{G}')$. As $\beta_*c \in \mathrm{Im}(\delta_2)$, $\exists$ a morphism of gerbes $\mathcal{G}' \xrightarrow[t]{} TORS(Y_{ét},\varphi_*GL(d))$ over the morphism of liens coming from λ_2. Also $\exists$ a morphism $\mathcal{G} \xrightarrow[s]{} \mathcal{G}'$ over β. Then $t \circ s$ is a λ_1-morphism, so $c \in \mathrm{Im}\ \delta_3$.

2nd Proof (using only the functoriality of δ).

Let $C = G_{m_Y}$, $C' = \varphi_*G_m$, and $G = \varphi_*GL(d)$.

Consider the following morphisms of central extensions :

$$
\begin{array}{ccccccccccc}
 & & & & 0 & & & & & & \\
 & & & & \downarrow & & & & & & \\
(1) & & 0 & \longrightarrow & C & \xrightarrow{\lambda_1} & G & \longrightarrow & G/C & \longrightarrow & 1 \\
\downarrow & & & & \downarrow{\scriptstyle\beta} & & \downarrow & & \downarrow{\scriptstyle\beta_2} & & \\
(2) & & 0 & \longrightarrow & C' & \xrightarrow{\lambda_2} & G & \longrightarrow & G/C' & \longrightarrow & 1 \\
\downarrow & & & & \downarrow{\scriptstyle\alpha} & & \downarrow & & \downarrow & & \\
(3) & & 0 & \longrightarrow & C'/C & \longrightarrow & G/C & \longrightarrow & G/C' & \longrightarrow & 1 \\
 & & & & \downarrow & & & & & & \\
 & & & & 0 & & & & & &
\end{array}
$$

$$
(4) \qquad 0 \longrightarrow C \xrightarrow{\beta} C' \xrightarrow{\alpha} C'/C \longrightarrow 1
$$

$$
\begin{array}{ccccccccc}
(1)\times(4)=(5) & 0 & \longrightarrow & C\times C & \longrightarrow & G\times C' & \rightarrow & (G/C)\times(C'/C) & \longrightarrow 1 \\
\downarrow{\scriptstyle s} & & & \downarrow{\scriptstyle\text{sum}} & & \downarrow{\scriptstyle\text{sum}} & & \downarrow{\scriptstyle\text{sum}} & \\
(1) & 0 & \longrightarrow & C & \longrightarrow & G & \longrightarrow & G/C & \longrightarrow 1
\end{array}
$$

We also have $pr_1, pr_2 : (5) \to (1), (4)$ resp. . $s = pr_1 + j \circ pr_2$, where $j : (4) \to (1)$ is defined using id_C on the first terms and λ_2 on the middle terms.

We are given $c \in H^2(C)$ s.t. $\beta_x c \in \operatorname{Im}\delta_2$. We have a commutative diagram

$$
\begin{array}{ccc}
 & & H^1(C'/C) \\
 & & \downarrow{\scriptstyle\delta_4} \\
H^1(G/C) & \xrightarrow{\delta_1} & H^2(C) \\
\downarrow{\scriptstyle\beta_{2*}} & & \downarrow{\scriptstyle\beta_*} \\
H^1(G/C') & \xrightarrow{\delta_2} & H^2(C') \\
 & \searrow{\scriptstyle\delta_3} & \downarrow{\scriptstyle\alpha_*} \\
 & & H^2(C'/C)
\end{array} \quad .
$$

So if $\beta_* c = \delta_2 x$, then $0 = \alpha_*(\beta_* c) = \delta_3 x$ so $\exists\, y \in H^1(G/C)$ s.t. $\beta_{2*} y = x$. Then $\beta_* \delta_1 y = \beta_* c$ so $z \overset{\text{def}}{=\!=} c - \delta_1(y) = \delta_4(w)$ for some $w \in H^1(C'/C)$.

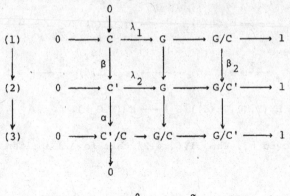

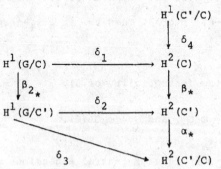

Consider

$$(y,w) \in H^1((G/C) \times (C'/C)) \xrightarrow[(pr_{1_*},pr_{2_*})]{\sim} H^1(G/C) \times H^1(C'/C) .$$

By functoriality w.r.t. pr_1 and pr_2, $\delta_5(y,w) = (\delta_1 y, \delta_4 w) = (\delta_1 y, z)$ so

$$c = sum_*(\delta_1 y, z) = sum_*(\delta_5(y,w)) = \delta_1(sum_*(y,w))$$

as required.

We now use the morphism

$$\begin{array}{ccccccccc}
(I) & 0 \longrightarrow & G_{m_Y} & \longrightarrow & \varphi_* GL(d) & \longrightarrow & \varphi_* GL(d)/G_m & \longrightarrow & 1 \\
\downarrow & & \downarrow & & \downarrow \gamma & & \downarrow & & \\
(IV) & 0 \longrightarrow & G_m & \longrightarrow & GL_{\mathcal{O}_Y}((\varphi_*\mathcal{O}_X)^d) & \longrightarrow & PGL_{\mathcal{O}_Y}((\varphi_*\mathcal{O}_X)^d) & \longrightarrow & 1
\end{array}$$

($GL_{\mathcal{O}_Y}(E) \overset{def}{=\!=} \underline{Aut}_{\mathcal{O}_Y}(E)$ for a vector bundle E) where γ is defined by $\varphi_* GL(d) = \varphi_* GL_{\mathcal{O}_X}(\mathcal{O}_X^d) \cong GL_{\varphi_*\mathcal{O}_X}(\varphi_*\mathcal{O}_X^d) \subset GL_{\mathcal{O}_Y}(\varphi_*\mathcal{O}_X^d)$. Using the assertion we obtain

$$c \in \mathrm{Im}[\delta_4 : H^1(Y,PGL(E)) \longrightarrow H^2(Y,G_m)]$$

where $E = \varphi_*\mathcal{O}_X^d$. If E is free (which is the only case we will need) this finishes the proof of Lemma 4. In general we have to show the existence of a commutative diagram

$$\begin{array}{ccc}
H^1(Y,PGL(E)) & \xrightarrow{\sim} & H^1(Y,PGL(s)) \\
& {\scriptstyle\delta_4}\searrow \quad \swarrow{\scriptstyle\delta} & \\
& H^2(Y,G_m) &
\end{array} \qquad s = rd = rk\ E$$

.

We consider the right $GL(s)$-torseur on $Y_{\acute{e}t}$

$$P = \underline{Isom}_{\mathcal{O}_Y}(\mathcal{O}_Y^s, E)$$

and $\zeta : GL(E) \xrightarrow[\text{left composition}]{\sim} \underline{Aut}_{GL(s)}(P)$.

We have a morphism of gerbes

$$\mathrm{TORS}(Y_{\text{ét}}, GL(s)) \xrightarrow{\varphi} \mathrm{TORS}(Y_{\text{ét}}, PGL(s)).$$
$$P \longmapsto P \overset{GL(s)}{\wedge} PGL(s)$$

($Y_{\text{ét}}$ denotes the étale site of Y.)

One checks that ζ induces $\bar{\zeta}$ s.t. the following diagram commutes.

$$\begin{array}{ccccc}
G_{m_Y} & \longrightarrow & GL(E) & \longrightarrow & PGL(E) \\
& \searrow^{h} & \wr\downarrow\zeta & & \wr\downarrow\bar{\zeta} \\
& & \underline{Aut}_{GL(s)}(P) & \longrightarrow & \underline{Aut}_{PGL(s)}(\varphi(P))
\end{array}.$$

h is defined using the right action of $G_{m_Y} \subset GL(s)$ on P. By [3, III 2.2.6] P and $\varphi(P)$ give equivalences ξ_1 and ξ_2 of gerbes

$$\begin{array}{ccc}
G = \mathrm{TORS}(Y_{\text{ét}}, GL(s)) & \xrightarrow[\text{equiv.}]{\xi_1} & \mathrm{TORS}(Y_{\text{ét}}, GL(E)) = G' \\
\downarrow\varphi & \overset{\alpha}{\swarrow\!\!\swarrow} & \downarrow\varphi' \\
\bar{G} = \mathrm{TORS}(Y_{\text{ét}}, PGL(s)) & \xrightarrow[\text{equiv.}]{\xi_2} & \mathrm{TORS}(Y_{\text{ét}}, PGL(E)) = \bar{G}'
\end{array}$$

such that ξ_1 is given together with an isomorphism

$\xi_1(P) \xrightarrow[\eta_1]{\sim} GL(E)_r$ (= the object of G'_Y obtained from GL(E) with its right action on itself)

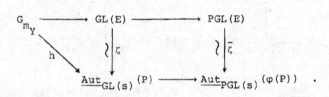

s.t. $\underline{Aut}(P) \xrightarrow[\text{by } \xi_1]{} \underline{Aut}(\xi_1(P)) \xrightarrow[\underline{Int}(\eta_1)]{\sim} \underline{Aut}(GL(E)_r) = GL(E)$ is ζ^{-1} and similarly for ξ_2. (The pair (ξ_i, η_i) is defined up to a unique isomorphism.) There is a unique isomorphism $\alpha : \varphi'\xi_1 \longrightarrow \xi_2\varphi$ compatible with η_1 and η_2. ξ_1 is compatible with the G_m-actions. Considering isomorphism classes of cartesian sections, ξ_2 gives a bijection $H^1(Y,PGL(s)) \xrightarrow[\Gamma\xi_2]{\sim} H^1(Y,PGL(E))$ (see also [3, Remark III 2.6.3]).

Moreover the connecting homomorphism is defined in [3, IV 2.5 and IV 4.2] by associating to each $s \in CART_{Y_{ét}}(Y_{ét},\bar{G})$ a G_m-gerbe $K_\varphi(s)$

$$Ob(K_\varphi(s)_\Omega) = \{(a,\alpha) \mid a \in Ob(G_\Omega),\ \alpha \in Isom_\Omega(\varphi(a),s(\Omega))\}\ \forall \Omega \in Y_{ét}$$

and letting $\delta(cl(s)) = cl(K_\varphi(s))$. Then one chekcs that ξ_1, ξ_2 and α induce a G_m-equivalence $K_\varphi(s) \longrightarrow K_{\varphi'}(\xi_2(s))$ which proves that $\delta_4 \cdot \Gamma\xi_2 = \delta$. A more general fact is in [3, Prop. IV 4.3.4.].

<u>LEMMA 5</u>. Let Y be a locally and commutatively ringed site, $c \in H^2(Y,G_m)$,

$$I = \{n \in \mathbb{Z}_+ \mid \exists\ A \in Az_n(Y) = \{\text{isomorphism classes of } \mathcal{O}_Y\text{-algebras locally} \simeq M_n(\mathcal{O}_Y)\} \text{ s.t. } \delta(cl(A)) = c\}.$$

Then I is closed under addition. If $I \neq \emptyset$ then $\exists\, d \in \mathbb{Z}_+$ s.t. $I \subset d\mathbb{Z}_+$ and $d\mathbb{Z}_+ - I$ is finite.

<u>Proof</u>. Let A,B be Azumaya algebras of ranks n^2, m^2 respectively s.t. $\delta(cl(A)) = \delta(cl(B)) = c$. Consider $x = (cl(A),cl(B)) \in H^1(Y,PGL(n) \times PGL(m))$. Then $\delta_2(x) = (c,c) = \Delta_* c \in H^2(Y,G_m \times G_m) \xrightarrow{\sim} H^2(Y,G_m) \times H^2(Y,G_m)$

where we consider the diagram

$$\begin{array}{cccccccccc}
(1) & 0 & \longrightarrow & G_{m_Y} & \longrightarrow & GL(n)\times GL(m) & \longrightarrow & H & \longrightarrow & 1 \\
\downarrow & & & \downarrow{\scriptstyle\Delta} & & \downarrow{\scriptstyle id} & & \downarrow & & \\
(2) & 0 & \longrightarrow & G_m \times G_m & \longrightarrow & GL(n)\times GL(m) & \longrightarrow & PGL(n)\times PGL(m) & \longrightarrow & 1 .
\end{array}$$

It now follows from the 1st or 2nd proof of the assertion on page 167 that $c \in \mathrm{Im}[\delta_1 : H^1(Y,H) \longrightarrow H^2(Y,G_m)]$. Consider the homomorphism

$$GL(n) \times GL(m) \longrightarrow GL(n+m) .$$

$$(A\,,\,B) \longmapsto \begin{pmatrix} A & 0 \\ 0 & B \end{pmatrix}$$

This defines a morphism of sequences

$$\begin{array}{ccccccccc}
(1) & 0 & \longrightarrow & G_m & \longrightarrow & GL(n)\times GL(m) & \longrightarrow & H & \longrightarrow & 1 \\
\downarrow & & & \downarrow{\scriptstyle id} & & \cap\!\downarrow & & \downarrow & & \\
(3) & 0 & \longrightarrow & G_m & \longrightarrow & GL(n+m) & \longrightarrow & PGL(n+m) & \longrightarrow & 1 .
\end{array}$$

Hence $c \in \mathrm{Im}\ \delta_1 \Rightarrow c \in \mathrm{Im}\ \delta_3 \Rightarrow n+m \in I$.

To prove the second assertion of Lemma 5 suppose $I \neq \emptyset$. Let $d \in \mathbb{Z}_+$ be such that $d\mathbb{Z} = I\mathbb{Z}$. $\exists\ k \geq 1$ and $x_1,\ldots,x_k \in I$, $y_1,\ldots,y_k \in \mathbb{Z}$ s.t. $d = \sum y_i x_i$. Let $s_i = \frac{x_i}{d}$. Suppose $\mathbb{Z}_+ \ni s \geq (s_1-1)(s_2+\ldots+s_k)$. We will show that $sd \in I$.

$sd = \sum_1^k (sy_i)x_i$. Write for each $i > 1$ $sy_i = q_i s_1 + r_i$ s.t. $0 \leq r_i < s_i$ $r_i, q_i \in \mathbb{Z}$. Then $sd = \sum_2^k r_i x_i + Nx_1$ where

$$N = sy_1 + \sum_2^k q_i s_i = sy_1 + \sum_{i=2}^k \frac{(sy_i - r_i)}{s_1} s_i \geq$$

$$s\cdot\left(\sum_1^k \frac{y_i s_i}{s_1}\right) - \sum_{i=2}^k \frac{(s_1-1)s_i}{s_1} = \frac{s}{s_1} - \frac{(s_1-1)\sum_2^k s_i}{s_1} \geq 0 .$$

Thus $sd \in \mathbb{Z}_0 \cdot I - \{0\} = I$.

<u>LEMMA 1</u>. Let A be an Azumaya algebra of rank d^2 over a local ring R. Then there exists a finite étale R subalgebra $S \subset A$ with rk S = d. For all such S, $A \otimes_R S \simeq M_d(S)$.

<u>Proof</u>. This almost follows from [2, I, Theorems 5.4, 5.7]. Note that A is necessarily a projective S-module as the left S-module homomorphism $(S \otimes_R S) \otimes_S A = S \otimes_R A \twoheadrightarrow A = S \otimes_S A$ splits since the S-bimodule map $S \otimes_R S \twoheadrightarrow S$ splits.

<u>Proof of Lemma 1</u> (stated on page 164).

(We use the "Sylow" idea as in the proof for fields in [10, Lemma 44.5].) If c = 0 we represent c by $M_1(\mathcal{O})$, so assume $n = \text{ord}(c) > 1$. Let I,d be as in Lemma 5. We know that mc = 0 for every $m \in I$ so $n|d$. Clearly $d^k \in I$ for $k >> 0$. Thus to prove Lemma 1 we have to check that for every $p \in P$, $p \nmid n \Rightarrow p \nmid d$. Suppose on the contrary that $\exists p$ s.t. $p|d$, $p \nmid n$. Write $c = \delta\, \text{cl}(\tilde{A}), [A] \in Az_m(\mathcal{O})$. By Lemma L $\exists R \overset{j}{\hookrightarrow} S$ étale finite s.t. $A \otimes_R S \simeq M_m(S)$. Hence $\pi^* c = \delta\, \text{cl}(\widetilde{A \otimes_R S}) = 0$ where π = Spec j. We can find a Galois cover $T \xrightarrow[\pi']{} \text{Spec } R$ which factorizes through Spec S (e.g., if $s = \deg \pi$ let $T \subset \text{sym}_R^s(\text{Spec } S)$ be the open and closed subscheme whose geometric points are systems of s distinct points in the geometric fibers of π). Thus $\pi'^* c = 0$. So we can assume π is Galois. Let G = Gal(S/R). Let $H \subset G$ be a p-sylow subgroup. Then we get

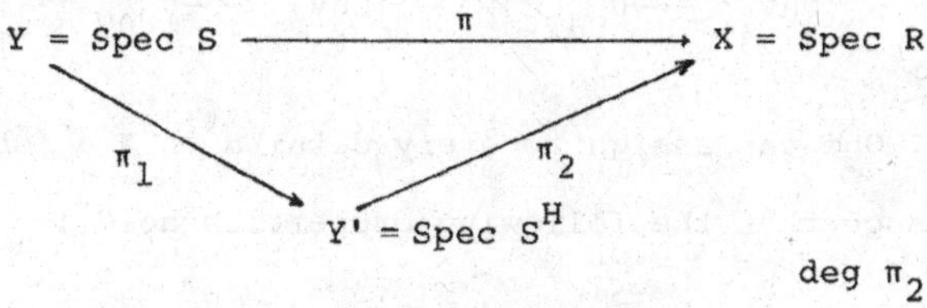

$\deg \pi_2 = |G|/|H|$

all maps being étale covers. It follows from

$$
\begin{array}{ccccccc}
 & & & H^2(Y,G_m) & & & \\
 & \nearrow^{\pi^*} & \nearrow^{\pi_1^*} & \uparrow \wr & & & \\
c\in H^2(X,G_m) & \xrightarrow[\pi_2^*]{} & H^2(Y',G_m) & \longrightarrow & H^2(Y',\pi_{1*}G_m) & \xrightarrow{\text{norm}_*} & H^2(Y',G_m) \\
 & & \lfloor & \multicolumn{3}{c}{\text{mult. by } \deg(\pi_1) = |H| = p^\alpha} & \uparrow
\end{array}
$$

that $|H|\cdot(\operatorname{Ker}\pi_1^*) = 0$, hence $0 = (|H|\mathbb{Z} + \operatorname{ord}(c)\mathbb{Z})\pi_2^*c = \mathbb{Z}\cdot\pi_2^*c \ni \pi_2^*c$. So by Lemma 4, $\alpha = \delta(\operatorname{cl}(B))$ for some $B \in Az_{|G|/|H|}(X)$ so $|G|/|H| \in I$ so $p|d|(|G|/|H|)$ which is a contradiction.

We now turn to the main lemma.

<u>LEMMA 2</u>. If $X = U \cup V$ where $U, V, U\cap V$ are affine schemes, $c \in H^2(X_{ét}, G_m)$, $\infty > \operatorname{ord}(c) = n \in \mathbb{Z}_+$, $\Sigma \subset \Sigma(n)$ and $\exists$ Azumaya algebras A, B on U, V respectively s.t. $\delta(\operatorname{cl} A) = c|_U$, $\delta(\operatorname{cl} B) = c|_V$, $\operatorname{rk} A = s^2$, $\operatorname{rk} B = t^2$, $\Sigma \not| s$, $\Sigma \not| t$, then $c \in \delta \operatorname{cl} Az_m(X)$ for some m s.t. $\Sigma \not| m$.

<u>Proof</u>.

<u>Definition</u>. A datum on a scheme (or a locally ringed topos) $X = U \cup V$ (U, V open) is a 5-tuple (A, B, P, Q, ζ) where A, B are Azumaya algebras on U, V respectively, Q, P vector bundles on $U \cap V$ and ζ is an isomorphism of $\mathcal{O}_{U\cap V}$ algebras

$$\zeta : A|_{U\cap V} \otimes \underline{\operatorname{End}}_{\mathcal{O}_{U\cap V}}(P) \xrightarrow{\sim} B|_{U\cap V} \otimes \underline{\operatorname{End}}_{\mathcal{O}_{U\cap V}}(Q) .$$

<u>PROPOSITION P</u>. One can assign to every datum α on X a cohomology class $c_\alpha \in H^2(X, G_m)$ such that the following properties hold :

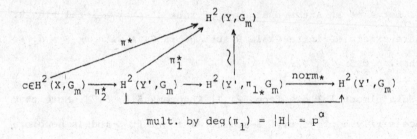

(1) $\alpha \simeq \alpha' \Rightarrow c_\alpha = c_{\alpha'}$.

(2) If we change the roles of U and V, c_α remains the same.

(3) $c_{\alpha\otimes\alpha'} = c_\alpha + c_{\alpha'}$ where $\alpha\otimes\alpha'$ is defined by taking $\otimes$ product on each of the components.

(4) $c_\alpha|_U = \delta(cl(A))$, $c_\alpha|_V = \delta(cl(B))$.

(5) If $P = Q = \mathcal{O}_{U\cap V}$, we obtain via ζ a glued Azumaya algebra C on X with isomorphisms $C|_U \xrightarrow[\varphi_1]{\sim} A$, $C|_V \xrightarrow[\varphi_2]{\sim} B$ s.t. $\zeta\circ\varphi_1|_{U\cap V} = \varphi_2|_{U\cap V}$. Then $c_\alpha = \delta(cl(C))$.

(6) If $A = \mathcal{O}_U$, $B = \mathcal{O}_V$ and $\zeta : \underline{End}(P) \xrightarrow{\sim} \underline{End}(Q)$ is induced from an isomorphism $P \xrightarrow[\varphi_3]{\sim} Q$ then $c_\alpha = 0$.

(7) Suppose $P = P_1 \otimes R$ and R extends to a bundle $\bar{R}$ on U. Then $c_{(A\otimes\underline{End}(\bar{R}),B,P_1,Q,\zeta)} = c_{(A,B,P,Q,\zeta)}$. Similarly when $Q = Q_1 \otimes R$ and R extends to $\bar{R}$ on V.

(8) If $A = \mathcal{O}_U$, $B = \mathcal{O}_V$, $P = \mathcal{O}_{U\cap V}$, rk Q = 1, then $c_\alpha = \delta_{MV}cl(Q)$ where $cl(Q) \in H^1(U\cap V, G_m)$ and δ_{MV} is the connecting map in the Mayer-Vietoris sequence

$$H^1(U,G_m) \oplus H^1(V,G_m) \to H^1(U\cap V,G_m) \xrightarrow{\delta} H^2(X,G_m)$$

which is obtained from the sequence of functors

$$\Gamma(X,F) \longrightarrow \Gamma(U,F) \oplus \Gamma(V,F) \longrightarrow \Gamma(U\cap V,F).$$

$$\alpha \longmapsto (\alpha|_U, \alpha|_V) \quad (\alpha,\beta) \longmapsto (\alpha|_{U\cap V} - \beta|_{U\cap V}).$$

Proof of the Proposition.

Given a datum α and $\Omega \in X_{ét}$ we define $Ob(\mathcal{G}_\Omega)$ to be the set of quintuples $(E,F,\varphi_1,\varphi_2,\varphi_3)$ s.t. E is a vector bundle on $\Omega \times U$,[1] F is

---

(1) The cartesian products are taken in the category $X_{ét}$.

a vector bundle on $\Omega \times V$,

$$\varphi_1 : A|_{\Omega\times U} \xrightarrow{\sim} \underline{\mathrm{End}}(E)$$

$$\varphi_2 : B|_{\Omega\times V} \xrightarrow{\sim} \underline{\mathrm{End}}(F)$$

$$\varphi_3 : E|_{\Omega\times U\times V} \otimes P|_{\Omega\times U\times V} \xrightarrow{\sim} F|_{\Omega\times U\times V} \otimes Q|_{\Omega\times U\times V}$$

such that the following diagram commutes. (In general, if $F \xrightarrow[u]{} G$ is an isomorphism of $\mathcal{O}$-modules, we denote by $\underline{\mathrm{End}}(u)$ the deduced isomorphism $\underline{\mathrm{End}}_{\mathcal{O}}(F) \longrightarrow \underline{\mathrm{End}}_{\mathcal{O}}(G)$ of their endomorphism sheaves.)

$$\begin{array}{ccc}
p_2^*A \otimes \underline{\mathrm{End}}(p_{23}^*P) & \xrightarrow[\sim]{p^*_{23}\zeta} & p_3^*B \otimes \underline{\mathrm{End}}(p_{23}^*Q) \\
\wr\downarrow p_{12}^*\varphi_1 \otimes \mathrm{id} & & \wr\downarrow p_{13}^*\varphi_2 \otimes \mathrm{id} \\
p_{12}^*\,\underline{\mathrm{End}}(E) \otimes \underline{\mathrm{End}}(p_{23}^*P) & & p_{13}^*\,\underline{\mathrm{End}}(F) \otimes \underline{\mathrm{End}}(p_{23}^*Q) \\
\wr\downarrow \mathrm{can.} \otimes \mathrm{id} & & \wr\downarrow \mathrm{can.} \otimes \mathrm{id} \\
\underline{\mathrm{End}}(p_{12}^*E) \otimes \underline{\mathrm{End}}(p_{23}^*P) & & \underline{\mathrm{End}}(p_{13}^*F) \otimes \underline{\mathrm{End}}(p_{23}^*Q) \\
\mathrm{can.}\ \downarrow\wr & & \wr\downarrow \mathrm{can.} \\
\underline{\mathrm{End}}(p_{12}^*E \otimes p_{23}^*P) & \xrightarrow[\underline{\mathrm{End}}(\varphi_3)]{\sim} & \underline{\mathrm{End}}(p_{13}^*F \otimes p_{23}^*Q)
\end{array}$$

where p_i, p_{ij} on $\Omega \times U \times V$ are the projections to factors or partial products in the category $X_{\text{ét}}$.

If $\Omega \xrightarrow{u} \Omega'$, $x \in \mathrm{Ob}(G_\Omega)$ $x' \in \mathrm{Ob}(G_\Omega)$, define

$\mathrm{Hom}_u(x,x') = \{(f_1,f_2) \mid f_1 : E \xrightarrow{\sim} u^*E',\ f_2 : F \xrightarrow[\sim]{} u^*F'$ such that

$$\underline{\mathrm{End}}(f_i)\varphi_i = u^*\varphi_i',\ i=1,2;\ (f_2 \otimes \mathrm{id}_Q)\circ\varphi_3 = u^*\varphi_3'\circ(f_1 \otimes \mathrm{id}_P)$$

on $\Omega \times U \times V\}$.

One checks that with the obvious definition of composition of arrows we get a category G with a functor $G \xrightarrow{\theta} X_{\text{ét}}$ s.t. $\mathrm{Ob}(G_\Omega) = \mathrm{Ob}(\theta^{-1}(\Omega))$ and $\mathrm{Hom}_u(x,y) = \{f \in \mathrm{Hom}_G(x,y) \mid \theta(f) = u\}$. Define $G_\Omega = \theta^{-1}(\Omega)$. Moreover G is clearly a champ. Define $G_\alpha = G$.

<u>ASSERTION</u>. G is a G_m-gerbe where G_m acts on G by the homomorphisms

$$\lambda \in G_m(\Omega) \longmapsto (\lambda,\lambda) \in \mathrm{Aut}_{\mathrm{id}_\Omega}(x) \qquad \text{for every } x \in G_\Omega.$$

<u>Proof</u>. (See a better proof in a remark on page 205.)

A champ G is defined to be a gerbe iff it has the following properties : (see [3, III 2.1.1])

(i) each fiber G_Ω is a groupoid.

(ii) G is locally non-empty, i.e., there exists a covering family $\Omega_i \in X_{\text{ét}}$ s.t. $\mathrm{Ob}(G_{\Omega_i}) \neq \emptyset \quad \forall_i$.

(iii) given $x,y \in \mathrm{Ob}(G_\Omega) \exists$ a covering family $\Omega_i \xrightarrow{u_i} \Omega$ s.t. $u_i^* x \simeq u_i^* y$ in G_{Ω_i}.

We check these properties for our G.

<u>Property (i)</u> holds by definition.

<u>Property (ii)</u>. There exist covering families

$\{U_i \to U\}_{i \in I}$, $\{V_j \to V\}_{j \in J}$ s.t. $\forall_i\ A|_{U_i} \simeq M_{n_i}(\mathcal{O}_{U_i})$ and $\forall_j\ B|_{V_j} \simeq M_{n_j}(\mathcal{O}_{V_j})$

for some n_i's, n_j's $\in \mathbb{Z}_+$. Then $\{U_i\}_{i\in I} \coprod \{V_j\}_{j \in J}$ is a covering family of X. We will prove that $G_{U_i} \neq \emptyset$ and $G_{V_j} \neq \emptyset$. By symmetry it is enough to check it for each U_i. We can also assume by restricting to U_i that $X = U$ and that $A \simeq M_n(\mathcal{O}_U)$ for some $n \in \mathbb{Z}_+$. Actually it suffices that there exists an isomorphism $A \xrightarrow[\varphi_1]{\sim} \underline{\mathrm{End}}(E)$ for some vector bundle E.

Then we have $\underline{\text{End}}(E)\big|_V \otimes \underline{\text{End}}(P) \simeq B \otimes \underline{\text{End}}(Q)$ so $[B] = 0$ in $\text{Br}(V)$. Then it is known that $B \xrightarrow[\varphi_2]{\sim} \underline{\text{End}}(F)$ for some vector bundle F on V (e.g., using $\text{Br}(V) \xhookrightarrow[\alpha_V]{} H^2(Y, G_m)$ and the nonabelian exact cohomology sequence in the statement of Theorem 1). Looking at the diagram on page 177 in which all sides are defined except $\underline{\text{End}}(\varphi_3)$ we get an isomorphism

$$\underline{\text{End}}(E\big|_V \otimes P) \xrightarrow[\xi]{\sim} \underline{\text{End}}(F \otimes Q) .$$

Our problem is to express ξ as $\underline{\text{End}}(\varphi_3)$ for some isomorphism φ_3. It is known (see [2] I 5.10) that such φ_3 exists locally in the Zariski topology on V, and that it is unique up to multiplication by a unit. We consider local $\varphi_{3,i}$'s on an open cover $V = \bigcup_i V_i$ and write $\varphi_{3,i} = \lambda_{ij} \cdot \varphi_{3,j}$, $\lambda_{ij} \in G_m(V_j \cap V_i)$. If L is a line bundle defined by the cocycle $\{\lambda_{ij}\}$, ξ is induced by an isomorphism $E\big|_V \otimes P \xrightarrow[\varphi_3]{\sim} F \otimes L \otimes Q$ so we have to replace F by $F \otimes L$.

Property (iii). Suppose we are given

$x = (E, F, \varphi_1, \varphi_2, \varphi_3)$, $y = (E', F', \varphi_1', \varphi_2', \varphi_3') \in \text{Ob}(G_\Omega)$.

Localizing, we can assume $X = U$ or $X = V$. By symmetry, we may assume $X = U$. On Ω we get an isomorphism $\underline{\text{End}}(E) \xrightarrow[\varphi_1'\varphi_1^{-1}]{\sim} \underline{\text{End}}(E')$. Covering Ω we can assume

$$\exists\, f_1 : E \xrightarrow{\sim} E' \quad \text{s.t.} \quad \varphi_1' = \underline{\text{End}}(f_1)\varphi_1 .$$

On $\Omega \times V$ we have $\varphi_2'\varphi_2^{-1} : \underline{\text{End}}(F) \xrightarrow{\sim} \underline{\text{End}}(F')$, so $\exists$ covering family $\Omega_i \longrightarrow \Omega \times V$ s.t. $\varphi_2'\varphi_2^{-1}\big|_{\Omega_i} = \underline{\text{End}}(f_{2,i})$, $f_{2,i} : F\big|_{\Omega_i} \xrightarrow{\sim} F'\big|_{\Omega_i}$. Now the diagram on page 177 determines $\underline{\text{End}}(\varphi_3)$ in terms of φ_1, φ_2. It follows that the diagram

$$\begin{array}{ccc} E|_{\Omega_i} \otimes P|_{\Omega_i} & \xrightarrow[\varphi_3|_{\Omega_i}]{\sim} & F|_{\Omega_i} \otimes Q|_{\Omega_i} \\ \wr\big\downarrow f_1|_{\Omega_i} \otimes id & & \wr\big\downarrow f_{2,i} \otimes id \\ E'|_{\Omega_i} \otimes P|_{\Omega_i} & \xrightarrow[\varphi'_3|_{\Omega_i}]{\sim} & F'|_{\Omega_i} \otimes Q|_{\Omega_i} \end{array}$$

is such that going 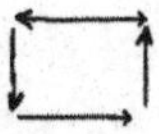 gives multiplication by some $\lambda_i \in G_m(\Omega_i)$ since it commutes after taking End. Then we replace $f_{2,i}$ by $\lambda_i \cdot f_{2,i}$ and so the diagram will commute. It follows from the commutativity that $f_{2,i}|_{\Omega_{ij}} = f_{2,j}|_{\Omega_{ij}}$ where $\Omega_{ij} = \Omega_i \underset{\Omega}{\times} \Omega_j$. Hence $\{f_{2,i}\}$ patch to $f_2 : F \xrightarrow{\sim} F'$. Then $(f_1, f_2) \in \mathrm{Isom}_{id_\Omega}(x,y)$.

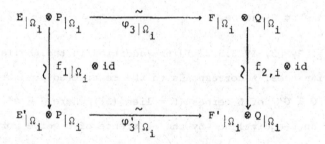

Property (iv). We have to check that

$$G_m(\Omega) \xrightarrow{\sim} \mathrm{Aut}_{G_\Omega}(x) \qquad \forall\, x \in G_\Omega .$$

This is obvious because if $(f_1, f_2) \in \mathrm{Aut}(E_1, E_2, \varphi_1, \varphi_2, \varphi_3)$ then $\underline{\mathrm{End}}(f_i)\varphi_i = \varphi_i$ so f_i is multiplication by $\lambda_i \in G_m(U_i \times \Omega)$ $(U_1 \overset{\text{def}}{=} U,\ U_2 \overset{\text{def}}{=} V)$ and $\lambda_1|_{U\times V\times\Omega} = \lambda_2|_{U\times V\times\Omega}$ because $(f_2 \otimes id_Q) \circ \varphi_3 = \varphi_3 \circ (f_1 \otimes id_P)$ on $U \times V \times \Omega$.

Now for a datum α define $c_\alpha = cl(G_\alpha) \in H^2(X_{\text{ét}}, G_m)$. Using the Giraud definition of non abelian H^2 [3, IV 3] and the comparison with abelian H^2 [3, IV 3.4] we verify now properties (1)-(8) in Proposition P.

Proof of (1). (1) follows because G_α is G_m-isomorphic to $G_{\alpha'}$.

Proof of (2). By the construction, if you change the order of U,V you get an isomorphic G_m-gerbe.

Proof of (3). By [3, IV 3.3.2(i)] the addition in the abelian $H^2(A)$ for an abelian sheaf A corresponds to the contracted product $[G],[G'] \longrightarrow [G \overset{L}{\wedge} G']$ of L-gerbes (L = lien (A)) where $G \overset{L}{\wedge} G'$ is determined up to an L-equivalence by the existence of a gerbe morphism $G \underset{X_{\text{ét}}}{\times} G' \to G \overset{L}{\wedge} G'$ equivariant w.r.t. $A \times A \xrightarrow[\text{sum}]{} A$.

In our case we define a morphism $G_\alpha \underset{X_{\text{ét}}}{\times} G_{\alpha'} \to G_{\alpha\otimes\alpha'}$ whose action on objects is

$$((E,F,\varphi_1,\varphi_2,\varphi_3),\ (E',F',\varphi_1',\varphi_2',\varphi_3')) \longmapsto (E\otimes E', F\otimes F', \varphi_1\otimes\varphi_1', \varphi_2\otimes\varphi_2', \varphi_3\otimes\varphi_3')$$

and similarly for morphisms, and one can check it has the necessary properties.

Proof of (4). For an Azumaya algebra A on a locally ringed site X (in [3] X is assumed to be a topos) Giraud defines [3, V 4.2] a gerbe d(A) s.t. $\mathrm{Ob}(d(A)_\Omega) = \{(E,u) \mid E$ vector bundle on $X_{/\Omega}, u : A_{|\Omega} \xrightarrow{\sim} \underline{\mathrm{End}}(E)\}$ with a natural definition of morphisms. This is a G_m-gerbe via the action of $G_m(\Omega)$ on E by scalar multiplication. Moreover, by [3, V, Remark 4.5],

$$\mathrm{cl}(d(A)) = \delta(\mathrm{cl}(A)) \text{ (defined via } H^1(X,PGL(n)) \xrightarrow{\delta} H^2(X,G_m)).$$

By [3, V 1.5] the pull back maps $H^2(E,A) \to H^2(E',f^*A)$ for a morphism of sites $E' \to E$ (A abelian) defined by homological algebra are equal to those defined by a suitable notion of pull-back of gerbes from E to E'. If $E' = E_{/U}$, $U \in \mathrm{Ob}\, E$, one checks that (up to an equivalence respecting the action of the structural lien) the pull-back of G is just $G \underset{E}{\times} (E_{/U})$. Now $(E,F,\varphi_1,\varphi_2,\varphi_3) \mapsto (E,\varphi_1)$, $(f_1,f_2) \mapsto f_1$

defines a morphism and hence an equivalence of G_m-gerbes $G_\alpha \underset{X_{ét}}{\times} U_{ét} = G_{\alpha}|_U \to d(A)$ so (4) follows. Similarly for V (or use symmetry (2)).

Proof of (5). Suppose $P = Q = \mathcal{O}_{U\cap V}$ and C,φ_1,φ_2 are as in the statement. Now we define a G_m-morphism $d(C) \to G_\alpha$ by

$$\text{Ob } d(C)_\Omega \ni (E,\varphi) \mapsto (E_{|\Omega\times U}, E_{|\Omega\times V}, \varphi_{|\Omega\times U}\varphi_1^{-1}{}_{|\Omega\times U}, \varphi_{|\Omega\times V}\varphi_2^{-1}{}_{|\Omega\times V}, \text{id}) \in \text{Ob } G_{\alpha_\Omega}$$

$$\text{Mor } d(C) \ni f \mapsto (f_{|\Omega\times U}, f_{|\Omega\times V}) \in \text{Mor } G_\alpha .$$

Proof of (6). $(\mathcal{O}_U,\mathcal{O}_V,\text{id},\text{id},\varphi_3) \in \text{Ob } G_{\alpha_X}$ so $cl(G_\alpha)$ is trivial.

Proof of (7). Let $\alpha = (A,B,P,Q,\zeta)$, $\alpha' = (A\otimes \underline{\text{End}}(\bar{R}),B,P_1,Q,\zeta)$; then one can define a G_m-morphism of gerbes $G_\alpha \to G_\alpha$ by

$$\text{Ob } G_{\alpha_\Omega} \ni (E,F,\varphi_1,\varphi_2,\varphi_3) \mapsto (E\otimes\bar{R},F,\varphi_1\otimes \text{id}_{\underline{\text{End}}(\bar{R})},\varphi_2,\varphi_3) \in \text{Ob } G_{\alpha'_\Omega},$$

where here $\bar{R}$ is restricted to $\Omega\times U$,

$$\text{Mor}_u\, G_\alpha \ni (f_1,f_2) \mapsto (f_1 \otimes \text{id}_{\bar{R}|_{\Omega\times U}}, f_2) \in \text{Mor}_u\, G_{\alpha'},$$

so $cl(G_\alpha) = cl(G_{\alpha'})$. Similarly for U replaced by V.

Proof of (8). Assume A,B,P,Q are as in the statement. $Y \overset{\text{def}}{=} U\cap V$. Now in the definition of Ob G_α φ_1 and φ_2 are uniquely determined so G_α is isomorphic to a gerbe G' s.t. $\text{Ob } G'_\Omega = \{(L_1,L_2,\varphi) \mid L_1$ line bundle on $\Omega \underset{X}{\times} U$, L_2 line bundle on $\Omega\times V$, $\varphi : L_{1|\Omega\times Y} \xrightarrow{\sim} Q_{|\Omega\times Y} \otimes L_{2|\Omega\times Y}\}$. Now $L \overset{\xi}{\mapsto} \underline{\text{Isom}}_{\mathcal{O}}(\mathcal{O},L)$ is a G_m-morphism from the G_m-gerbe of line bundles to the G_m-gerbe of right G_m-torseurs. Moreover, we have an obvious map $\underline{\text{Isom}}(\mathcal{O},L_1) \times \underline{\text{Isom}}(\mathcal{O},L_2) \to \underline{\text{Isom}}(\mathcal{O},L_1\otimes L_2)$ which is functorial, commuting with restriction and equivariant w.r.t. $G_m \times G_m \xrightarrow[\text{sum}]{} G_m$. We recall that if P,P' are right A-torseurs for an abelian sheaf A, $P \overset{A}{\wedge} P'$ is the A-torseur obtained from the A×A-torseur P×P' by "extending the structural group" via $A \times A \xrightarrow{\text{sum}} A$. Thus one gets an isomorphism

$$\underline{\mathrm{Isom}}(\mathcal{O},L_1) \overset{G_m}{\wedge} \underline{\mathrm{Isom}}(\mathcal{O},L_2) \xrightarrow[\eta(L_1,L_2)]{} \underline{\mathrm{Isom}}(\mathcal{O},L_1 \otimes L_2) .$$

Define a G_m-gerbe $\mathcal{G}''$ over $X_{\acute{e}t}$ by

(i) $\mathrm{Ob}\ \mathcal{G}''_\Omega = \{(P_1,P_2,\varphi) \mid P_1$ a G_m-torseur on $(U \times \Omega)_{\acute{e}t}$,

P_2 a G_m-torseur on $(\Omega \times V)_{\acute{e}t}$,

$\varphi : P_1|_{\Omega\times Y} \xrightarrow{\sim} P|_{\Omega\times Y} \overset{G_m}{\wedge} P_2|_{\Omega\times Y}\}$ where $P \overset{\text{def}}{=\!=} \mathrm{Isom}_{\mathcal{O}}(\mathcal{O},Q)$

(ii) for $s : \Omega \to \Omega'$ and $(x,x') \in \mathrm{Ob}\ \mathcal{G}''_\Omega \times \mathrm{Ob}\ \mathcal{G}''_{\Omega'}$,

$\mathrm{Mor}_s(x,x') = \{(f_1,f_2) \mid f_1 : P_1 \xrightarrow{\sim} s^*P'_1,\ f_2 : P_2 \xrightarrow{\sim} s^*P'_2$

s.t. $((\mathrm{id}_P) \wedge f_2) \circ \varphi = s^*\varphi' \circ f_1$ on $\Omega \times Y\}$.

The G_m-action is :

$$\lambda \in G_m(\Omega) \longmapsto (\lambda,\lambda) \in \underline{\mathrm{Aut}}_{\mathcal{G}''_\Omega}(P_1,P_2,\varphi) .$$

Define a G_m-morphism $\mathcal{G}' \to \mathcal{G}''$ by

$$\mathrm{Ob}\ \mathcal{G}'_\Omega \ni (L_1,L_2,\varphi) \longmapsto (\xi(L_1),\xi(L_2),\ \eta(Q|_{\Omega\times Y},L_2|_{\Omega\times Y})^{-1} \circ \xi(\varphi)) \in \mathrm{Ob}\ \mathcal{G}''_\Omega$$

$$\mathrm{Mor}\ \mathcal{G}' \ni (f_1,f_2) \longmapsto (\xi(f_1),\xi(f_2)) \in \mathrm{Mor}\ \mathcal{G}'' .$$

$\mathcal{G}''$ makes sense for any abelian sheaf A on $X_{\acute{e}t}$ and a right $A|_Y$-torseur P, or more generally on a site S given together with two subsheaves U,V of the final sheaf e on S such that U,V cover e. By [3, III 3.6] we have a commutative diagram of isomorphisms

$$\begin{array}{ccc} H^1_{\text{non ab}}(X_{\text{ét}},A) & \xrightarrow[\alpha_1]{\sim} & H^1_{\text{ab}}(X_{\text{ét}},A) \\ & \underset{\alpha_3}{\searrow\!\sim} \qquad \underset{\alpha_2}{\wr\!\swarrow} & \\ & H^1(X_{\text{ét}},A) & \end{array} .$$

α_1 and α_2 are defined by commutativity with connecting homomorphisms and α_3 is defined without assuming A is abelian by associating to a torseur P with sections $p_i \in P(\Omega_i)$ over a covering family $\{\Omega_i\}_{i\in I}$ of $X_{\text{ét}}$ the cocycle $\{u_{ij}\}$ defined by $p_i|_{\Omega_i\times\Omega_j} \cdot u_{ij} = p_j|_{\Omega_i\times\Omega_j}$.

For the case $P = \underline{\text{Isom}}_{\mathcal{O}}(\mathcal{O},L)$ (and in similar cases) we get $p_i^{-1}|_{\Omega_{ij}} = u_{ij}p_j^{-1}|_{\Omega_{ij}} : L|_{\Omega_{ij}} \xrightarrow{\sim} \mathcal{O}_{\Omega_{ij}}$, thus $\alpha_1([P])$ is the class usually associated with L. Now it is enough to prove for any A,P that

$$\alpha^{(2)}([G''_P]) = \delta_{MV}(\alpha_1([P]))$$

where $\alpha^{(2)} : H^2_{\text{non ab}}(X_{\text{ét}},A) \xrightarrow{\sim} H^2_{\text{ab}}(X_{\text{ét}},A)$ is the isomorphism ζ^{-1} of page 174. Notice that $\alpha_1([P]) = \alpha_2(\alpha_3([P]))$ is what is denoted by cl(Q) in the statement of Proposition P(8). We denote $\alpha_1([P])$ also by cl(P). $\alpha^{(2)}$ is compatible with the connecting homomorphisms. Let $0 \to A \to J_0 \to J_1 \to \dots$ be an injective resolution of A. As $H^1(Y,J_0) = 0$, we can choose an isomorphism $P \overset{A}{\wedge} J_0 \xrightarrow[\xi]{\sim} J_0$ of J_0-torseurs. Then we get an A-embedding $P \overset{}{\underset{\xi'}{\hookrightarrow}} J_0$. Let $\pi : J_0 \to J_0/A$ be the projection. Then $\text{Im}(\pi\xi')$ is the subsheaf generated by some $s \in \Gamma(Y,J_0/A)$, i.e., $P \xrightarrow[\xi]{\sim} \pi^{-1}(s)$. By definition $[P] = \delta_{\text{non ab}}(s)$ so $\alpha_1([P]) = \delta_{\text{ab}}(s)$. (The δ's come from the exact sequence $0 \to A|_Y \to J_0|_Y \to (J_0/A)|_Y \to 0$.) The Mayer-Vietoris sequence is constructed from the exact sequence of complexes

$$0 \to J_i(X) \to J_i(U) \oplus J_i(V) \to J_i(U \cap V) \to 0 .$$

$$\alpha \longmapsto (\alpha_{|U}, \alpha_{|V}) \qquad (\alpha, \beta) \longmapsto (\alpha_{|Y} - \beta_{|Y})$$

Using [10, IV 2.1] if $0 \to A' \to A \to A'' \to 0$ then the diagram

$$\begin{array}{ccc} H^p(Y,A'') & \xrightarrow{\delta_{MV}} & H^{p+1}(X,A'') \\ \downarrow \delta_{ab} & & \downarrow \delta_{ab} \\ H^{p+1}(Y,A) & \xrightarrow{\delta_{MV}} & H^{p+2}(X,A) \end{array}$$ anti commutes.

Thus $\delta_{MV}(cl(P)) = -\delta_{ab}\delta_{MV}(s)$. We first give a torseur interpretation of $\delta_{MV}(a)$ for any $a \in \Gamma(Y,A)$, A an abelian sheaf. Let $0 \to A \overset{\varepsilon}{\hookrightarrow} J_0 \xrightarrow{d_0} J_1 \xrightarrow{d_1} \ldots$ be an injective resolution. Write :

$$\varepsilon a = s_1|_Y - s_2|_Y$$

$$s_1 \in J_0(U), \qquad s_2 \in J_0(V) .$$

By definition $\delta_{MV}(a)$ is represented by a section $\sigma \in J_1(X)$ s.t. $\sigma_{|U} = d_0 s_1$, $\sigma_{|V} = d_0 s_2$. By [10, V 7.1] (adapted to injective resolutions with $n = 0$, $i = 0$) $\delta_{MV}(a) = -\delta_{ab}(\sigma)$. ($\delta_{ab}$ comes from the exact sequence $0 \to A \xrightarrow{\varepsilon} J_0 \xrightarrow{d_0} \underline{Ker}(d_1) \to 0$.) By definition $\delta_{non\ ab}(\sigma) = [d_0^{-1}\sigma]$. In general, if G is a sheaf of groups on $X_{ét}$ and $g \in G(Y)$ then there is a triple (Q_g, s_1, s_2) where Q_g is a G-torseur, $s_1 \in Q_g(U)$, $s_2 \in Q_g(V)$, $s_{1|Y} = (s_{2|Y}) \cdot g$, and it is unique up to a unique isomorphism. $(d_0^{-1}\sigma, s_1, s_2)$ is such a triple for A and $a \in A(Y)$. So we obtain that $-\delta_{MV}(a) = \alpha_1([Q_a])$. We now return to P. We get

$$\delta_{MV}(\alpha_1([P])) = \delta_{ab}(-\delta_{MV}(s))) = \delta_{ab}(\alpha_1[Q_s]) = \alpha^{(2)}\ \delta_{non\ ab}([Q_s]) .$$

By definition $\delta_{\text{non ab}}([Q_s])$ (w.r.t. the exact sequence $0 \to A \to J_0 \xrightarrow[\pi]{} J_0/A \to 0$) is represented by the A-gerbe $K_\pi(Q_s) = K$ defined by

(i) $\quad \operatorname{Ob} K_\Omega = \{(R,\varphi) \mid R \text{ a } J_{0|\Omega}\text{- torseur}, \ \varphi : R \overset{J_0}{\wedge} (J_0/A) \xrightarrow{\sim} Q_{s|\Omega}\}$

(ii) $\quad$ for $u : \Omega \to \Omega'$,

$$\operatorname{Hom}_u((R,\varphi),(R',\varphi')) = \{f : R \xrightarrow{\sim} u^*R' \mid u^*\varphi' \circ (f \overset{J_0}{\wedge} (J_0/A)) = \varphi\}.$$

Hence the proof will be finished if we give an A-morphism $G'' \to K_\pi(Q_s)$. Now given $(P_1,P_2,\varphi) \in \operatorname{Ob} G''_\Omega$ we have a canonical isomorphism of $J_{0|\Omega\times Y}$-torseurs (proven by universal properties) $(P \overset{A}{\wedge} P_2) \overset{A}{\wedge} J_0 \xrightarrow[w]{\sim} (P \overset{A}{\wedge} J_0) \overset{J_0}{\wedge} (P_2 \overset{A}{\wedge} J_0)$. (All sheaves here and in the definition of ψ below are to be restricted to $\Omega \times Y$.) Define

$$\psi = \text{can.} \circ (\xi \overset{J_0}{\wedge} \text{id}_{P_2 \overset{A}{\wedge} J_0}) w(\varphi \overset{A}{\wedge} J_0) : (P_1 \overset{A}{\wedge} J_0)|_{\Omega\times Y} \xrightarrow{\sim} (P_2 \overset{A}{\wedge} J_0)|_{\Omega\times Y} .$$

Thus $P_1 \overset{A}{\wedge} J_0$ and $P_2 \overset{A}{\wedge} J_0$ patch to a $J_{0|\Omega}$-torseur P' on $\Omega_{\text{ét}}$ with isomorphisms

$$\zeta_U : P'|_{U\times\Omega} \xrightarrow{\sim} P_1 \overset{A}{\wedge} J_0 \ , \quad \zeta_V : P'|_{V\times\Omega} \xrightarrow{\sim} P_2 \overset{A}{\wedge} J_0$$

s.t. $\quad \psi \cdot (\zeta_U|_{Y\times\Omega}) = \zeta_V|_{Y\times\Omega}$. Consider

$$\zeta_U^{-1} \overset{J_0}{\wedge} (J_0/A) : P_1 \overset{A}{\wedge} J_0 \overset{J_0}{\wedge} (J_0/A) \xrightarrow{\sim} (P' \overset{J_0}{\wedge} (J_0/A))|_{U\times\Omega}.$$

As the LHS is canonically $\simeq J_0/A$ as a J_0/A-torseur we obtain

$$\zeta'_U : (J_0/A)|_{U\times\Omega} \xrightarrow{\sim} (P' \overset{J_0}{\wedge} (J_0/A))|_{U\times\Omega} .$$

Let $\sigma_1 = \zeta'_U(0) \in (P' \overset{J_0}{\wedge} (J_0/A))(U \times \Omega)$. Similarly we define ζ'_V and $\sigma_2 = \zeta'_V(0) \in (P' \overset{J_0}{\wedge} (J_0/A))(V \times \Omega)$.

We have a commutative diagram (with a triangle of J_0/A-torseurs in its RHS) in which all sheaves are understood to be restricted to $\Omega \times Y$:

$$\begin{array}{ccccc} P_1 & \xrightarrow{\quad o \quad} & J_0/A & \xrightarrow{\quad \zeta'_U \quad} & P' \overset{J_0}{\wedge} (J_0/A) \\ \Big\downarrow \varphi & & \psi \overset{J_0}{\wedge}(J_0/A) \ \wr\Big\downarrow & & \Big\uparrow \zeta'_V \\ & & J_0/A & \longrightarrow & \\ & & \Big\uparrow \pi \text{ essentially} & & \\ P \overset{A}{\wedge} P_2 & \xrightarrow{\quad \xi' \overset{A}{\wedge} P_2 \quad} & J_0 \overset{A}{\wedge} P_2 & . & \end{array}$$

As $\pi\xi'$ "=" $s \in (J_0/A)(Y)$ we deduce $\psi \overset{J_0}{\wedge} (J_0/A)$ = action of $s|_{\Omega \times Y}$. Hence (if we write the action multiplicatively)

$$\zeta'_U(0) = \sigma_1|_{\Omega \times Y} = \sigma_2|_{\Omega \times Y} \cdot s|_{\Omega \times Y} .$$

We define $\psi_1 : P' \overset{J_0}{\wedge} (J_0/A) \xrightarrow{\sim} Q_{s|\Omega}$ such that $\psi_1(\sigma_1) = s_1$, $\psi_1(\sigma_2) = s_2$. So $(P', \psi_1) \in \mathrm{Ob}\, K_\Omega$. One checks that the construction is functorial w.r.t. restrictions and isomorphisms, so we get a map of gerbes $G'' \xrightarrow{\Phi} K_\pi(Q_s)$. To check the equivariance of Φ w.r.t. the A-actions note that $a \in A(\Omega)$ acts on $(P_1, P_2, \varphi) \in \mathrm{Ob}\, G''_\Omega$ by the right A-actions on the A-torseurs P_1, P_2 and that ζ_U, ζ_V are equivariant w.r.t. the action of any $a \in J_0(\Omega)$.

This finishes the proof of Proposition P.

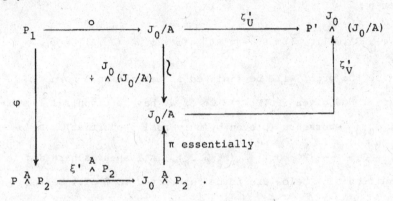

We now continue the proof of Lemma 2 from page 174. If $Y = \emptyset$, define $[C] \in Az_{st}(X)$ by $C|_U = M_t(A)$ $C|_V = M_s(B)$. Then $\delta(cl(C)) = c$. Hence we may assume that $Y \neq \emptyset$. Using the monomorphism $Br(Y) \hookrightarrow H^2(Y,G_m)$ we see that $[A|_Y] = [B|_Y]$ in $Br(Y)$. Hence $[A \otimes B^o] = 0$. Hence $A|_Y \otimes B^o|_Y \simeq \underline{End}(E)$ for some E on Y s.t. $rk\,E = st$. So $A|_Y \otimes \underline{End}(B|_Y) \simeq A|_Y \otimes B^o|_Y \otimes B|_Y \simeq B|_Y \otimes \underline{End}(E)$. Define the composite isomorphism to be ζ. Define $P = B|_Y$, $Q = E$. Then $\alpha = (A,B,P,Q,\zeta)$ is a datum on X. Moreover $\Sigma \not|\, rk\,A$, $rk\,B$, $rk\,P$, $rk\,Q$. We will exclude in the proof the trivial case $n = 1$. By Proposition P(4) $c_\alpha|_U = \delta(cl(A)) = c|_U$, $c_\alpha|_V = c|_V$. Then by the Mayer-Vietoris sequence $\exists\, x \in H^1(Y,G_m)$ s.t. $c = c_\alpha + \delta_{MV}(x)$. Let $[P] \in Pic(Y)$ be s.t. $\alpha_1([P]) = x$. Then by Proposition P(8) if $\alpha' = (\mathcal{O}_U, \mathcal{O}_V, \mathcal{O}_Y, P, \text{can.})$ then $c_{\alpha'} = \delta_{MV}(x)$. Hence by P(3) $c_{\alpha\otimes\alpha'} = c_\alpha + \delta_{MV}(x) = c$. Replacing α by $\alpha \otimes \alpha'$ we can assume $c_\alpha = c$. By P(3) $0 = nc = c_{\alpha_n}$, where $\alpha_n = \alpha \otimes \alpha \ldots \otimes \alpha$ (n times). As $\delta(cl(A^{\otimes n})) = n\cdot(c|_U) = 0$ $\exists$ vector bundle A_1 on U s.t. $A^{\otimes n} \simeq \underline{End}(A_1)$. Similarly $\exists\, B_1$ s.t. $B^{\otimes n} \simeq \underline{End}(B_1)$. Applying P(7) once for U and one for V we find

$$0 = c_{\alpha_n} = c_{(\mathcal{O}_U, \mathcal{O}_V, A_1 \otimes P^{\otimes n}, B_1 \otimes Q^{\otimes n}, \zeta^{\otimes n})} .$$

The isomorphism $\zeta^{\otimes n} : \underline{End}(A_1|_Y \otimes P^{\otimes n}) \xrightarrow{\sim} \underline{End}(B_1|_Y \otimes Q^{\otimes n})$ is (cf. page 174) induced from an isomorphism $A_1|_Y \otimes P^{\otimes n} \otimes L \xrightarrow{\sim} B_1|_Y \otimes Q^{\otimes n}$ for some $[L] \in Pic(Y)$. Thus by P(1)

$$0 = c_{(\mathcal{O}_U, \mathcal{O}_V, A_1 \otimes P^{\otimes n}, A_1 \otimes P^{\otimes n} \otimes L, \text{can.})} \underset{\overline{\overline{P(3)}}}{} \quad (\text{if we define } A_2 = A_1|_Y \otimes P^{\otimes n})$$

$$c_{(\mathcal{O}_U, \mathcal{O}_V, A_2, A_2, \text{id})} + c_{(\mathcal{O}_U, \mathcal{O}_V, \mathcal{O}_Y, L, \text{can.})} \overset{P(6),P(8)}{=\!=\!=\!=} 0 + \delta_{MV} cl(L) .$$

By the Mayer-Vietoris sequence ∃ line bundles M_1 on U and M_2 on V s.t. $L \simeq M_1|_Y \otimes M_2|_Y$. Let $A' = A_1 \otimes M_1$, $B' = B_1 \otimes M_2^{-1}$. Then A', B' are vector bundles on U,V respectively of ranks prime to Σ and

$A'|_Y \otimes P^{\otimes^n} \simeq B'|_Y \otimes Q^{\otimes^n}$.

For a scheme X let K_0X be the Grothendieck group of the exact category of locally free $\mathcal{O}_X$-modules of finite type. This is a commutative ring with 1 s.t. $[E]\cdot[E'] = [E \otimes E']$. Let $K_0A = K_0$ Spec A for a commutative ring A.

<u>LEMMA K</u>. Consider the homomorphism $K_0X \xrightarrow{rk} C(X)$, $C(X) \overset{def}{=}$ the ring of locally constant functions $X \to \mathbb{Z}$. Suppose X is affine = Spec A. Then:

(i) If $x \in K_0X$, $rk(x) = 0$, then x is nilpotent.

(ii) If $x \in K_0X$, $rk(x) > 0$, then $\exists\, n_0 \in \mathbb{Z}_+$ s.t. $\forall\, n \geq n_0$ ∃ a vector bundle E_n on X s.t. $[E_n] = n\cdot x$.

(iii) If E,F are vector bundles on X s.t. $[E] = [F]$ in K_0X, then $\exists\, n_0$ s.t. $\forall\, n \geq n_0$ $E \underset{\mathcal{O}_X}{\otimes} \mathcal{O}_X^n \simeq F \underset{\mathcal{O}_X}{\otimes} \mathcal{O}_X^n$.

<u>Proof</u>. Write $A = \varinjlim A_i$ where the A_i's are the subrings of A which are finitely generated over $\mathbb{Z}$. Then $K_0A \xleftarrow{\sim} \varinjlim K_0A_i$, $C(A) \xleftarrow{\sim} \varinjlim C(A_i)$ [where $C(A) \overset{def}{=} C(\text{Spec } A)$] etc.. Then one can reduce the proof to the case that A is finitely generated over $\mathbb{Z}$ and in particular max Spec A is a noetherian space of finite dimension d. Then by "Serre's theorem" and the cancellation theorem [14, Prop. 10.7] (ii) is true for $n_0\cdot rk\, E \geq d$ and (iii) is true for $n_0 \cdot rk\, E \geq d+1$. In (i) we actually have $x^{d+1} = 0$. This follows from [14; § 15, Cor. 15.2 and Prop. 15.3] This can be shown as follows : If $x = [E] - [\mathcal{O}_X^n]$, rk E = n, we consider complexes $K_i = \{0 \to \mathcal{O}_X^n \xrightarrow[\phi_i]{} E \to 0\}$ concentrated in degrees 0,1 $(1 \leq i \leq d+1)$ which are chosen inductively s.t. $\dim(V(\phi_1) \cap \ldots \cap V(\phi_i) \cap \max \text{Spec } A) \leq d-i$ where $V(\phi_i)$ denotes the

complement of the maximal open set over which ϕ_i is an isomorphism. Then $K_1 \otimes \ldots \otimes K_{d+1}$ is exact so $0 = [K_1 \otimes \ldots \otimes K_{d+1}] = [K_1]\cdot \ldots \cdot [K_{d+1}] = x^{d+1}$, where for a bounded complex $K = \{K^i\}_{i\in\mathbb{Z}}$ of vector bundles $[K] \overset{def}{=} \sum(-1)^n[K^n]$ and one checks that $[K\otimes K'] = [K]\cdot[K']$.

We have a section (in the category of rings) (cf. [14, page 40]) $s : C(X) \to K_0(X)$ of rk defined by the requirement that $s(\varphi) = [E_\varphi]$ if $\varphi : X \to \mathbb{Z}_0$ is locally constant where E_φ is defined by $E_\varphi|_{\varphi^{-1}(m)} = \mathcal{O}^m|_{\varphi^{-1}(m)}\ \forall\ m \in \mathbb{Z}_0$. If Σ is a set of prime numbers define $\mathbb{Z}_\Sigma$ to be the ring of those rational numbers whose denominator is prime to each $p\in\Sigma$. Define $K_0'(X) = K_0X \otimes_{\mathbb{Z}} \mathbb{Z}_\Sigma$, $C'(X) = C(X) \otimes_{\mathbb{Z}} \mathbb{Z}_\Sigma$, $rk' = rk \otimes \mathbb{Z}_\Sigma$, $s' = s \otimes \mathbb{Z}_\Sigma$. Now suppose X is affine. $K_0'(X) \xrightarrow{rk'} C'(X)$ is a surjective ring homomorphism (use s') with a kernel consisting of nilpotents (use Lemma K(i)). It follows that $rk'(x) \in C'(X)^*$ implies $x \in (K_0'X)^*$, and by Hensel's lemma that if $n \in \mathbb{Z}_\Sigma^* \cap \mathbb{Z}_+$ and $rk'x = 1$ then $\exists_1\ y \in K_0'X$ s.t. $rk'y = 1$ and $y^n = x$. For $x \in (K_0'X)^*$ define $Sx = x/s'(rk'(x)) \in (K_0'X)_{rk'=1}$.

We now return to the proof of Lemma 2. From the isomorphism $A'|_Y \otimes P^{\otimes n} \simeq B'|_Y \otimes Q^{\otimes n}$ we get $S[A'|_Y](S[P])^n = S[B'|_Y](S[Q])^n \overset{def}{=} x$. Let $z = x^{-1/n} \in K_0'(Y)_{rk'=1}$. Then $z\cdot S[P] = (S[A'|_Y])^{-1/n} = S[A']^{-1/n}|_Y \in \mathrm{Im}(K_0'(U)_{rk'=1} \to K_0'(Y))$. Similarly $z\cdot S[Q] \in \mathrm{Im}(K_0'(V)_{rk'=1} \to K_0'Y)$. $\exists\ k \in \mathbb{Z}_+$ s.t, $k\cdot z \in \mathrm{Im}(K_0Y \to K_0'Y)$ and $\Sigma \not| k$. Since $\mathbb{Z} \neq \mathbb{Z}_\Sigma$ (as $n \neq 1$) k can be chosen sufficiently large and then by Lemma K(ii) $\exists$ vector bundle R on Y s.t. $[R]' = k\cdot z$. So $rk\ R = k \in \mathbb{Z}_\Sigma^*$. $[R\otimes P]' \in \mathrm{Im}(K_0'U \to K_0'Y)$. Enlarging k multiplicatively in $\mathbb{Z}_+ \cap \mathbb{Z}_\Sigma^*$ and replacing R by $R \otimes_{\mathcal{O}_Y} \mathcal{O}_Y^{k'} = R^{k'}$ when k is changed to kk' we can assume first that $[R\otimes P] \in \mathrm{Im}(K_0U \to K_0Y)$, then that $\exists$ a vector bundle E on U of constant rank s.t. $[R\otimes P] = [E]|_Y$ (use Lemma K(ii), then that $E|_Y \simeq R\otimes P$ (by Lemma K(iii)). If we repeat those arguments with V instead of U we see that we can further assume that there exists a vector bundle F of rank k rk(Q) on U s.t. $F|_Y \simeq R\otimes Q$. By Proposition P(6) $0 = c_{(\mathcal{O}_U,\mathcal{O}_V,R,R,id)}$ so

$$c = c_\alpha \overset{P(3)}{=\!=} c_{(\mathcal{O}_U,\mathcal{O}_V,R,R,id)\otimes\alpha} = c_{(A,B,E|_Y,F|_Y,\xi)} =$$

$$\overset{P(7)_U}{=\!=} c_{(A\otimes \underline{End}(E),B,\mathcal{O}_Y,F|_Y,\xi)} \overset{P(7)_V}{=\!=} c_{(A\otimes\underline{End}(E),B\otimes\underline{End}(F),\mathcal{O}_Y,\mathcal{O}_Y,\xi)}.$$

Now we are in the situation of Proposition P(4). So we get a glued Azumaya algebra C s.t. $\delta(cl(C)) = c$ and $rk\ C = rk\ A\cdot(rk\ P\cdot k)^2 \in \mathbb{Z}^*_\Sigma$. This completes the proof of Lemma 2.

<u>COROLLARY</u>. If $X = U\cup V$, $U,V,U\cap V$ are affine, $c\in H^2(X,G_m)_{tors}$, $n = ord(c)$, $\Sigma\subset\Sigma(n)$, then $(\forall_{x\in X}(1\Sigma)(c|_{Spec(\mathcal{O}_{X,x})})) \Rightarrow (1\Sigma)c$.

<u>Proof</u>. By Lemma 2 $(1\Sigma)(c|_U)$ and $(1\Sigma)(c|_V)$ imply $(1\Sigma)c$. Hence we reduce to prove the corollary when X is affine = Spec A. $\forall\ x\in X\ \exists$ by assumption an Azumaya algebra B_x over $\mathcal{O}_{X,x}$ s.t. $rk\ B_x\in\mathbb{Z}^*_\Sigma$ and $\delta(cl(B_x)) = c|_{Spec(\mathcal{O}_{X,x})}$. By standard limit arguments $\exists\ f_x\in A - x$ (x is a prime ideal in A) s.t. A_x extends to an Azumaya algebra $\tilde{A}_x$ over $A[\frac{1}{f_x}]$ (of the same rank) and $\delta(cl(\tilde{A}_x)) = c|_{Spec(A[\frac{1}{f_x}])}$. Let us call a subset U of X good iff $U = Spec(A[\frac{1}{\varphi}])$ for some $\varphi\in A$ and $(1\Sigma)(c|_U)$. Thus it follows that X is covered by its good subsets (since Spec $A[\frac{1}{f_x}]$ is good and contains x). We prove the corollary by induction on the minimal $k\in\mathbb{Z}_+$ s.t. X is the union of k good subsets. (A posteriori k = 1.) If k = 1 we are done. If k = 2 use Lemma 2. Suppose $k > 2$. Then we can find $f_1,\dots,f_k\in A$ s.t. $U_i \overset{def}{=\!=}$ Spec $A[\frac{1}{f_i}]$ is good and $\sum_1^k f_iA = A$. Write $1 = \sum_1^k g_if_i$, $g_i\in A$. Define $W = Spec\ A[\frac{1}{1-g_1f_1}]$. Then $X = U_1\cup W$ and $W = \bigcup_{i=2}^k(W\cap U_i)$. By induction on k we get $(1\Sigma)(c|_W)$, and then apply Lemma 2.

LEMMA 3. α_X is surjective when $X = \operatorname{Spec} A$ and A is local.

Proof. Clearly there exists an étale A-algebra B s.t. $\operatorname{Spec} B \xrightarrow{\pi} \operatorname{Spec} A$ is surjective and $\pi^* c = 0$. Choose $p \in \operatorname{Spec} B$ s.t. πp = the maximal ideal of A. By a local structure theorem for étale morphisms [9, IV 18.4.6] $B_p \simeq (A[x]/(f(x)))_{m_1}$ where $f \in A[x]$ is monic and $f'(x) \notin m_1$ and m_1 is maximal. If $f(x) = x^N + \sum_{i=1}^{N} a_i x^{N-i}$ $(a_i \in A)$ define

$$C \overset{\text{def}}{=} A[T_1,\ldots,T_N]/(s_i(T_1,\ldots,T_N) - (-1)^i a_i)_{1 \le i \le N}$$

where $\quad s_i(X_1,\ldots,X_N) = \sum_{1 \le \lambda_1 < \ldots < \lambda_i \le N} X_{\lambda_1} \cdot \ldots \cdot X_{\lambda_i}$.

Let $\bar{T}_i$ be the image of T_i in C.

LEMMA S. C is a free A-module with basis $\bar{T}_1^{\alpha_1} \cdot \ldots \cdot \bar{T}_N^{\alpha_N}$, $0 \le \alpha_i \le N-i$. Let S_N act on C by $\sigma(\bar{T}_i) = \bar{T}_{\sigma(i)}, \sigma|_A = id_A$. Then S_N acts transitively on the fibers of the morphism $\operatorname{Spec} C \xrightarrow{h} \operatorname{Spec} A$.

Proof. We define by induction A-algebras C_i and monic polynomials $g_i \in C_i[t]$ of degree N-i for $0 \le i \le N$. Define $C_0 = A$, $g_0 = f$. For $N \ge i > 0$ define $C_i = C_{i-1}[T_i]/(g_{i-1}(T_i))$. Then in $C_i[t]$ we can write $g_{i-1}(t) = g_i(t) \cdot (t - \tilde{T}_i)$ where $\tilde{T}_i$ is the image of T_i in C_i. $C_N = A[T_1,\ldots,T_N]/J$ for some J. In $C_N[t]$ we have $f(t) = \prod_{i=1}^{N} (t - \tilde{T}_i)$. Thus $s_i(T_1,\ldots,T_N) - (-1)^i a_i \in J$. So if $C = A[T_1,\ldots,T_N]/J_0$ then $J_0 \subset J$. Using the factorization $f(t) = \prod_{i=1}^{N} (t - \bar{T}_i)$ in C we get by induction maps $C_i \to C$ which send $\tilde{T}_j$ to $\bar{T}_j$ for $j \le i$. The map $C_N \to C$ is then inverse to the projection $C = A[T]/J_0 \to A[T]/J = C_N$ so $C \simeq C_N$. The freeness assertion is easily verified in C_N. To check the second statement of the lemma we notice that if Ω is a field, then the fiber of the map $\operatorname{Hom}(C,\Omega) \longrightarrow \operatorname{Hom}(A,\Omega)$ [Hom is in the category of rings] over $\lambda : A \to \Omega$ is isomorphic to the set of N-tuples $(\xi_1, \ldots, \xi_N)$ in Ω s.t.

$$t^N + \sum_{i=1}^{N} \lambda(a_i) t^{N-i} = \prod_{i=1}^{N} (t - \xi_i)$$

and thus S_N acts on this fiber transitively. If $x, y \in \mathrm{Spec}(C)$ and $h(x) = h(y) = z \in \mathrm{Spec}(A)$ take Ω to be a compositum of $k(x)$ and $k(y)$ over $k(z)$. We get homomorphisms $C \xrightarrow{e_x} k(x) \xrightarrow{j_1} \Omega$ and $C \xrightarrow{e_y} k(y) \xrightarrow{j_2} \Omega$ with the same restriction to A. By the above $\exists\ \sigma \in S_N$ s.t. $j_2 e_y = j_1 e_x \sigma$. Hence $y = \mathrm{Ker}\ (j_2 e_y) = \sigma^{-1}(x) = \mathrm{Spec}(\sigma)(x)$.

Suppose $m \in \mathrm{Max}\ C$. We choose $m' \in \mathrm{Max}\ C$ s.t. $m' \cap C_1 = m_1$. By Lemma S $\exists\ \sigma \in S_N$ s.t. $\sigma m = m'$. Then we have a commutative diagram :

$$\begin{array}{ccccc}
\mathrm{Spec}(C_m) & \xrightarrow[\mathrm{Spec}(\sigma^{-1})]{\sim} & \mathrm{Spec}(C_{m'}) & \longrightarrow & \mathrm{Spec}(C_{1\,m_1}) \\
\downarrow{\scriptstyle\alpha} & & & & \wr\downarrow \\
\mathrm{Spec}(A) & \xleftarrow[\pi]{} & \mathrm{Spec}(B) & \longleftarrow & \mathrm{Spec}(B_p)\ .
\end{array}$$

Because $\pi^* c = 0$ we obtain $\alpha^* c = 0$. So $\alpha^* c$ is the cohomology class of the trivial Azumaya algebra $\mathcal{O}_{\mathrm{Spec}(C_m)}$. The same is true when m is replaced by a non maximal ideal. Hence by the corollary to Lemma 2 $h^* c$ is represented by an Azumaya algebra. Hence by Lemma 4 c is represented by an Azumaya algebra. This finishes the proof of Lemma 3 and hence the proof of Theorem 1.

PART 2. Additional Results

Using Theorem 1 it is possible to compute $H^2(X, \mathbb{G}_m)_{\mathrm{tors}}$ for Brauer-Severi schemes. $X \xrightarrow[\pi]{} S$ is called a Brauer-Severi scheme (cf. [2, I 8]) iff $\exists$ étale covering family $S_i \to S$ s.t. $X \times_S S_i \xrightarrow[\varphi_i]{\sim} \mathbb{P}^{n_i}_{S_i}$.

The n_i's patch to $n \in H^0(S,\mathbb{Z}_0)$. We recall that $\underline{\mathrm{Aut}}_S(\mathbb{P}^n_S) = \Gamma(S, PGL(n+1))$ [12, Chap 0 § 5] and (cf. [2, I 8]) that every descent datum on $X' \to S'$ where $X' \simeq \mathbb{P}^n_{S'}$, relative to an f.p.q.c. morphism $S' \to S$ is effective because $(\Omega^n_{X'/S'})^{-1}$ is a relatively very ample line bundle "equivariant" under the descent. Hence $H^1_{ét}(S, PGL(n+1))$ classifies the isomorphism classes of Brauer-Severi schemes of relative dimension n over S. Explicitly, if in the notation of the definition

$\varphi_i|_{S_i \times_S S_j} = g_{ij} \circ \varphi_j|_{S_i \times_S S_j}$, then we correspond to X the cohomology class of the 1-cocycle $\{g_{ij}\} \in Z^1(\{S_i \to S\}, PGL(n+1))$. We denote this cohomology class by cl(X).

<u>THEOREM 2</u>. Let $X \xrightarrow{\pi} S$ be a Brauer-Severi scheme. Then

$H^2(S,G_m)_{tors} \xrightarrow[\pi^*_{tors}]{} H^2(X,G_m)_{tors}$ is surjective and

$\mathrm{Ker}(\pi^*) = \delta(\mathrm{cl}(X)) \cdot H^0(S,\mathbb{Z})$.

<u>Remark</u>. The "hard" case is for p-torsion when p is not invertible in $\mathcal{O}_S$.

<u>COROLLARY</u>. $R^2\pi_{fl_*}\mu_n \xleftarrow{\sim} \mathbb{Z}/n\mathbb{Z}\ \forall\ n \in \mathbb{Z}_+$ where π_{fl} is the morphism of sites $X_{fppf} \to S_{fppf}$ defined by π.

<u>Proof of the corollary (assuming Theorem 2)</u>. Using the Kummer sequence $0 \to \mu_n \to G_m \xrightarrow{n} G_m \to 0$ on X_{fppf} we get an exact sequence

$$(*)\quad 0 \to (R^1\pi_{fl_*} G_m) \otimes_{\mathbb{Z}} (\mathbb{Z}/n) \to R^2\pi_{fl_*}\mu_n \to (R^2\pi_{fl_*} G_m)_n \to 0 .$$

$(R^2\pi_{fl_*}G_m)_n$ is the fppf sheafification of $(S' \xrightarrow[p.f.]{} S) \to H^2_{fl}(X \times_S S', G_m)_n$. But for any scheme Z, $H^2_{ét}(Z,G_m) \xrightarrow{\sim} H^2_{fl}(Z,G_m)$ by a theorem of Grothendieck [2, III Thm. 11.7] comparing étale and fppf cohomology of smooth group schemes. Thus by Theorem 2

$H^2_{fl}(S',G_m)_{tors} \twoheadrightarrow H^2_{fl}(X \times_S S',G_m)_{tors}$ so the étale and hence the fppf sheafification of the RHS is 0. Also it is known (cf. page 202) that $\mathrm{Pic}(\mathbb{P}^N_R) = \mathbb{Z}\cdot[\mathcal{O}(1)]$ if R is local. Thus $\mathbb{Z} \xrightarrow{\sim} R^1\pi_{fl_*}G_m$. Substituting in the exact sequence (*) we get the corollary. We note also that it actually follows from the argument that $\mathbb{Z}/n\mathbb{Z} \xrightarrow{\sim} H^2_{fl}(\mathbb{P}^N_R, \mu_n)$ when R is a strictly henselian local ring.

<u>Proof of Theorem 2</u>.

<u>LEMMA 1'</u>. If Theorem 2 is true when S = Spec $\mathcal{O}$ and $\mathcal{O}$ is strictly henselian, then it is true in general.

<u>Proof</u>. For any S the assumption gives us $R^2\pi_*G_m = 0$. We know $\mathbb{Z} \xrightarrow{\sim} R^1\pi_*G_m$ and $\pi_*G_m = G_m$. Hence by the Leray spectral sequence we get an exact sequence

$$(*) \qquad H^0(S,\mathbb{Z}) \xrightarrow{d_2} H^2(S,G_m) \xrightarrow{\pi^*} H^2(X,G_m) \to H^1(S,\mathbb{Z}).$$

$H^1(S,\mathbb{Z})_{tors} = 0$ since if $n \in H^0(S,\mathbb{Z}_+)$ we have the exact sequence

$$H^0(S,\mathbb{Z}) \xrightarrow{\alpha} H^0(S,\mathbb{Z}/n\mathbb{Z}) \xrightarrow{\delta} H^1(S,\mathbb{Z})_n \to 0$$

and α is clearly surjective. By (*) it follows that π^*_{tors} is surjective provided we show that $\mathrm{Ker}(\pi^*) = \mathrm{Im}(d_2)$ is torsion. So it remains to prove the second part of Thm. 2, i.e., that $\mathrm{Ker}(\pi^*) = H^0(S,\mathbb{Z})\cdot\delta\,\mathrm{cl}(X)$. [Notice that $\delta(\mathrm{cl}(X))$ is annihilated by $\dim(X/S)+1$.] Since d_2 is $H^0(S,\mathbb{Z})$-linear it suffices to show that $d_2(1) = \pm\delta\,\mathrm{cl}(X)$. This is proven in [3, V 4.8.3]. (It is stated there with the sign $+1$. However the correct sign is (I believe) -1 using [3, Lemma V 4.8.1 and Prop. V 3.2.1].)

Using Lemma 1' we can reduce Theorem 2 to the case $X = \mathbb{P}(E)$ where E is a vector bundle on S of constant rank $N > 0$. In this case we have to prove :

$$(*)_N : \qquad H^2(S,G_m)_{tors} \xrightarrow[\pi^*]{\sim} H^2(X,G_m)_{tors}$$

under the above assumptions. We will prove it by induction on N.

It is clear from the proof of Lemma 1' that we have :

<u>LEMMA $1'_N$</u>. If $(*)_N$ is true for S strictly local then $(*)_N$ is true in general.

Assume $N > 0$ and that $(*)_{N'}$ holds if $0 < N' < N$. We have (by Lemma $1'_N$) to show that $H^2_{et}(\mathbb{P}_{\mathcal{O}}^{N-1}, G_m)_{tors} = 0$ for $\mathcal{O}$ strictly henselian. If $N = 1$ then $\mathbb{P}_{\mathcal{O}}^{N-1} \simeq \mathrm{Spec}(\mathcal{O})$ so the statement is trivial. If $N = 2$ we will prove more generally

<u>LEMMA 2'</u>. $H^2(X,G_m)_{tors} = 0$ if $\mathcal{O}$ is strictly local and $\pi : X \longrightarrow S \overset{def}{=\!=} \mathrm{Spec}(\mathcal{O})$ is a proper morphism of relative dimension ≤ 1.

<u>Remarks</u>. Lemma 2' is proven in [2, III Cor. 3.2] if X is regular and $\mathcal{O}$ is an excellent discrete valuation ring. Our proof of Lemma 2' will be similar to that in ibid., using Thm. 1 and Artin's approximation theorem which were not available in [2]. Also the part of Lemma 2' dealing with ℓ-torsion where $\ell \in \mathcal{O}^*$ can be done using an argument in [2, III pages 100-101]. The properness is necessary even when S is a point as can be seen by considering $\mathrm{Br}(\mathbb{A}_k^1)$ when $k = k_{sep} \neq \bar{k}$. See also a Remark on page 205 for a complement.

<u>Proof of Lemma 2'</u>. We first recall that under the assumptions of Lemma 2' (with $\mathcal{O}$ required only to be henselian) π is projective.

We give some details how one proves it. One proves first that every curve (separated and of finite type) over a field is quasi-projective. Choose an ample line bundle L_1 on $X_1 = X \times_S \mathrm{Spec}(k)$ $(k = \mathcal{O}/m)$. It is enough to prove that L_1 can be extended to a line bundle L' on X because then L' will be ample by [9, III 4.7.1] (see also [9, IV 8.10.5.2] for elimination of the noetherian hypothesis) so some power of L' will embed X in a projective space over S. By a limit argument we can reduce to the case when $\mathcal{O} = (B_p)^h$ is the henselization of a localisation of a finitely generated ring B. Define $X_n = X \otimes_{\mathcal{O}} (\mathcal{O}/m^n)$. From the exact sequence

$$0 \longrightarrow m^{n-1}\mathcal{O}_{X_n} \longrightarrow \mathcal{O}^*_{X_n} \longrightarrow \mathcal{O}^*_{X_{n-1}} \longrightarrow 0 \quad \text{on } X_{1\,\mathrm{Zar}}$$
$$\alpha \longmapsto 1+\alpha$$

we get an exact cohomology sequence :

$$(*) \qquad \mathrm{Pic}\, X_n \xrightarrow{\ \alpha_n\ } \mathrm{Pic}\, X_{n-1} \longrightarrow H^2(X_n, m^{n-1}\mathcal{O}_{X_n}) \ .$$

The last term in (*) is zero because $\dim X_n \le 1$. Hence α_n is surjective. So we can lift successively L_1 to line bundles L_n on X_n with isomorphisms $L_n \otimes_{\mathcal{O}_{X_n}} \mathcal{O}_{X_{n-1}} \xrightarrow{\sim} L_{n-1}$. This determines an invertible sheaf $\hat{L} \overset{\mathrm{def}}{=\!=} \varprojlim L_n$ on $\hat{X} = (X_1, \varprojlim \mathcal{O}_{X_n})$. By Grothendieck's theorem on algebraization of formal sheaves [9, III 5.1.3, 5.1.4] $\hat{L}$ comes from a line bundle L on $X \otimes_{\mathcal{O}} \hat{\mathcal{O}}$. By [16, Thm. 3.5] the map $\mathrm{Pic}\, X \longrightarrow \mathrm{Pic}(X \otimes_{\mathcal{O}} \hat{\mathcal{O}})$ has a dense image and thus each L_n can be extended to a line bundle on X. Now we know that there exists an embedding $X \overset{i}{\hookrightarrow} \mathbb{P}^n_S = P$. Since $\dim X_1 \le 1$, there exists $d \ge 1$ ($d = 1$ if k is infinite) and $F, G \in H^0(P, \mathcal{O}_P(d))$ s.t. $X \cap V(F) \cap V(G) = \emptyset$. The open subsets $X_F = X - V(F)$ and $X_G = X - V(G)$ of X are affine and cover X .

We now return to the proof of Lemma 2'. We first notice that it is equivalent to saying that $(R^2\pi_*G_m)_{tors} = 0$ if $\pi : X \to S$ is proper of relative dimension ≤ 1. To prove it one can make a reduction by a projective limit argument to the case when S is of finite type over Spec($\mathbb{Z}$). Then the closed points of S are very dense and thus give a conservative family of fiber functors on $\tilde{S}_{ét}$ (= the étale topos of S). Thus it suffices to look at the stalks of $R^2\pi_*G_m$ at the closed points of S (where the residue field is finite, hence perfect). This will be done in case (iv) below. Before we do that we give some cases in which Theorem 2' is true.

Case (i). $S = \mathrm{Spec}(k)$, $k = \bar{k}$. Then we do not need the properness assumption on X. We only need X to be of finite type and of dimension ≤ 1. Define $N = \underline{\mathrm{Ker}}(\mathcal{O}_X \to \mathcal{O}_{X_{red}})$. We first argue by induction on the index of nilpotency of N. Suppose $N^k \neq 0$ and $N^{k+1} = 0$ where $k > 0$. Then

$$0 \to N^k \to G_{m_X} \to G_{m_{V(N^k)}} \to 0$$

$$\alpha \longmapsto 1+\alpha$$

is exact. Since $H^2(X,N^k) = 0$ we get that $H^2_{ét}(X,G_m) = 0$ provided we know it for $V(N^k)$. It follows that we may assume that X is reduced. Let $\{x_i\}$ be its generic points, with morphisms $\mathrm{Spec}\, k(x_i) \xrightarrow[j_i]{} X$. Consider the exact sequence

$$0 \longrightarrow G_{m_X} \longrightarrow \bigoplus_i j_{i_*}G_m \longrightarrow C \longrightarrow 0 .$$

C is concentrated at the closed points of X so $H^1_{ét}(C) = 0$, so by the exact cohomology sequence we reduce to show that $H^2(j_{i_*}G_m) = 0$. The Leray spectral sequence for j_i gives an exact sequence

$$(*):\quad H^0(X, R^1 j_{i_*} G_m) \xrightarrow{d_2} H^2(X, j_{i_*} G_m) \longrightarrow H^2(k(x_i), G_m)\ .$$

The 3rd term in (*) is 0 by Tsen's theorem and the 1st term is 0 because $R^1 j_{i_*} G_m = 0$.

Case (ii). $\mathcal{O}$ is artinian with an algebraically closed residue field. The proof given in Case (i) applies also to this case.

Case (iii). $\mathcal{O}$ is noetherian and complete and $\mathcal{O}/m$ is algebraically closed. Suppose $\alpha \in H^2(X, G_m)_{tors}$ and we wish to show $\alpha = 0$. We recall from page 196 that X can be covered by two affine open subsets. Hence by Theorem 1 $\alpha = \delta(cl(A))$ for some Azumaya algebra A on X. If $n \geq 1$ then by case (ii) $[A \otimes_{\mathcal{O}_X} \mathcal{O}_{X_n}]$ is trivial in $Br(X_n)$ so we can choose a vector bundle E_n over $\mathcal{O}_{X_n}$ and an isomorphism $A \otimes_{\mathcal{O}_X} \mathcal{O}_{X_n} \xrightarrow[\varphi_n]{\sim} \underline{End}(E_n)$. We get

$\underline{End}(E_{n+1} \otimes_{\mathcal{O}_{X_{n+1}}} \mathcal{O}_{X_n}) \simeq \underline{End}(E_{n+1}) \otimes_{\mathcal{O}_{X_{n+1}}} \mathcal{O}_{X_n} \xrightarrow[\varphi_n \varphi_{n+1}^{-1}]{\sim} \underline{End}(E_n)$ so there exists a line bundle L_n on X_n with an isomorphism $E_{n+1} \otimes_{\mathcal{O}_{X_{n+1}}} L_n \xrightarrow[\xi_n]{\sim} E_n$ s.t. $\underline{End}(\xi_n) = \varphi_n \varphi_{n+1}^{-1}$. We will now show that the E_n's can be chosen by induction s.t. $L_n = \mathcal{O}_{X_n} \forall_n$. By page 196 we can extend L_n to a line bundle $\bar{L}_n$ on X_{n+1}. Thus, replacing E_{n+1} by $E_{n+1} \otimes \bar{L}_n$ we can assume that $\varphi_n \varphi_{n+1}^{-1}$ is induced by an isomorphism $E_{n+1} \otimes_{\mathcal{O}_{X_{n+1}}} \mathcal{O}_{X_n} \xrightarrow[\xi_n]{\sim} E_n$. $\hat{E} \overset{def}{=} \varprojlim E_n$ is a locally free coherent sheaf on $\hat{X}$. So by Grothendieck's theorem $\hat{E}$ is the completion of a locally free coherent E on X and the isomorphism $\varprojlim \varphi_n : \hat{A} \xrightarrow{\sim} \underline{End}(\hat{E})$ comes from an isomorphism $A \simeq \underline{End}(E)$. So $[A] = 0$ in $Br(X)$ and applying α_X we get that $\alpha = 0$. Compare with Lemma 7 on page 135.

<u>Case (iv)</u>. $\mathcal{O} = (B_m)^{st}$ where B is a finitely generated ring, $m \subset B$ a maximal ideal and "st" denotes the strict henselization (which is unique over B_m up to a non unique isomorphism). Write $\hat{\mathcal{O}} = \varinjlim B_i$ where the B_i's run over all finitely generated $\mathcal{O}$-subalgebras of $\hat{\mathcal{O}}$ ordered by inclusion. Using [16, 1.10] one shows that for every i there exists an $\mathcal{O}$-algebra map $B_i \xrightarrow{s_i} \mathcal{O}$ (if B_i is given over $\mathcal{O}$ by generators T_j $(1 \le j \le m)$ and relations $F_\alpha(T_1,\ldots,T_m) = 0$ $(\alpha \in I)$, then the inclusion $B_i \to \hat{\mathcal{O}}$ gives a solution of the system of equations $\{F_\alpha(\vec{T}) = 0\}_{\alpha\in I}$ in $\hat{\mathcal{O}}$ and by ibid. we can approximate it by a solution with values in $\mathcal{O}$). If F is the covariant functor on the category of $\mathcal{O}$-algebras defined by $F(B) = H^2(X \otimes_{\mathcal{O}} B, G_m)_{tors}$ then we have by Case (iii) $0 = F(\hat{\mathcal{O}}) \xleftarrow{\sim} \varinjlim F(B_i)$. The map $F(\mathcal{O}) \to F(B_i)$ is injective because $F(s_i)$ is a left inverse of it. Hence the map $F(\mathcal{O}) \to \varinjlim F(B_i)$ is injective so $F(\mathcal{O}) = 0$.

<u>Remark on Case (iv)</u>. In order to apply Thm. 1.10 of [16] we have to check that $\mathcal{O}$ is of the form $(C_p)^h$ where C is an algebra of finite type over an excellent discrete valuation ring R. This can be done as follows : let $p = \mathrm{char}(\mathcal{O}/m) > 0$. Take $R = \mathbb{Z}_{(p)}^{st}$ (where $\mathbb{Z}_{(p)}$ is the localization of $\mathbb{Z}$ at the prime $p\mathbb{Z}$) and $C = B \otimes_{\mathbb{Z}} R$. Since the prime ideal $p\mathbb{Z}$ is contracted from R, $m \subset B$ is contracted from some prime ideal $\mathfrak{p}$ of C. $k(\mathfrak{p})$ is a composition of $k(m)$ and $R/pR = \overline{\mathbb{F}}_p$ and thus $R/pR \xrightarrow{\sim} k(\mathfrak{p})$ so $C_{\mathfrak{p}}^{st} \xleftarrow{\sim} C_{\mathfrak{p}}^h$. One checks that the map $B_m \to C_{\mathfrak{p}}$ induces an isomorphism on the strict henselizations.

We now return to the inductive proof of $(*)_N$ (see page 195). We have to consider the case $N \ge 3$ and prove that $H^2(P,G_m)_{tors} = 0$ if $\mathcal{O}$ is strictly henselian and $P = \mathbb{P}_{\mathcal{O}}^{N-1}$. Let $(x_0,x_1,\ldots,x_{N-1})$ be the standard basis of $H^0(P,\mathcal{O}_P(1))$, and $W \overset{\mathrm{def}}{=} V(x_1,\ldots,x_{N-1})$.

W projects isomorphically onto $S = \mathrm{Spec}(\mathcal{O})$. The sections $(x_1,\dots,x_{N-1})$ of $\mathcal{O}_P(1)$ define a morphism $v: P-W \to \mathbb{P}_{\mathcal{O}}^{N-2} = P'$. Let $\Gamma \subset P \times P'$ be the graph of v. The schematic closure Z of Γ in $P \times P'$ is, via the projection to P, the blowing up of P along W. It is well known that as a scheme over P' Z is isomorphic to $\mathbb{P}(\mathcal{O}_{P'} \oplus \mathcal{O}_{P'}(1))$. Hence using $(*)_2$ and $(*)_{N-2}$ we get $0 = H^2(P',G_m)_{tors} \xrightarrow{\sim} H^2(Z,G_m)_{tors}$. W lies in the open set $P-V(x_0)$ of P and is defined there by the regular sequence $\frac{x_1}{x_0},\dots,\frac{x_{N-1}}{x_0}$. (Recall (cf. [19, VII, Def. 1.1]) that if A is a not necessarily noetherian ring a sequence of elements of A $(f_1,\dots,f_n)$ is called regular iff the Koszulcomplex $K_*(f_1,\dots,f_n)$ is acyclic in positive degrees.) It remains to apply the following lemma.

<u>LEMMA 3'</u>. Suppose that $Y \subset X$ is a closed embedding of schemes s.t. Y is locally defined by a non empty regular $\mathcal{O}_X$-sequence. Then the map $H^2(X,G_m) \to H^2(Z,G_m)$ is injective where $Z = bl_Y(X) \stackrel{\text{def}}{=} \mathrm{Proj}(\bigoplus_{n\geq 0} I^n_{X,Y})$.

<u>Proof</u>. We consider the projection $\pi: Z \to X$, the exceptional divisor $E = \pi^{-1}(Y) \subset Z$, and the line bundle $\mathcal{O}_Z(1)$ (defined on Proj of any homogeneous algebra). One knows that there is a canonical isomorphism $\mathcal{O}_Z(1) \simeq \mathcal{O}_Z(-E)$. (This is true for any blow-up and can be shown as follows : recall first that if $A = \bigoplus_{n\geq 0} A_n$ is a graded ring generated by A_0 and A_1 then $\mathrm{Proj}(A) = P$ has an open affine cover $U_a = \mathrm{Spec}(A[\frac{1}{a}]_{deg=0})$ $(a \in A_1)$, and $\mathcal{O}_P(1)$ is defined s.t. there are isomorphisms

$$\mathcal{O}_P(1)\Big|_{U_a} \xrightarrow[\varphi_a]{\sim} \mathcal{O}_{U_a} \quad \text{and} \quad \varphi_a = \left(\frac{a}{b}\right)^{-1} \varphi_b \text{ on } U_{ab}\,.$$

Now in the case of blowing up $A_n = I^n$ where I is an ideal in A_0.

If $x \in A$ we denote by $[x]_n$ the image of x in A as a graded element of degree n, provided that $x \in I^n$. The exceptional divisor E is defined in U_a by the ideal $[I]_0\Gamma(U_a,\mathcal{O}_{U_a}) = [f]_0\Gamma(U_a,\mathcal{O}_{U_a})$ if $a = [f]_1$.

Define $\psi_a : \mathcal{O}_P(1)|_{U_a} \xrightarrow{\sim} J_{P,E}|_{U_a} = \mathcal{O}_P(-E)|_{U_a}$ by $\psi_a(s) = \varphi_a(s)\cdot[f]_0$, and check that $\psi_a|_{U_{ab}} = \psi_b|_{U_{ab}}$.) We now state the following :

<u>LEMMA 4'</u>. Under the assumptions of Lemma 3'

$$G_m \xrightarrow{\sim} \pi_{ét_*} G_m \quad \text{and} \quad \mathbb{Z}_Y \xrightarrow{\sim} R^1 \pi_{ét_*} G_m .$$

$$\mathbb{Z} \ni n \longmapsto cl(\mathcal{O}_Z(n)) = cl(\mathcal{O}_Z(-nE)) .$$

We assume Lemma 4' for the moment. Then from the Leray spectral sequence for $\pi_{ét}$ we get an exact sequence

$$(*)\ \mathrm{Pic}(Z) \xrightarrow[\alpha]{\text{edge homom.}} H^0(X,R^1\pi_*G_m) \longrightarrow H^2_{ét}(X,\pi_*G_m) \xrightarrow{\beta} H^2_{ét}(Z,G_m) .$$

The second term in (*) is $\simeq H^0(X,\mathbb{Z}_Y) \simeq H^0(Y,\mathbb{Z}_Y)$ by Lemma 4'. If $n \in H^0(Y,\mathbb{Z}_Y)$ then $[\mathcal{O}_Z(-nE)] \in \mathrm{Pic}(Z)$ is mapped by α to "n" so α is surjective and hence β is injective. As $\pi_*G_m \xleftarrow{\sim} G_m$ by Lemma 4' this is what was to be shown in Lemma 3'.

<u>Proof of Lemma 4'</u>. By [19, VII, 3.5 (ii)] we have that $\pi_*\mathcal{O}_Z \xleftarrow{\sim} \mathcal{O}_X$ (in the Zariski topology). By standard arguments (π_* commutes with étale base changes) this holds also in the étale topology. Taking the sheaves of invertible elements we get that $G_{mX} \xrightarrow{\sim} \pi_{ét_*} G_{mZ}$. We now show that the map $\mathbb{Z}_Y \xrightarrow[\gamma]{} R^1\pi_{ét_*} G_m$ is an isomorphism. Let π' be the projection of E into Y. $E = \mathrm{Proj}(\bigoplus_{n\geq 0} (I^n/I^{n+1}))$ where $I \subset \mathcal{O}_X$ is the ideal defining Y. Using [19, VII 1.3 (ii), (iii)], we see that $E = \mathbb{P}(I/I^2)$ and I/I^2 is locally free of finite (nowhere 0) rank over $\mathcal{O}_Y$.

Hence by [17, Lecture 13] the map $\mathbb{Z}_Y \to R^1\pi'_{\text{ét}_*} G_m$ sending 1 to the cohomology class of $\mathcal{O}_E(1)$ is an isomorphism. It follows that γ is injective. To show that γ is surjective it is enough to show that (after any étale base change) if $L \in \mathrm{Pic}(Z)$ then locally in the Zariski topology on X $L \simeq \mathcal{O}_Z(n)$ for some $n \in \mathbb{Z}$. Since the map π is an isomorphism above X-Y we can restrict ourselves to considering neighborhoods of a point $y \in Y$. We can assume that X is affine and that $L \otimes \mathcal{O}_E \simeq \mathcal{O}_E(n)$ for some $n \in \mathbb{Z}$. We will then prove that $L \simeq \mathcal{O}_Z(n)$ over a neighborhood of Y. If we replace L by L(-n) we see that it is enough to prove that if we assume that $L \otimes \mathcal{O}_E \simeq \mathcal{O}_E$ then $L \simeq \mathcal{O}_Z$ over a neighborhood of Y. As the map π is proper it is enough to show that the nowhere vanishing section $\sigma \in \Gamma(E, L \otimes \mathcal{O}_E)$ extends to $\bar{\sigma} \in \Gamma(Z, L)$. (Use that $\pi(V(\bar{\sigma}))$ is closed.) For $n \geq 1$ let nE be the subscheme of Z defined by J_E^n. We have exact sequences

$$(*)_n : \qquad 0 \longrightarrow J_E^n L / J_E^{n+1} L \longrightarrow L \otimes \mathcal{O}_{(n+1)E} \longrightarrow L \otimes \mathcal{O}_{nE} \longrightarrow 0 .$$

In $(*)_n$ the first term is isomorphic to

$$L \otimes \mathcal{O}_E \otimes (J_E^n / J_E^{n+1}) \simeq (J_E^n / J_E^{n+1}) \simeq \mathcal{O}_E(-nE) \simeq \mathcal{O}_E(n) .$$

So the cohomology sequence associated to $(*)_n$ gives an exact sequence

$$H^0(L/J_E^{n+1}L) \xrightarrow{r_n} H^0(L/J_E^n L) \longrightarrow H^1(E, \mathcal{O}_E(n)).$$

It is well known that the last term is 0 and hence r_n is surjective. Hence for any $n \geq 1$ the map $H^0(L/J_E^n L) \to H^0(L/J_E L)$ is surjective. Hence to show the existence of $\bar{\sigma}$ in the above it is enough to show that if n is sufficiently large then the map $H^0(L) \longrightarrow H^0(L/J_E^n L)$ is surjective. For this it is enough to show that

$(*)$: $\exists\, n_0$ s.t. $0 = H^1(J_E^n L) (= H^1(L(n)))$ if $n \geq n_0$.

$(*)$ is well known if X is noetherian (Theorem B). If X is not necessarily noetherian we can proceed as follows : we recall from [19, I] that an $\mathcal{O}_X$-module F (where $(X,\mathcal{O}_X)$ is a scheme, or even a ringed topos) is called pseudo-coherent if $\forall\, n \geq 0$ there exists a covering $\{U_\alpha\}$ of X and exact sequences of $\mathcal{O}_{U_\alpha}$-modules

$$\mathcal{O}_{U_\alpha}^{m_n} \xrightarrow{d_n} \cdots\cdots \xrightarrow{d_2} \mathcal{O}_{U_\alpha}^{m_1} \xrightarrow{d_1} \mathcal{O}_{U_\alpha}^{m_0} \xrightarrow{\varepsilon} F|_{U_\alpha} \longrightarrow 0 .$$

So L is a pseudo-coherent $\mathcal{O}_Z$-module. As X is affine, $H^1(L(n)) = 0 \iff R^1\pi_*(L(n)) = 0$. Hence to show $(*)$ we can localize further on X and assume that $X = \mathrm{Spec}(A)$ and Y is defined by a regular sequence $f_1,\ldots,f_s \in A$. By assumption $s \geq 1$. The epimorphism $\mathcal{O}_X^s \xrightarrow[u]{} J_{X,Y}$ defined by the f's induces an embedding $Z \xhookrightarrow[j]{} \mathbb{P}_X^{s-1} = P$. By [19, VII 1.8] j is a regular immersion. By [19, III 1.1.1, 1.1.2] j_*L is a pseudo-coherent $\mathcal{O}_P$-module. By [19, II 2.2.9(a)] it follows that j_*L admits an (infinite) resolution

$$\text{(R)} \qquad \cdots \longrightarrow E_3 \xrightarrow{d_3} E_2 \xrightarrow{d_2} E_1 \xrightarrow{d_1} E_0 \xrightarrow{d_0} j_*L \longrightarrow 0$$

s.t. $\forall\, i \geq 0\ \exists\, m_i, k_i \in \mathbb{Z}_+$ s.t. $E_i \simeq \mathcal{O}_P(-m_i)^{k_i}$.

We define (see (R)) F_i = Image (d_i). Thus $F_0 = j_*L$. From (R) we get for each $i \geq 0$ an exact sequence

$$(R_i) \qquad 0 \longrightarrow F_{i+1} \hookrightarrow E_i \longrightarrow F_i \longrightarrow 0 .$$

If $i \geq 0$ and $n \in \mathbb{Z}$ we get from $(R_i) \otimes \mathcal{O}_P(n)$ an exact sequence

$$(e_{i,n}) \qquad H^{i+1}(E_i(n)) \longrightarrow H^{i+1}(F_i(n)) \xrightarrow{\delta_i} H^{i+2}(F_{i+1}(n)) .$$

One knows $H^p(\mathbb{P}_A^{s-1},\mathcal{O}(t)) = 0$ if either $0 < p \neq s-1$ or $0 < p = s-1$ and $t > -s$. It follows that if

$$(+) \qquad s = 1, \text{ or } s > 1 \text{ and } n \geq m_{s-2} + 1 - s ,$$

then the first term in $(e_{i,n})$ is zero for all $i \geq 0$ and thus the maps δ_i are all injective then. We can prove by decreasing induction on i that $(+) \Rightarrow H^{i+1}(F_i(n)) = 0$. (It is true for $i \geq s-1$ because P is covered by s affine open subsets.) The case $i = 0$ gives that

$$(+) \Rightarrow 0 = H^1(F_0(n)) = H^1((j_*L)(n)) \simeq H^1(Z,L(n))$$

which proves (*).

<u>Remark</u>. It is possible to prove Theorem 2 without using Theorem 1. To show that the p-torsion in $(R^2\pi_{\text{ét}_*}G_m)_\xi$ (where ξ is a geometric point of characteristic p) is zero one can make a reduction to the case of a perfect reduced base scheme of characteristic p and deal with the last case using the exact sequence

$$0 \longrightarrow G_{m_X} \longrightarrow fr_*G_{m_X} \longrightarrow fr_*Z_X^1 \xrightarrow{C-I} \Omega_X^1 \longrightarrow 0$$

of [13, Thm. 1.3]. I also want to mention that using Lemma 2' and the next Remark, I can prove the following theorem which generalizes Theorem 2 and Lemma 4'.

<u>THEOREM</u>. Let $\pi : X \to Y$ be a morphism s.t. locally in the étale topology on Y we can embed $X \underset{j}{\hookrightarrow} \mathbb{P}_Y^N = P$ s.t. $\mathcal{J}_{P,X}(1)$ is a regular sheaf on P [in the sense that $R^q p_* \mathcal{J}_{P,X}(1-q) = 0 \ \forall \ q \geq 1$ where $p : P \to X$ is the projection]. Then $G_{m_S} \xrightarrow{\sim} \pi_{\text{ét}_*}G_m$, $\mathbb{Z}_T \underset{\alpha}{\xrightarrow{\sim}} R^1\pi_{\text{ét}_*}G_m$ and $R^q\pi_{\text{ét}_*}G_m$ is uniquely divisible for $q \geq 2$. Here S is the schematic image of π and

$T = \{y \in Y | \dim \pi^{-1}(y) \geq 1\}$ and if j is as in the assumptions on π then the map α sends 1 to the cohomology class of $j^*\mathcal{O}_P(1)$.

I can show that if $Z \subset Y$ is a regular immersion then $bl_Z(Y) \longrightarrow Y$ satisfies the assumptions of the theorem.

Remark on Lemma 2'. It is stated in [15, p. 125] that if $X \xrightarrow{f} Y$ is proper of relative dimension ≤ 1 then the étale sheaves $R^q f_* G_m$ are uniquely divisible for $q \geq 2$. Lemma 2' is the part of the last statement saying that $R^2 f_* G_m$ is torsion free. The other parts, i.e., that $R^q f_* G_m$ is divisible if $q \geq 2$ and is torsion free if $q \geq 3$ are easier and can be shown (cf. [2, III Lemma 3.2.1]) by considering for each prime p the exact sequence

$$0 \longrightarrow K \longrightarrow G_m \longrightarrow G_m \longrightarrow L \longrightarrow 0$$
$$x \longmapsto x^p$$

on $X_{ét}$ and using that the proper base change theorem and the cohomological dimension theorems [1, X 4.3, 5.2] imply that $R^2 f_* K = 0$ if $q \geq 3$ and $R^q f_* L = 0$ if $q \geq 2$ (notice that L is concentrated on the locus $p = 0$).

A remark on the assertion on page 177.

1) Recall ([3, V 4.2]; cf. Proof of Proposition P(4), page 175) that if B is an Azumaya algebra on a scheme X we have a G_m-gerbe d(B) over $X_{ét}$ s.t. Ob d(B) = $\{(\Omega, e), (E, \varphi) | \Omega \xrightarrow{e} X$ is étale, E is a vector bundle on Ω, and $\varphi : e^*B \xrightarrow{\sim} \underline{\mathrm{End}}(E)\}$ (in the definition of d(B) on page 175 E is a locally free sheaf of finite type over the étale site $\Omega_{ét}$ of Ω. However by descent theory for quasi-coherent sheaves the

category of such E is equivalent (via restriction to Ω_{Zar}) to the category of vector bundles on $(\Omega, \mathcal{O}_\Omega)$).

2) If, furthermore, Q is a vector bundle on X, we get a functor

$$d(B) \xrightarrow{v} d(B \otimes \underline{End}(Q)) \text{ s.t.}$$

$$v((\Omega,e),(E,\varphi)) = ((\Omega,e),\ (E \otimes e^*Q, \varphi \otimes id_{\underline{End}(e^*Q)})) .$$

v is compatible with the projections to $X_{ét}$ and with the G_m-actions and hence by [3, IV 2.2.7] it is an equivalence of gerbes. Thus the functors $d(B)_\Omega \xrightarrow{v_\Omega} d(B \otimes \underline{End}(Q))_\Omega$ are equivalences of categories.

3) Given categories C_1, C_2, C_3 and functors $C_1 \xrightarrow{u} C_2$, $C_3 \xrightarrow{v} C_2$, we define the "loose" fibered product $D = C_1 \bar{\times}_{C_2} C_3$ by

$$Ob(D) = \{(x,y,\zeta) \mid x \in Ob\ C_1,\ y \in Ob\ C_2,\ \zeta : u(x) \xrightarrow{\sim} v(y)\},$$

$$Mor_D((x,y,\zeta),(x',y',\zeta')) =$$

$$= \{(\alpha,\beta) \mid \alpha \in Mor_{C_1}(x,x'),\ \beta \in Mor_{C_3}(y,y'), \zeta' u(\alpha) = v(\beta)\zeta\}.$$

4) In (3), if v is an equivalence of categories, then the functor $D \to C_1$ is an equivalence of categories.

5) The category G_Ω (see pages 175-177) is of the form $C_1 \bar{\times}_{C_2} C_3$ where $C_1 = d(A)_{\Omega\times U}$, $C_3 = d(B)_{\Omega\times V}$, and $C_2 = d(B|_{U\cap V} \otimes \underline{End}(Q))_{\Omega\times U\times V}$.

6) If image $(\Omega \to X) \subset U$ then by (2) the functor $C_3 \to C_2$ (the C's are as in (5)) is an equivalence of categories so by (4) we get an equivalence of categories $G_\Omega \xrightarrow{\psi_\Omega} d(A)_\Omega$. ψ_Ω is the functor on the fibers defined by a G_m-morphism of champs $G|_{U_{ét}} \xrightarrow{\psi} d(A)$

($\downarrow$ is defined already in the proof of Proposition P(4) on page 175). It follows that $G|_{U_{ét}}$ is a G_m-gerbe since it is G_m-equivalent to d(A). Similarly $G|_{V_{ét}}$ is a G_m-gerbe and hence we get that G is a G_m-gerbe as desired.

REFERENCES

[1] Théorie des Topos et Cohomologie Etale des Schemas (SGA 4), dirigé par M. Artin, A. Grothendieck, J.L. Verdier, "Lecture Notes in Mathematics", Vols. 269, 270, 305, Springer Verlag, 1972, 1973.

[2] A. Grothendieck, "Le Groupe de Brauer I, II, III" in Dix Exposés Sur la Cohomologie des Schémas, by J. Giraud (et al), North Holland Publ. Co., 1968.

[3] J. Giraud, Cohomologie non Abélienne, Springer Verlag, 1971.

[4] A. Grothendieck, Cohomologie Locale des Faisceaux Cohérents et Théorèmes de Lefschetz Locaux et Globaux (SGA 2), North Holland Publ. Co., 1968.

[5] M. Raynaud, Anneaux Locaux Henséliens, "Lecture Notes in Mathematics", Vol. 169, Springer Verlag, 1970.

[6] R. Elkik, "Equations à Coefficients dans un Anneau Hensélien", Annales Scientifiques de l'Ecole Normale Supérieure, Vol. 6 (1973), pp. 553-604.

[7] Revêtements Etales et Groupe Fondamental (SGA 1), dirigé par A. Grothendieck, "Lecture Notes in Mathematics", Vol. 224, Springer Verlag, 1970.

[8] A. Altman and S. Kleiman, Introduction to Grothendieck Duality Theory, "Lecture Notes in Mathematics", Vo. 146, Springer Verlag, 1970.

[9] A. Grothendieck and J. Dieudonné, "Eléments de Géométrie Algébrique", Publications Mathématiques de l'I.H.E.S., Vols. 4, 8, 11, 17, 20, 24, 28, 32, 1960-1967.

[10] I.N. Herstein, Non Commutative Rings, MAA, 1968.

[11] H. Cartan and S. Eilenberg, Homological Algebra, Princeton University Press, 1956.

[12] D. Mumford, Geometric Invariant Theory, Springer Verlag, 1965.

[13] R. Hoobler, "Cohomology of Purely Inseparable Galois Coverings", Jour. für die Reine und Angew. Math., vol. 266 (1974), 183-199.

[14] H. Bass, "K theory and stable algebra", Publications Math. I.H.E.S., Vol. 22 (1964), 5-60.

[15] M. Artin and J.S. Milne, "Duality in the Flat Cohomology of Curves", Inventiones Math., Vol. 35 (1976), 111-129.

[16] M. Artin, "Algebraic Approximation of Structures Over Complete Local Rings", Publ. Math. I.H.E.S., vol. 36 (1969), 23-58.

[17] D. Mumford, Lectures on Curves on an Algebraic Surface, Harvard University, 1964.

[18] R. Hoobler, "A Cohomological Interpolation of Brauer Groups of Rings", preprint.

[19] P. Berthelot, A. Grothendieck, L. Illusie, Théorie des Intersections et Théorème de Riemann-Roch (SGA 6). Lecture Notes in Mathematics, vol. 225, Springer Verlag, 1971.

COHOMOLOGIE ETALE ET GROUPE DE BRAUER

M.-A. KNUS & M. OJANGUREN

Introduction

Soit R un anneau commutatif. Une R-algèbre associative A est dite algèbre d'Azumaya sur R si

1) A est un R-module projectif de type fini

2) A/mA est centrale simple sur K = R/m pour tout idéal maximal m de R.

On définit le groupe de Brauer de R, Br(R), comme en théorie classique des algèbres : deux algèbres d'Azumaya A et A' sont dites semblables s'il existe des R-modules fidèlement projectifs (projectifs de type fini et fidèles) P et P' tels que $A \otimes_R \mathrm{End}_R P \simeq A' \otimes_R \mathrm{End}_R P'$. C'est une relation d'équivalence et le produit tensoriel induit la structure de groupe cherchée sur l'ensemble des classes d'équivalence. Soit G_m le foncteur "groupe des unités". Le groupe de Brauer Br(R) se plonge dans le groupe de cohomologie étale $H^2_{\text{ét}}(\mathrm{Spec}\, R, G_m)$. Comme Br(R) est de torsion, on peut conjecturer que

$$\mathrm{Br}(R) = H^2(\mathrm{Spec}\, R, G_m)_{\text{torsion}} .$$

Une démonstration de ce résultat se trouve dans la thèse de Gabber [Ga]. Gabber montre même le résultat équivalent pour un schéma X réunion de deux ouverts affines U et V tels que l'intersection $U \cap V$ est affine. La démonstration utilise des techniques de cohomologie non-abélienne.

Le but de ce travail est de donner une version plus élémentaire de la démonstration de Gabber, dans le cas d'un anneau R. Un outil essentiel est la construction explicite de deux suites de Mayer-Vietoris, l'une en cohomologie étale et l'autre pour Pic(R) et Br(R).

Nous noterons FP(R) la catégorie des modules fidèlement projectifs, Pic(R) la catégorie des modules projectifs de rang un sur R et Az(R) la catégorie des algèbres d'Azumaya sur R. Les produits tensoriels ne seront pas toujours indéxés, l'index étant souvent donné par le contexte. On trouvera les définitions manquantes et les résultats de base dans [KO].

1. Cohomologie d'Amitsur

Nous rappelons quelques résultats. Soit R un anneau commutatif. On dit qu'une R-algèbre S est étale si

(1.1)
- S est une R-algèbre de présentation finie
- S est plate sur R
- S est séparable sur R (S est un $S \otimes S$-module projectif)

Exemples (1.2). Un localisé R_f, $f \in R$, est étale sur R. Une extension séparable finie de corps est étale. Si S/R est étale, $S/S \otimes S$ est étale (pour la structure donnée par la multiplication).

Proposition (1.3).

(a) Si S/R est étale, alors $S \otimes T/T$ est étale pour toute R-algèbre T.

(b) Si S/R et T/S sont étales, T/R est étale. Inversement, si T/R et S/R sont étales, T/S est étale.

(c) Si S/R et T/R sont étales, alors $S \otimes T$ et $S \times T$ sont étales sur R.

On dit que S est un <u>recouvrement étale</u> de R si S est étale et fidèlement plate sur R. Ces recouvrements définissent une topologie sur X = Spec R. Le foncteur "groupe des unités" G_m est alors un faisceau et on peut définir des groupes de cohomologie de Čech $\check{H}^n(R) = \check{H}^n(X, G_m)$. Rappelons cette définition (comme nous nous limitons au cas affine, nous n'utiliserons pas le langage des schémas). Pour tout recouvrement étale S/R notons $(S/R)^{(n)} = S^{(n)}$ le produit tensoriel de n copies de S sur R, et définissons

$$(1.4) \qquad \partial_n : G_m(S^{(n)}) \to G_m(S^{(n+1)})$$

par $\partial_n u = u_1 u_2^{-1} \ldots u_{n+1}^{(-1)}$ où, si $u = a_1 \otimes \ldots \otimes a_n$, $u_i = a_1 \otimes \ldots \otimes 1 \otimes \ldots \otimes a_n$, 1 introduit en i-ème position.

On vérifie que $\partial_{n+1}\partial_n = 1$ et on définit $\check{H}^n(S/R) = \operatorname{Ker} \partial_{n+1}/\operatorname{Im}\partial_n$. Tout homomorphisme de R-algèbres $S \to S'$ induit un homomorphisme $\check{H}^n(S/R) \to \check{H}^n(S'/R)$ et on peut montrer que deux homomorphismes $S \to S'$ induisent des applications $G_m(S^{(n)}) \to G_m(S'^{(n)})$ homotopes ([KO] prop. 1.7, p. 123). L'application induite $\check{H}^n(S/R) \to \check{H}^n(S'/R)$ est donc indépendante de l'homomorphisme particulier $S \to S'$. Cela permet de définir

$$(1.5) \qquad \check{H}^n(R) = \varinjlim_S H^n(S/R), \text{ S recouvrement étale de R.}$$

Ces groupes de cohomologie sont aussi appelés les groupes de <u>cohomologie d'Amitsur</u> de R. D'après un résultat d'Artin ([A] Cor. 4.2) ces groupes sont isomorphes aux groupes de cohomologie étale $H^n_{ét}(\operatorname{Spec} R, G_m)$ définis comme foncteurs dérivés. Le lemme suivant est une conséquence d'un résultat fondamental pour le théorème de comparaison d'Artin.

<u>Lemme</u> (1.6). Soient S_i/R, $i = 1,\ldots,n$ des recouvrements étales. Alors si $T/S_1 \otimes \ldots \otimes S_n$ est un recouvrement étale, il existe des recouvrements étales S_i'/S_i, $i = 1,\ldots,n$ tels que $S_1' \otimes \ldots \otimes S_n'/S_1 \otimes \ldots \otimes S_n$ se factorise par T.

<u>Démonstration</u>. Le cas $S_1 = \ldots = S_n$ est un résultat d'Artin ([A] Thm. 4.1). Le cas général s'obtient en appliquant le résultat d'Artin à $(S_1 \times \ldots \times S_n)^{(n)}$.

Dorénavant nous noterons $\check{H}^n(R) = H^n(R)$.

<u>Interprétation des groupes H^0, H^1 et H^2</u>

(1.7) On vérifie immédiatement que $H^0(R) = G_m(R)$.

(1.8) Il est bien connu que $H^1(R) \cong \mathrm{Pic}(R)$. L'isomorphisme se construit de la façon suivante : soit $I \in \underline{\underline{\mathrm{Pic}}}(R)$; il existe alors un recouvrement étale S/R tel que

$$\sigma : I \otimes S \tilde{\to} S$$

(S peut être choisi de la forme $R_{f_1} \times \ldots \times R_{f_n}$, $f_1,\ldots,f_n \in R$ et $Rf_1 + \ldots + Rf_n = R$).

Soit $D(\sigma) : S \otimes S \to S \otimes S$ l'isomorphisme défini par

$$(1.9) \qquad \begin{array}{ccc} S \otimes I \otimes S & \xrightarrow{1 \otimes \sigma} & S \otimes S \\ {\scriptstyle \tau \otimes 1}\downarrow & & \downarrow{\scriptstyle D(\sigma)} \\ I \otimes S \otimes S & \xrightarrow[\sigma \otimes 1]{} & S \otimes S \end{array}$$

où τ est le switch. Alors $D(\sigma)$ définit une unité de $S \otimes S$ (aussi notée $D(\sigma)$) telle que $\partial_2 D(\sigma) = 1$. Soit $[D(\sigma)]$ sa classe dans $H^1(R)$. L'isomorphisme $\mathrm{Pic}(R) \tilde{\to} H^1(R)$ est induit par $I \mapsto [D(\sigma)]$.

(1.10) Montrons comment $Br(R)$ se plonge dans $H^2(R)$. Soit $A \in \underline{\underline{Az}}(R)$. Sans trop de restrictions, on peut supposer que A est de rang constant n^2. L'algèbre A peut être <u>neutralisée</u> par un recouvrement étale S/R :

(1.11) $$\alpha : A \otimes S \tilde{\to} M_n(S)$$

où $M_n(S)$ est l'algèbre des $n \times n$-matrices à coefficients dans S. Soit $\varphi : M_n(S \otimes S) \to M_n(S \otimes S)$ l'automorphisme défini par le diagramme analogue à (1.9). L'automorphisme φ n'est pas nécessairement intérieur. L'obstruction est donnée par un élément de $Pic(S \otimes S)$ (voir [KO] p. 107). Il existe donc un recouvrement étale $T/S \otimes S$ qui neutralise cette obstruction (voir (1.8)). Quitte à remplacer S par un recouvrement S', on peut d'après (1.6) se ramener au cas où φ est intérieur. Notons $D(\alpha)$ un automorphisme de $(S \otimes S)^n$ qui induit φ, $\varphi = End(D(\alpha))$. Il suit de la relation $\varphi_2 = \varphi_3 \varphi_1$ (où φ_i dénote φ tensorisé avec 1 en i-ème position) que l'élément

(1.12) $$\partial_{na}(D(\alpha)) = D(\alpha)_2^{-1}\, D(\alpha)_3\, D(\alpha)_1$$

(attention à l'ordre des facteurs!) de $G_m(M_n(S^{(3)}))$ appartient au centre de $M_n(S^{(3)})$ donc à $G_m(S^{(3)})$. On vérifie de plus que c'est un 2-cocycle et on construit un plongement

(1.13) $$\beta : Br(R) \to H^2(R)$$

en associant à l'algèbre A la classe de $\partial_{na}(D(\alpha))$. La construction explicite de β donnée ci-dessus permet de vérifier facilement que $Br(R)$ est de torsion. De façon plus précise :

<u>Théorème</u> (1.15). Soit A une R-algèbre d'Azumaya de rang n^2. Alors $n[A] = 1$ dans $Br(R)$.

<u>Démonstration</u>. Soit $c = \partial_{na}(D(\alpha))$ un cocycle associé à une neutralisation $\alpha : A \otimes S \tilde{\to} M_n(S)$ de A. A l'algèbre $A^{(n)} = A \otimes \ldots \otimes A$ (n fois) correspond le cocycle $c(\alpha)^n$. On a $c(\alpha)^n = [\partial_{na}(D(\alpha))]^n = \det(\partial_{na}(D(\alpha))) = \partial_2(\det(D(\alpha)))$ où det est le déterminant. L'élément c^n est donc un cobord.

2. <u>Les suites de Mayer-Vietoris</u>

Soient $f,g \in R$ tels que $Rf + Rg = R$. Par un argument de suites spectrales, on sait associer une suite de Mayer-Vietoris au recouvrement $R_f \times R_g$ de R ([M] p. 110). Nous aurons besoin d'une construction explicite de la partie

(2.1) $\quad H^1(R_f) \oplus H^1(R_g) \to H^1(R_{fg}) \xrightarrow{d_{MV}} H^2(R) \to H^2(R_f) \oplus H^2(R_g) \to H^2(R_{fg})$

de cette suite. La première et la dernière application sont induites par la différence, la troisième par la diagonale. Les deux lemmes suivants rassemblent les résultats techniques nécessaires à la définition de d_{MV} et à la vérification de l'exactitude.

<u>Lemme</u> (2.2). Pour calculer $H^n(R_{fg}) = \varinjlim_S H^n(S/R_{fg})$, il suffit de considérer des recouvrements S de la forme $S_1 \otimes S_2$, S_1 recouvrement étale de R_f et S_2 recouvrements étale de R_g.

<u>Démonstration</u>. C'est une conséquence immédiate de (1.6) puisque $R_{fg} = R_f \otimes R_g$.

<u>Lemme</u> (2.3). Soient S_1 un recouvrement étale de R_f et S_2 un recouvrement étale de R_g. Pour tout triplet $\lambda = (u_1,v,u_2)$ de $G_m(S_1^{(3)} \times (S_1 \otimes S_2)^{(2)} \times S_2^{(3)})$, soit $M(\lambda) = M(u_1,v,u_2)$ l'élément de

$(S_1 \times S_2)^{(3)} = S_1^{(3)} \times S_1^{(2)} \otimes S_2 \times S_1 \otimes S_2 \otimes S_1 \times S_1 \otimes S_2^{(2)} \times S_2 \otimes S_1^{(2)} \times$

$S_2 \otimes S_1 \otimes S_2 \times S_2^{(2)} \otimes S_1 \times S_2^{(3)}$ défini par la formule suivante :

si $v = \Sigma\alpha \otimes \beta \otimes \gamma \otimes \delta \otimes \in S_1^{(2)} \otimes S_2^{(2)}$, alors

$M(\lambda) = (u_1, \Sigma\alpha \otimes \beta \otimes \gamma\delta, (\Sigma\alpha \otimes \gamma\delta \otimes \beta)^{-1}, (\Sigma\alpha\beta \otimes \gamma \otimes \delta)^{-1}, \Sigma\gamma\delta \otimes \alpha \otimes \beta,$

$\Sigma\gamma \otimes \alpha\beta \otimes \delta, (\Sigma\gamma \otimes \delta \otimes \alpha\beta)^{-1}, u_2)$.

Cette construction vérifie les propriétés suivantes :

1) $M(\lambda_1\lambda_2^{-1}) = M(\lambda_1)M(\lambda_2)^{-1}$

2) a) Si $z_i \in G_m(S_i^{(2)})$, $i = 1,2$, sont des cochaînes normalisées (c'est-à-dire telles que l'image dans S_i par la multiplication est 1), alors $M(1, z_1 \otimes z_2^{-1}, 1) = \partial(z_1, 1, 1, z_2) \cdot (\partial z_1^{-1}, 1, \ldots, 1, \partial z_2^{-1})$

b) $M(\partial z_1, z_1^{-1} \otimes z_2, \partial z_2) = \partial(z_1^{-1}, 1, 1, z_2^{-1})$

3) Si $z \in G_m(S_1 \otimes S_2)$, alors $M(1, \partial z, 1) = \partial(1, z^{-1}, \partial z, 1)$ où τ est le switch.

4) Si $u_i \in G_m(S_i^{(3)})$ sont des cocycles normalisés et si $u_1 \otimes 1 = (1 \otimes u_2)\partial v$, alors $M(u_1, v, u_2)$ est un cocycle.

5) Si $S_1 = S_2 = S$, si $u = \Sigma a \otimes b \otimes c \in G_m(S^{(3)})$ est un cocycle normalisé et si $v = (\Sigma a \otimes 1 \otimes b \otimes c)(\Sigma a \otimes b \otimes 1 \otimes c)^{-1}$, alors $u^{-1} \otimes u = \partial v$ et $M(u,v,u)$ est cohomologue à l'image diagonale de u dans $(S \times S)^{(3)}$.

<u>Démonstration</u>. Les propriétés 1), 2) et 3) se vérifient directement facilement. Pour démontrer 4) notons $u_1 = \Sigma a \otimes b \otimes c$ et $u_2 = \Sigma u \otimes y \otimes z$. En contractant différents produits tensoriels (par multiplication) et en utilisant que u_1 et u_2 sont normalisés, on tire de la formule $u_1 \otimes u_2^{-1} = \partial v$ les relations suivantes :

a) $(\Sigma a\otimes b\otimes c\otimes 1)(\Sigma\alpha\otimes 1\otimes\beta\otimes\gamma\delta) = (\Sigma 1\otimes\alpha\otimes\beta\otimes\gamma\delta)(\Sigma\alpha\otimes\beta\otimes 1\otimes\gamma\delta)$.

b) $(\Sigma\alpha\beta\otimes\gamma\otimes 1\otimes\delta) = (\Sigma 1\otimes w\otimes y\otimes z)(\Sigma\alpha\beta\otimes 1\otimes\gamma\otimes\delta)(\Sigma\alpha\beta\otimes\gamma\otimes\delta\otimes 1)$.

c) $\Sigma\alpha\beta\otimes\gamma\delta = 1$, $\Sigma w\otimes yz = \Sigma wy\otimes z = 1$

$\Sigma ab\otimes c = \Sigma a\otimes bc = 1$.

d) $\Sigma b\otimes ac\otimes 1 = (\Sigma\alpha\otimes\beta\otimes\gamma\delta)(\Sigma\beta\otimes\alpha\otimes\gamma\delta)$, d'où par symétrie

$\Sigma b\otimes ac\otimes 1 = \Sigma ac\otimes b\otimes 1$.

e) $(\Sigma ac\otimes b\otimes 1\otimes 1)(\Sigma\alpha\beta\otimes 1\otimes\gamma\otimes\delta) = (\Sigma\beta\otimes\alpha\otimes\gamma\otimes\delta)(\Sigma\alpha\otimes\beta\otimes\gamma\delta\otimes 1)$

$(\Sigma b\otimes ac\otimes 1\otimes 1)(\Sigma 1\otimes\alpha\beta\otimes\gamma\otimes\delta) = (\Sigma\alpha\otimes\beta\otimes 1\otimes\gamma\delta)(\Sigma\beta\otimes\alpha\otimes\gamma\otimes\delta)$.

De d) et e) on tire

f) $(\Sigma\alpha\beta\otimes 1\otimes\gamma\otimes\delta)(\Sigma\alpha\otimes\beta\otimes 1\otimes\gamma\delta) = (\Sigma 1\otimes\alpha\beta\otimes\gamma\otimes\delta)(\Sigma\alpha\otimes\beta\otimes\gamma\delta\otimes 1)$.

Le fait que $M(u_1,v,u_2)$ soit un cocycle suit alors de a), b) et f).

5) se vérifie en utilisant certaines des formules ci-dessus.

(2.4) <u>Construction de d_{MV} dans (2.1)</u>.

Soit v un 1-cocycle sur R_{fg}. D'après (2.2) on peut supposer que v est un élément de $G_m((S_1\otimes S_2)^{(2)})$. D'après 4) $M(1,v,1)$ est un 2-cocycle sur R ($S_1\times S_2$ est un recouvrement de R) et on pose $d_{MV}[v] = [M(1,v,1)]$. Le fait que l'application est bien définie suit essentiellement de 1) et 3).

(2.5) <u>Exactitude de (2.1)</u>.

<u>En $H^1(R_{fg})$</u> : Si v est un cocycle sur R_{fg} de la forme $u_1\otimes u_2^{-1}$, alors d'après 2)a) $M(1,v,1) = \partial(u_1,1,1,u_2)$ donc $d_{MV}[v] = 1$. Inversement, si $d_{MV}[u] = \partial(u_1,v,w,u_2)$, alors u_1 et u_2 sont des cocycles et u est cohomologue à $u_1\otimes u_2^{-1}$.

En $H^2(R)$: Soit $u \in G_m(S^{(3)})$ un cocycle dont les images dans S_1 et S_2 sont des cobords ∂z_1, resp. ∂z_2. Notons φ_i les applications $S \to S_i$ $i = 1,2$. On vérifie alors à l'aide de (2.3) que $(\varphi_1 \times \varphi_2)^{(3)} u$ est cohomologue à $M(1,(\varphi_1^{(2)} \otimes \varphi_2^{(2)} v^{-1})(z_1^{-1} \otimes z_2),1)$ où v est défini comme dans 5).

En $H^2(R_f) \oplus H^2(R_g)$: c'est une conséquence de 4).

Nous construisons maintenant une suite de Mayer-Vietoris pour Pic et le groupe de Brauer. Nous définissons tout d'abord un groupe de Brauer "relatif" pour un diagramme

$$\begin{array}{ccc} & & R_1 \\ & & \downarrow \\ R_2 & \longrightarrow & R_3 \end{array} \quad .$$

Soient $A_i \in \underline{\underline{A_z}}(R_i)$, $i = 1,2$ des algèbres qui deviennent semblables sur R_3. Autrement dit, il existe P et $Q \in \underline{\underline{FP}}(R_3)$ et un isomorphisme

(2.6) $$\xi : A_1 \otimes_{R_1} R_3 \otimes \mathrm{End}_{R_3} P \xrightarrow{\sim} A_2 \otimes_{R_2} R_3 \otimes \mathrm{End}_{R_3} Q .$$

Appelons un quintuplet

(2.7) $\Lambda = (A_1, A_2, P, Q, \xi)$ où ξ vérifie (2.6) une donnée. Un isomorphisme de données $\Lambda \xrightarrow{\sim} \Lambda'$ est défini par un quadruplet $(\alpha_1,\alpha_2,\sigma,\rho)$, $\alpha_i : A_i \xrightarrow{\sim} A_i'$ $i = 1,2$, $\sigma : P \xrightarrow{\sim} P'$, $\rho : Q \xrightarrow{\sim} Q'$ tels que

(2.8) $$(\alpha_2 \otimes \mathrm{End}\rho) \circ \xi = \xi' \circ (\alpha_1 \otimes \mathrm{End}\sigma) \quad .$$

On dira qu'une donnée $(\mathrm{End}_{R_1} P_1, \mathrm{End}_{R_2} P_2, N, M, \mathrm{End}\varphi)$, où $P_i \in \underline{\underline{FP}}(R_i)$ $i = 1,2$, N et $M \in \underline{\underline{FP}}(R_3)$ et

(2.9) $$\varphi : P_1 \otimes_{R_1} R_3 \otimes N \xrightarrow{\sim} P_2 \otimes_{R_2} R_3 \otimes M$$

est <u>triviale</u> et qu'une donnée $\Delta = (A_1, A_2, P, Q, \xi)$ est <u>neutralisée</u> par les recouvrements S_i/R_i, $i = 1,2,3$,

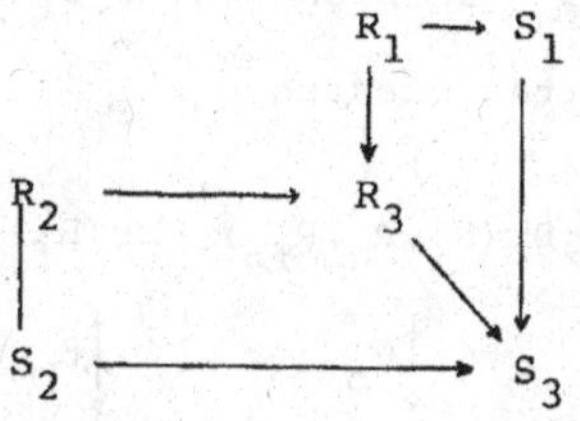

si la donnée obtenue de façon évidente par extension des scalaires est triviale.

<u>Proposition</u> (2.10). Soient $f, g \in R$ tels que $R_f + R_g = R$ et soient $R_1 = R_f$, $R_2 = R_g$ et $R_3 = R_{fg}$. Toute donnée $\Delta = (A_1, A_2, P, Q, \xi)$ peut être neutralisée.

<u>Démonstration</u>. On prend pour S_i, $i = 1,2$ des neutralisations de A_i et pour S_3 un recouvrement de $S_1 \otimes S_2$ qui rend $\xi \otimes 1_{S_1 \otimes S_2}$ intérieur. Remarquons que d'après (1.6) on peut choisir S_3 de la forme $S_1' \otimes S_2'$.

Le produit "tensoriel" de deux données se définit de façon évidente et on dira que deux données Δ et Δ' sont <u>semblables</u> s'il existe deux données triviales ε et ε' et un isomorphisme de données $\Delta \otimes \varepsilon \xrightarrow{\sim} \Delta' \otimes \varepsilon'$. Notons $Br(R_1, R_2, R_3)$ l'ensemble des classes d'équivalence pour cette relation. C'est un groupe avec la multiplication induite par le produit tensoriel. L'élément neutre est la classe de $(R_1, R_2, R_3, R_3, 1)$ et l'inverse de (A_1, A_2, P, Q, ξ) est $(A_1^o, A_2^o, P^*, Q^*, \xi^o)$ où N^* est le dual de N.

Les projections $(A_1, A_2, P, Q, \xi) \to A_i$ $i = 1,2$ induisent des homomorphismes de groupes

$$\pi_i : Br(R_1,R_2,R_3) \to Br(R_i) \qquad i = 1,2 .$$

Théorème (2.11). Soient $R_1 = R_f$, $R_2 = R_g$, $R_3 = R_{fg}$, $Rf + Rg = R$. Alors le groupe $Br(R_f,R_g,R_{fg})$ se plonge dans $H^2(R)$ et il existe un diagramme commutatif de suites exactes

$$\begin{array}{ccccccccccc}
\mathrm{Pic}\, R_f \oplus \mathrm{Pic}\, R_g & \to & \mathrm{Pic}\, R_{fg} & \xrightarrow{d} & Br(R_f,R_g,R_{fg}) & \to & Br(R_f) \oplus Br(R_g) & \to & Br(R_{fg}) \\
\downarrow\wr & & \downarrow\wr & & \downarrow\beta_o & & \downarrow\beta_o & & \downarrow\beta_o \\
H^1(R_f) \oplus H^1(R_g) & \to & H^1(R_{fg}) & \xrightarrow{d_{MV}} & H^2(R) & \longrightarrow & H^2(R_f) \oplus H^2(R_g) & \to & H^2(R_{fg})
\end{array}$$

La deuxième suite est la suite (2.3).

Démonstration. Nous nous limiterons à définir β_o et d. La commutativité et l'exactitude sont des vérifications longues, mais de routine. Le fait que β_o soit injectif suit finalement du lemme des 5.

Définition de β_o : Soit $\Delta = (A_1,A_2,P,Q,\xi)$ une donnée. Sans restriction, on peut supposer que A_1,A_2,P et Q sont de rang constant. Soient S_i/R_i, $i = 1,2,3$ des recouvrements qui neutralisent Δ. Soient $\alpha_i : A_i \otimes S_i \tilde{\to} M_{n_i}(S_i)$ $i = 1,2$. Comme dans la construction de (1.14), on peut supposer que φ_i défini par

$$\begin{array}{ccc}
S_i \otimes A_i \otimes S_i & \xrightarrow{1\otimes\alpha_i} & M_{n_i}(S_i \otimes S_i) \\
\tau\otimes 1 \downarrow & & \downarrow \varphi_i \\
A_i \otimes S_i \otimes S_i & \xrightarrow[\alpha_i\otimes 1]{} & M_{n_i}(S_i \otimes S_i)
\end{array} \qquad (2.11)$$

est intérieur, $\varphi_i = \mathrm{End}(D(\alpha_i))$. D'autre part on peut choisir S_3 de la forme $S_1 \otimes S_2$ et des isomorphismes $\sigma : P \otimes S_3 \tilde{\to} S_3^{r_1}$, $\rho : Q \otimes S_3 \tilde{\to} S_3^{r_2}$. Soient $D(\sigma)$ et $D(\rho)$ les applications $(S_3 \otimes S_3)^{r_i} \to (S_3 \otimes S_3)^{r_i}$ définies par des diagrammes commutatifs correspondant à (2.11) pour P et Q. Par défintion de la neutralisation d'une donnée, il existe un homomorphisme

$$\psi : S_3^{n_1} \otimes_{S_3} P \otimes S_3 \to S_3^{n_2} \otimes_{S_3} Q \otimes S_3$$

tel que le diagramme

$$\begin{array}{ccc} A_1 \otimes S_1 \otimes S_2 \otimes_{S_3} \mathrm{End}_{S_3}(P \otimes S_3) & \xrightarrow{\xi \otimes 1} & A_2 \otimes S_1 \otimes S_2 \otimes_{S_3} \mathrm{End}_{S_3}(Q \otimes S_3) \\ \downarrow {\scriptstyle (\alpha_1 \otimes 1) \otimes 1} & & \downarrow {\scriptstyle (\alpha_2 \otimes 1) \otimes 1} \\ M_{n_1}(S_3) \otimes_{S_3} \mathrm{End}_{S_3}(P \otimes S_3) & \xrightarrow[\mathrm{End}(\psi)]{} & M_{n_2}(S_3) \otimes_{S_3} \mathrm{End}_{S_3}(Q \otimes S_3) \end{array}$$

commute. Quitte à remplacer ψ par $1 \otimes \sigma \circ \psi \circ 1 \otimes \rho^{-1}$, on peut supposer $P \otimes S_3$ et $Q \otimes S_3$ libres et on a un isomorphisme

$$\mathrm{End}\,\psi : M_{n_1}(S_3) \otimes_{S_3} M_{r_1}(S_3) \to M_{n_2}(S_3) \otimes_{S_3} M_{r_2}(S_3) .$$

Par définition de End ψ le diagramme

$$\begin{array}{ccc} M_{n_1 r_1}(S_3^{(2)}) & \xrightarrow{\mathrm{End}(1 \otimes \psi)} & M_{n_2 r_2}(S_3^{(2)}) \\ \downarrow {\scriptstyle \mathrm{End}(D(\alpha_1) \otimes 1)} & & \downarrow {\scriptstyle \mathrm{End}(D(\alpha_2) \otimes 1)} \\ M_{n_1 r_1}(S_3^{(2)}) & \xrightarrow[\mathrm{End}(\psi \otimes 1)]{} & M_{n_2 r_2}(S_3^{(2)}) \end{array}$$

commute. Il existe donc $u \in G_m(S_3^{(2)})$ tel que

$$D(\alpha_1) \otimes 1 = (\psi \otimes 1)^{-1}(1 \otimes D(\alpha_2))(1 \otimes \psi) \cdot u$$

On vérifie alors que $\partial_{na}(D(\alpha_1)) \otimes 1 = (1 \otimes \partial_{na}(D(\alpha_2)))\partial u$ dans $G_m((S_1 \otimes S_2)^{(3)})$. On normalise alors $\partial_{na}(D(\alpha_i))$, $i = 1,2$ (et on change u en conséquence). D'après (2.3) 4) l'élément $M = M(\partial_{na}(D(\alpha_1)), u, \partial_{na}(D(\alpha_2)))$ est un 2-cocycle sur R et on définit $\beta_o[\Delta] = [W]$. La vérification que cette application est bien définie est laissée au lecteur.

<u>Définition de d.</u>

Soit $I \in \underline{\underline{Pic}}(R_{fg})$. Alors $\Delta = (R_f, R_g, R_{fg}, I, can)$ où can est l'isomorphisme canonique $R_{fg} \xrightarrow{\sim} End_{R_{fg}}(I)$ est une donnée et on pose $d[I] = [\Delta]$.

3. <u>Le théorème de Gabber</u>

Soit $\beta : Br(R) \to H^2(R)$ le plongement du groupe de Brauer construit en (1.14).

<u>Théorème</u> (3.1) (Gabber). Soit $c \in H^2(R)$ un élément de torsion. Alors il existe une R-algèbre d'Azumaya A telle que $\beta[A] = c$.

Le théorème est une conséquence des résultats suivants :

<u>Lemme</u> (3.2) (Recollement). Soit $c \in H^2(R)$ un élément de torsion et soient $f, g \in R$ tels que $Rf + Rg = R$. Soient c_f et c_g les images de c dans $H^2(R_f)$, resp. $H^2(R_g)$. Alors, s'il existe une R_f-algèbre d'Azumaya A et une R_g-algèbre d'Azumaya B telles que $\beta[A] = c_f$ et $\beta[B] = c_g$, il existe une R-algèbre d'Azumaya C telle que $\beta[C] = c$.

<u>Corollaire</u> (3.3). Soit $c \in H^2(R)$ un élément de torsion. Supposons que pour tout idéal maximal $\underline{m}$ de R, il existe une $R_{\underline{m}}$-algèbre d'Azumaya

$A(\underline{m})$ telle que $\beta[A(\underline{m})] = c_{\underline{m}}$ (où $c_{\underline{m}}$ est l'image de c dans $H^2(R_{\underline{m}})$.
Alors il existe une R-algèbre d'Azumaya A telle que $\beta[A] = c$.

<u>Démonstration du corollaire</u>. Soit $\alpha(\underline{m}) : A(\underline{m}) \otimes S(\underline{m}) \tilde{\to} M_n(S(\underline{m}))$ une neutralisation de $A(\underline{m})$ telle que la classe du cocycle associé $\partial_{na}(D(\alpha(\underline{m})))$ soit $c_{\underline{m}}$. Puisque les algèbres $\alpha(\underline{m})$, $S(\underline{m})$ et l'isomorphisme $\alpha(\underline{m})$ peuvent être décrits par un nombre fini de conditions algébriques sur $R_{\underline{m}}$, il existe $f \notin \underline{m}$, une R_f-algèbre étale $S(f)$, une R_f-algèbre d'Azumaya $A(f)$ et une neutralisation $\alpha(f) : A(f) \otimes S(f) \to M_n(S(f))$ tels que $[\partial_{na}(D(\alpha(f)))] = c_f$. Soit maintenant $\Sigma = \{f \in R \mid \exists A(f)$ telle que $\beta[A(f)] = c_f\}$. Il suffit alors de montrer que Σ est un idéal de R, car alors $\Sigma = R$ et $1 \in \Sigma$. Il est clair que si $a \in R$ et $f \in \Sigma$, alors $a \cdot f \in \Sigma$. D'autre part il suit du lemme 1 appliqué à R_{f+g} que si $f, g \in \Sigma$, alors $f + g \in \Sigma$.

Pour obtenir (3.1), il suffit maintenant de vérifier le cas local :

<u>Lemme</u> (3.4) (Version locale de 3.1). Soit R un anneau local. Alors pour tout élément $c \in H^2(R)_{torsion}$, il existe une R-algèbre d'Azumaya A telle que $\beta[A] = c$.

Nous démontrerons maintenant (3.2) puis (3.4).

<u>Démonstration de (3.2)</u>. Puisque $\beta[A_g] = \beta[B_f] = c_{fg}$, il existe une donnée Δ de la forme (A,B,P,Q,ξ). Sans restriction de la généralité, nous pouvons supposer que les rangs de A,B,P et Q sont constants. Il est clair que $(\beta_o[\Delta])_f = c_f$, $(\beta_o[\Delta])_g = c_g$. Par conséquent $\beta_o[\Delta]c^{-1} = d_{MV}(x)$, $x \in H^1(R_{fg}) = \mathrm{Pic}(R_{fg})$. Soit $I \in \underline{\underline{\mathrm{Pic}}}(R_{fg})$ un représentant de x. Quitte à remplacer Δ par $\Delta \otimes (R_f, R_g, R_{fg}, I, \mathrm{can})$, on peut donc supposer que $\beta_o[\Delta] = c$. Soit n l'ordre de c. De $1 = c^n = \beta_o[\Delta]^n = \beta_o[\Delta^{(n)}]$ et de l'injectivité de β_o, on déduit que la donnée

$\Delta^{(n)} = (A^{(n)}, B^{(n)}, P^{(n)}, Q^{(n)}, \xi^{(n)})$ est triviale, donc isomorphe à une donnée (End M, End N, E, F, η) où η est intérieur. En particulier, on a que $E \cong P^{(n)}$, $F = Q^{(n)}$. On peut donc supposer que $E = P^{(n)}$, $F = Q^{(n)}$ et $\eta = \mathrm{End}\,\omega$ où $\omega : M \otimes P^{(n)} \to N \otimes Q^{(n)}$ est un isomorphisme de R_{fg}-modules. D'après Bass [B, § 7], il existe un homomorphisme

$$\mathrm{Rang} : K_o \underline{\underline{FP}}(R) \to C^+(R)$$

où $C^+(R)$ est le groupe multiplicatif des fonctions continues sur Spec(R) à valeurs dans $\mathbb{Q}^+$, qui associe à la classe [P] d'un R-module fidèlement projectif son rang en chaque point de Spec(R). Cet homomorphisme admet une section σ. Pour tout $x \in K_o \underline{\underline{FP}}(R)$, l'élément $\rho(x) = x/\sigma(\mathrm{Rang}\ x)$ est de rang un. Il suit des résultats de ([B], Ch. IV, § 7) que tout élément de $K_o \underline{\underline{FP}}(R)$ de rang 1 admet une racine n-ième unique. Rappelons encore que pour tout $x \in K_o \underline{\underline{FP}}(R)$, nx est la classe d'un module projectif dès que n est multiple d'un certain entier $n_o(x)$. Soit maintenant x la classe de $M \otimes P^{(n)} \cong N \otimes Q^{(n)}$ dans $K_o \underline{\underline{FP}}(R)$ et soit $y = \rho(x)$. Soit z la racine n-ième de y^{-1}. On vérifie alors que $(z\rho([P]))^n = y^{-1}\rho([P])^n = \rho([M])^{-1}$ et par conséquent $z\rho([P]) = (\rho([M]))^{-1/n}$ appartient à l'image de $K_o \underline{\underline{FP}}(R_f)$ dans $K_o \underline{\underline{FP}}(R_{fg})$. Puisque les rangs de P et de Q sont constants, z[P] et z[Q] appartiennent également à l'image de $K_o \underline{\underline{FP}}(R_f)$, resp. $K_o \underline{\underline{FP}}(R_g)$ dans $K_o \underline{\underline{FP}}(R_{fg})$. Il existe donc des entiers h arbitrairement grands tels que hz = [H], $hz[P] = [P'_{fg}]$ et $hz[Q] = [Q'_{fg}]$, où $H \in \underline{\underline{FP}}(R_{fg})$, $P' \in \underline{\underline{FP}}(R_f)$ et $Q' \in \underline{\underline{FP}}(R_g)$. Les égalités $[H][P] = [P'_{fg}]$ et $[H][Q] = [Q'_{fg}]$ entraînent, d'après le théorème d'effacement de Bass ([B], Chap. IX, § 7), que $(H \otimes P)^m = {P'_{fg}}^m$ et $(H \otimes Q)^m \simeq {Q'_{fg}}^m$ pour tout entier m assez grand. Quitte à remplacer h par hm, on peut donc supposer que $H \otimes P \cong P'_{fg}$ et $H \otimes Q \cong Q'_{fg}$. La classe de $\Gamma = (A,B,P,Q,\xi) \otimes (R_f, R_g, H, H, \mathrm{Id})$

est $\beta_o[\Delta]$ et $\Gamma \cong (A,B,P'_{fg},Q'_{fg},\xi')$. En recollant $A \otimes \mathrm{End}\, P'$ et $B \otimes \mathrm{End}\, Q'$ à l'aide de ξ', on obtient l'algèbre C cherchée.

<u>Démonstration de (3.4)</u>. Soit R un anneau local et soit $c \in H^2(R)$ un cocycle de torsion. Remarquons tout d'abord qu'il existe un recouvrement étale S/R tel que l'image c_S de c dans $H^2(S)$ est triviale. En effet, si $u \in G_m(S^{(3)})$ est un cocycle qui représente c, on déduit de $\partial u = 1$ que $u \otimes 1_S$ est un cobord dans $G_m((S \otimes S/S)^{(3)})$. Soit $\underline{p}$ un idéal premier de S au-dessus de l'idéal maximal $\underline{m}$ de R. Alors $S_{\underline{p}}$ est de la forme $(R[t]/f(t))_{\underline{m}_1}$ où $f(t) = t^N + a_1 t^{N-1} + \ldots + a_N$ et $\underline{m}_1$ est un idéal maximal de $R[t]/(f(t))$ qui ne contient pas l'image de f'. Soit C un anneau neutralisant pour f : $C = R[t_1,\ldots,t_N]/I$ où I est l'idéal engendré par tous les éléments $s_i(T_1,\ldots,t_N) - (-1)^i a_i$, s_i étant la i-ème fonction symétrique. C est un R-module projectif de rang N! et le polynôme f a (au moins) N racines distinctes dans C qui sont permutées transitivement par le groupe symétrique S_N. De plus S_N permute transitivement les idéaux maximaux de C au-dessus de l'idéal maximal $\underline{m}$ de R. Notons $R[t]/(f(t)) = C_1$, donc $S_{\underline{p}} = (C_1)_{\underline{m}_1}$. Soit $\underline{m}'$ un idéal maximal de C au-dessus de $\underline{m}_1$ et soit $\underline{m}''$ un idéal maximal quelconque de C. On a un diagramme

(3.5)
$$\begin{array}{ccccc} R \longrightarrow & C_1 & \longrightarrow & C & \\ & \downarrow & & \downarrow & \\ & (C_1)_{\underline{m}_1} & \longrightarrow & C_{\underline{m}'} & \underset{\sigma}{\overset{\sim}{\rightarrow}} C_{\underline{m}''} \\ & \downarrow\wr & & & \\ & S_{\underline{p}} & & & \end{array}$$

où σ est induit par un élément de S_N. Puisque l'image c_S du cocycle c dans $H^2(S)$ est nulle, il suit de (3.5) que l'image de c dans tous les $H^2(C_{\underline{m}''})$ est nulle. Par conséquent, l'image de c dans $H^2(C)$ est représentée localement par une algèbre d'Azumaya (l'algèbre triviale!).

On en conclut d'après (3.3) que cette image est représentée par une algèbre d'Azumaya sur C. Le résultat est alors conséquence du lemme suivant :

<u>Lemme</u> (3.6). Soit C/R une R-algèbre commutative, fidèlement projective comme R-module. Soit $c \in H^2(R)$ un élément tel que son image dans $H^2(C)$ soit représentée par une algèbre d'Azumaya sur C. Alors c est représenté par une algèbre d'Azumaya sur R.

<u>Démonstration</u>. D'après (3.3) on peut supposer que R est local. Soit S/R un recouvrement étale et $u \in G_m(S^{(3)})$ un cocycle qui représente c. Soit A une C-algèbre d'Azumaya telle que $\beta[A] = 1 \otimes c \in H^2(C)$. Soit R^{hs} le hensélisé strict de R. Puisque C est une algèbre finie sur R, tout recouvrement étale de $C \otimes R^{hs}$ admet une section ([Gr], § 18). La $C \otimes R^{hs}$-algèbre $A \otimes R^{hs}$ est donc triviale et comme R^{hs} est limite inductive de recouvrements étales, il existe un recouvrement étale local T/R tel que $A \otimes T$ est triviale sur $C \otimes T$. Soit $\alpha : A \otimes T \xrightarrow{\sim} M_n(C \otimes T)$ une neutralisation. Quitte à remplacer T par un recouvrement de T, on peut supposer que l'automorphisme $(\alpha \otimes 1)\, \tau\, (1 \otimes \alpha^{-1})$ de $M_n(C \otimes T \otimes T)$ est induit par $D(\alpha) : ((C \otimes T/C)^{(2)})^n \to ((C \otimes T/C)^{(2)})^n$. Comme R^{hs} recouvre S et T, on peut remplacer S et T par R^{hs}. Le cocycle $\partial_{na}(D(\alpha))$ est cohomologue à $1 \otimes u$; on a donc $1 \otimes u = \partial_{na}(D(\alpha)) \cdot \partial w$ où $w \in G_m((W/C)^{(3)})$, W un recouvrement étale de C. Comme la $C \otimes R^{hs}$-algèbre étale $W \otimes R^{hs}$ admet une section, on peut remplacer W par $W \otimes R^{hs}$ d'abord et par $C \otimes R^{hs}$ ensuite. Nous pouvons donc choisir w dans $G_m((C \otimes R^{hs}/C)^{(3)})$. L'isomorphisme $D(\alpha)w : (C \otimes R^{hs} \otimes R^{hs})^n \to (C \otimes R^{hs} \otimes E^{hs})^n$ peut s'interpréter comme un isomorphisme de $R^{hs} \otimes R^{hs}$-modules libres de rang nd où d est le rang de C. Il induit alors un isomorphisme de $R^{hs} \otimes R^{hs}$-algèbres

$$\psi = \mathrm{End}(D(\alpha)w) : M_{nd}(R^{hs} \otimes R^{hs}) \to M_{nd}(R^{hs} \otimes R^{hs})$$

qui vérifie la condition de descente de R^{hs} à R : $\psi_2 = \psi_3\psi_1$. L'algèbre descendue $B = \{x \in M_n(R^{hs}) \mid \psi(1 \otimes x) = x \otimes 1\}$ est une R-algèbre d'Azumaya et par construction le cocycle associé à B est u.

Bibliographie

[A] M. Artin, On the joins of Hensel rings, Advances in Math. 7 (1971), 282-296.

[B] H. Bass, Algebraic K-theory, Benjamin, New York, 1968.

[Ga] O. Gabber, Some theorems on Azumaya algebras, ce volume

[Gr] A. Grothendieck, Eléments de géométrie algébrique IV, Publ. Math. I.H.E.S. Vol. 32, 1967.

[KO] M.-A. Knus et M. Ojanguren, Théorie de la descente et algèbres d'Azumaya, Springer LN 389, 1974.

[M] J.S. Milne, Etale cohomology, Princeton Mathematical Series, Vol. 33, Princeton, 1980.

SUR LE GROUPE DE BRAUER D'UN ANNEAU DE POLYNÔMES EN CARACTERISTIQUE P ET LA THEORIE DES INVARIANTS

W. Hürlimann *

Introduction.

Un cas simple d'anneau commutatif de caractéristique $p > 0$, dont le groupe de Brauer est encore mal connu, est l'anneau de polynômes à deux variables $R=F_q[X,Y]$ sur un corps fini à $q=p^r$ éléments. Nous savons que le groupe de Brauer $Br(R)$ est très grand : c'est une somme directe infinie dénombrable de copies de $\mathbb{Z}(p^\infty)$ ([KOS], p. 40). Notre travail contribue à une meilleure compréhension de la structure de $Br(R)$ et détermine en particulier le sous-groupe ${}_pBr(R)$ des éléments de $Br(R)$ d'ordre p. L'action naturelle du groupe linéaire $G=SL(2,q)$ sur R induit une action de G sur le groupe de Brauer $Br(R)$. L'étude de cette action a été suggérée par Amitsur.

Dans le premier chapitre, nous montrons que le groupe ${}_pBr(R)$ est muni naturellement d'une structure de F_qG-module. Nous construisons alors un plongement de R dans ${}_pBr(R)$ compatible avec l'action de G. Les résultats de Glover sur la structure de $R=F_p[X,Y]$ comme F_pG-module fournissent ainsi des renseignements sur la structure de F_pG-module de ${}_pBr(R)$. En particulier, nous obtenons que ${}_pBr(R)$ contient tous les types de F_pG-modules de dimension finie. A l'aide des résultats de Yuan, nous construisons ensuite une suite exacte pour les groupes ${}_pBr(F_q[X,Y])$ et ${}_pBr(F_q(X,Y))$. Nous en tirons un isomorphisme explicite de ${}_pBr(F_q[X,Y])$ avec p^2-1 copies du groupe additif de $F_q[X,Y]$.

L'étude de l'action de $SL(2,q)$ sur ${}_pBr(R)$ nous a conduit à considérer les invariants de R par l'action d'un sous-groupe quelconque G de $SL(2,q)$. En caractéristique zéro, les invariants de R sont bien connus ([Sp], chap. 4). En caractéristique p, nous n'avons trouvé dans la littérature que des résultats partiels. Dans le deuxième chapitre, nous construisons un système générateur minimal d'invariants de l'anneau $k[X,Y]^G$ pour tous les sous-groupes finis G de $SL(2,k)$, k un corps quelconque de caractéristique $p > 2$. Nous en tirons des conditions nécessaires et suffisantes pour que $k[X,Y]^G$ soit un anneau de polynômes. Ces résultats s'appliquent au groupe de Brauer. Le plongement de l'anneau $R=F_q[X,Y]$ dans ${}_pBr(R)$ induit un plongement de l'anneau des invariants R^G dans la partie invariante ${}_pBr(R)^G$ du groupe de Brauer. Il suit que $Br(R)^G = \bigoplus_\infty \mathbb{Z}(p^\infty)$. Si R^G est un anneau de polynômes, alors on a même

* Version améliorée d'une thèse présentée à l'EPFZ en juin 1980.

${}_p\mathrm{Br}(R^G)=(p^2-1)R^G$.

Dans le troisième chapitre, nous étudions l'homomorphisme $\mathrm{Br}(R^G) \longrightarrow \mathrm{Br}(R)^G$ induit par l'inclusion d'anneaux $R^G \subset R$. Si S est le corps des fractions de R, l'homomorphisme de p-groupes abéliens $\mathrm{Br}(S^G)_p \longrightarrow \mathrm{Br}(S)_p^G$ est surjectif pour tout sous-groupe G de $GL(2,q)$. Si p ne divise pas $\mathrm{card}(G)$, c'est même un isomorphisme. L'étude de l'homomorphisme $\mathrm{Br}(R^G) \longrightarrow \mathrm{Br}(R)^G$ est plus difficile puisque les extensions R/R^G ne sont pas galoisiennes. Nous démontrons que l'homomorphisme $\mathrm{Br}(R^G) \longrightarrow H^0(R/R^G,\mathrm{Br})$ (cohomologie d'Amitsur) est surjectif si R^G est régulier. Par contre, si G est un sous-groupe de $SL(2,q)$ dont l'ordre est divisible par p, alors l'homomorphisme $\mathrm{Br}(R^G) \longrightarrow \mathrm{Br}(R)^G$ n'est pas surjectif. Le conoyau de cet homomorphisme contient même une copie de R^G. Pour les corps, nous donnons également un critère suffisant pour décider de la surjectivité de l'homomorphisme $\mathrm{Br}(.^G) \longrightarrow \mathrm{Br}(.)^G$. Par exemple, l'homomorphisme $\mathrm{Br}(F_q(X)^G) \longrightarrow \mathrm{Br}(F_q(X))^G$ est surjectif pour tout sous-groupe parfait G de $PGL(2,q)$.

Mentionnons pour finir qu'une grande partie de nos résultats est formulée et démontrée pour des anneaux de polynômes et des corps de fonctions rationnelles à plusieurs variables sur un corps parfait de caractéristique p.

J'aimerais remercier M.-A. Knus pour son soutien généreux lors de la préparation de cette thèse ainsi que pour une suggestion qui est à l'origine de la version améliorée du premier chapitre.

CHAPITRE I. Groupe de Brauer d'un anneau de polynômes en caractéristique p.

Nous rappelons d'abord quelques résultats de la théorie des algèbres d'Azumaya en caractéristique $p > 0$. Puis nous résumons certains résultats de Yuan ([Y]) concernant le groupe de Brauer $\mathrm{Br}(S/R)$ pour une extension d'anneaux $R \subset S$ radicielle finie de hauteur un. Nous construisons ensuite une suite exacte contenant le groupe ${}_p\mathrm{Br}(R)$ formé des éléments de $\mathrm{Br}(R)$ annulés par p pour l'anneau de polynômes $R=K[X_1,\dots,X_n]$ et le corps des fonctions rationnelles $R=K(X_1,\dots,X_n)$ sur un corps parfait K de caractéristique p. Ce résultat donne en particulier une description de ${}_p\mathrm{Br}(R)$ en termes de générateurs et relations. Nous montrons ensuite que le groupe additif de l'anneau de polynômes à $n > 1$ variables $R=K[X_1,\dots,X_n]$ sur un corps quelconque de caractéristique p se plonge dans le groupe ${}_p\mathrm{Br}(R)$. Si le corps K est fini, nous nous intéressons aux plongements équivariants de R dans ${}_p\mathrm{Br}(R)$. Pour cela, nous considérons l'action naturelle de $G=GL(n,K)$

sur R ainsi que l'action induite par G sur ${}_p\mathrm{Br}(R)$. Nous construisons un homomorphisme injectif de $KU_n(K)$-module $R \longrightarrow {}_p\mathrm{Br}(R)$ où $U_n(K)$ est le sous-groupe unipotent de G. Si $R=K[X,Y]$, nous obtenons un homomorphisme injectif de $KSL(2,K)$-module $R \longrightarrow {}_p\mathrm{Br}(R)$. Nous appliquons ce résultat au cas particulier $R=F_p[X,Y]$ pour $p > 2$. D'après les résultats de Glover sur la structure du $F_pSL(2,p)$-module R, nous obtenons que ${}_p\mathrm{Br}(R)$ contient tous les types de $F_pSL(2,p)$-modules de dimension finie. Finalement, nous considérons l'anneau de polynômes à n variables $R=K[X_1,\ldots,X_n]$ sur un corps parfait de caractéristique p. Pour tout sous-groupe G de $GL(n,K)$, le groupe ${}_p\mathrm{Br}(R)$ est non seulement un KG-module, mais aussi un Δ-module où Δ est le produit croisé trivial de R avec G. Nous montrons alors que ${}_p\mathrm{Br}(R)$ est un R-module libre de rang $(n-1)(p^n-1)$ en construisant une base. En particulier, nous obtenons un isomorphisme explicite de ${}_p\mathrm{Br}(R)$ avec $(n-1)(p^n-1)$ copies du groupe additif de R. Il est alors possible de décrire de façon semblable la partie p-primaire $\mathrm{Br}(R)_p$. L'étude de la structure de Δ-module de ${}_p\mathrm{Br}(R)$ fera l'objet d'une recherche ultérieure. Le cas d'un anneau de polynômes à deux variables nous suggère que ${}_p\mathrm{Br}(R)$ est un Δ-module indécomposable et que ${}_p\mathrm{Br}(R)$ possède une série de composition dont les facteurs proviennent tous de KG-modules complètement réductibles de dimension finie.

§ 1. Algèbres d'Azumaya en caractéristique p.

Rappels (cf. [KOS])

Soit R un anneau commutatif de caractéristique $p > 0$.

(1.1) Définition. Une R-algèbre d'Azumaya A est dite cyclique de degré p^e si A est neutralisée par une sous-algèbre commutative de la forme $R[a^{1/p^e}]=R[T]/(T^{p^e}-a)$, $a \in R$.

(1.2) Une algèbre cyclique de degré p est de la forme

$$(a,b)_R=R\langle x,y\rangle/(x^p-a,y^p-b,yx-xy-1)$$

où $R\langle x,y\rangle$ est l'algèbre non commutative libre sur R engendrée par x et y. Nous écrirons également (a,b) pour cette algèbre lorsqu'aucune confusion n'est possible.

(1.3) Soient v et w les images de x et y dans $(a,b)_R$. Alors

a) $(a,b)_R$ est un R-module libre de base v^iw^j $(0 \leq i,j < p)$, et si S est une R-algèbre commutative, alors $(a,b)_R \otimes S \simeq (a,b)_S$.

b) $(a,b)_R \simeq M_p(R)$ si $a=c^p$ pour $c \in R$.

c) Si $c \in R[v]$, alors $(a,b)_R \simeq (a,b+c^p+d_w^{p-1}(c))_R$ où d_w est la dérivation intérieure déterminée par w. De même, si $c \in R[w]$, alors $(a,b)_R \simeq (a+c^p+d_v^{p-1}(c),b)_R$.

d) $(a,b)_R \simeq (b,-a)_R$, $(a,b)_R^o \simeq (a,-b)_R \simeq (-a,b)_R$.

e) $(a,b)_R \otimes (a,b')_R \simeq (a,b+b')_R \otimes (0,b)_R$ et $(a,b)_R \otimes (a,b')_R$ est semblable à $(a,b+b')_R$.

(1.4) Soit $q=p^e$ et K une extension de R telle que $K^q=R$, alors $Br(K/R)$ est l'ensemble $_q Br(R)$ des éléments de $Br(R)$ d'exposant q (ou annihilés par q).

(1.5) Le sous-groupe $_p Br(R)$ de $Br(R)$ de tous les éléments d'exposant p est engendré par les algèbres (a,b).

(1.6) Soit S une extension de R de la forme $S=R[a^{1/p}]$, soit d la R-dérivation de S telle que $d(t)=1$ où t est la classe de T dans $S=R[T]/(T^p-a)$, soit encore $E_a=\{d^{p-1}(c)+c^p,\ c \in S\}$. Supposons que $Pic(S)=0$. Alors

$$Br(S/R) \simeq H^2(S/R) \simeq R^+/E_a.$$

En particulier, $(a,b)=0$ si et seulement si $b \in E_a$.

(1.7) Si A est une R-algèbre d'Azumaya neutralisée par $C=R[a_1^{1/p},\dots,a_n^{1/p}]$, alors A est semblable à $A_1 \otimes A_2 \otimes \dots \otimes A_n$ où A_i est neutralisée par $R[a_i^{1/p}]$.

(1.8) Si A est une R-algèbre d'Azumaya contenant $C=R[a^{1/p}]$ comme sous-algèbre commutative maximale et si A est C-projectif, alors il existe $b \in R$ tel que $A \simeq (a,b)$.

Appliquons les résultats précédents à l'anneau de polynômes $R=K[X_1,\dots,X_n]$ où K est un corps parfait de caractéristique p. Nous obtenons

(1.9) Le groupe $_p Br(R)$ est engendré comme groupe abélien par les classes d'algèbres de la forme $\bigotimes_{i=1}^{n} (X_i,a_i)$ où $a_i \in R$.

<u>Démonstration</u>. D'après (1.4), $_p Br(R)=Br(K[X_1^{1/p},\dots,X_n^{1/p}]/R)$. Ainsi une p-algèbre d'Azumaya A est neutralisée par $C=K[X_1^{1/p},\dots,X_n^{1/p}]=(K[X_1^{1/p},\dots,X_n^{1/p}])^p[X_1^{1/p},\dots,X_n^{1/p}]=R[X_1^{1/p},\dots,X_n^{1/p}]$ (remarquer que $K^p=K$ puisque K est parfait). Elle est donc équivalente à une algèbre $\bigotimes_{i=1}^{n} A_i$

où A_i est neutralisée par $R[X_i^{1/p}]$ (1.7). De plus, si A_i est neutralisée par $R[X_i^{1/p}]$, alors d'après ([ChR], p. 29), A_i est semblable à A_i' avec A_i' contenant $R[X_i^{1/p}]$ comme sous-algèbre commutative maximale et A_i' est $R[X_i^{1/p}]$-projectif. D'après (1.8), on a $A_i' \simeq (X_i, a_i)$ pour $a_i \in R$. Par conséquent A est semblable à $\bigotimes_{i=1}^{n} (X_i, a_i)$, d'où l'affirmation.

§ 2. Extensions régulières et algèbres d'Azumaya.

Dans tout ce paragraphe, $R \subset S$ est une extension radicielle finie de hauteur un. Ceci signifie que R est un anneau commutatif de caractéristique p, que S est un R-module projectif de type fini et que l'anneau $\mathrm{End}_R(S)$ est engendré en tant que R-algèbre par S et la p-algèbre de Lie $\underline{g}=\underline{g}(S/R)$ des R-dérivations de S. Nous notons par $Br(S/R)$ le groupe de Brauer des R-algèbres d'Azumaya neutralisées par S. L'étude du groupe $Br(S/R)$ à l'aide du groupe de Hochschild $E(S,\underline{g})$ des extensions régulières de p-algèbres de Lie de S par $\underline{g}$ a été faite par Yuan. Nous résumons ici certains de ses résultats indispensables à notre travail.

Considérons S comme une p-algèbre de Lie abélienne avec la p-application $s \longrightarrow s^p$.

<u>Définition 2.1</u>. Une <u>extension régulière</u> de S par $\underline{g}$ est une suite exacte de p-algèbre de Lie

$$L : 0 \longrightarrow S \longrightarrow \Lambda \xrightarrow{\Psi} \underline{g} \xrightarrow{\phi} 0$$

telle que les conditions suivantes soient satisfaites :

(2.2) $\Psi^{-1}[x,\Psi s]=(\phi x)(s)$ pour tout $x \in \Lambda$ et $s \in S$

(2.3) Λ possède une structure de S-module pour laquelle ϕ et Ψ sont S-linéaires

(2.4) $[sx,s'x']=s(x.s')x' - s'(x'.s)x + ss'[x,x']$ pour tout $s,s' \in S$ et $x,x' \in \Lambda$, avec $x.s=\Psi^{-1}[x,\Psi s]$

(2.5) $(sx)^p=s^px^p + D_{sx}^{p-1}(s)x$ pour tout $s \in S$ et $x \in \Lambda$, et où D_x pour $x \in \Lambda$ désigne l'application $s \longrightarrow x.s$.

Deux extensions régulières L, L' sont <u>équivalentes</u> s'il existe un homomorphisme S-linéaire de p-algèbres de Lie $\xi : \Lambda \longrightarrow \Lambda'$ tel que le diagramme suivant soit commutatif :

(2.6)
$$\begin{array}{ccccc} & & \Lambda & & \\ & \overset{\Psi}{\nearrow} & & \overset{\phi}{\searrow} & \\ S & & \downarrow \xi & & \underline{g} \\ & \underset{\Psi'}{\searrow} & & \underset{\phi'}{\nearrow} & \\ & & \Lambda' & & \end{array}$$

L'ensemble des classes d'équivalence d'extensions régulières peut être muni d'une structure additive de groupe ([Y], p. 431). Nous obtenons le groupe de Hochschild $E(S,\underline{g})$ des extensions régulières de S par $\underline{g}$.

Considérons maintenant R comme un S-module via $s.r=s^p r$, $s \in S$, $r \in R$. Le groupe $E(S,\underline{g})$ s'identifie à un quotient de $\mathrm{Hom}_S(\underline{g},R)$ de la manière suivante (voir [Y] pour les détails). Pour tout $x \in S$, notons dx le S-homomorphisme

(2.7) $dx : \underline{g} \longrightarrow S$, $dx(\partial)=\partial x$ pour tout $\partial \in \underline{g}$.

Il est possible de prolonger l'application $d : S \longrightarrow \Omega^1 := \mathrm{Hom}_S(\underline{g},S)$ à une application R-linéaire $d : \Omega \longrightarrow \Omega$ où Ω est l'algèbre extérieure de $\mathrm{Hom}_S(\underline{g},S)$ sur S ([Y], lemme 11). Nous notons Z le noyau de $\Omega^1 \longrightarrow \Omega^2$. On montre que

(2.8) $Z=\{\omega=dx + \Sigma\, r_i x_i^{p-1} dx_i \mid x, x_i \in S,\ r_i \in R\}$ ([Y], lemme 13)

Pour tout $\omega \in Z$, nous avons une application $\Gamma\omega : \underline{g} \longrightarrow R$, appelée <u>opérateur de Cartier</u>, donnée par

(2.9) $$(\Gamma\omega)(\partial)=\omega(\partial^p) - \partial^{p-1}(\omega(\partial))$$

On vérifie que

(2.10) Γ est R-linéaire, $\Gamma(dx)=0$ et $\Gamma(x^{p-1}dx)(\partial)=(\partial(x))^p$

et que $\Gamma : Z \longrightarrow \mathrm{Hom}_S(\underline{g},R)$ est un homomorphisme de groupes additifs. Par conséquent, l'application

(2.11) $\delta : Z \longrightarrow \mathrm{Hom}_S(\underline{g},R)$, $(\delta\omega)(\partial)=(\Gamma\omega)(\partial) - \omega(\partial)^p$, $\partial \in \underline{g}$,

est aussi un homomorphisme de groupes additifs.

Soit maintenant $\theta \in \mathrm{Hom}_S(\underline{g},R)$. Le S-module somme directe $S \oplus \underline{g}$ est une p-algèbre de Lie Λ_θ relativement au crochet $[(x,\partial),(x',\partial')]=(\partial x'-\partial' x,[\partial,\partial'])$ et à la p-application $(x,\partial)^p=(x^p+\partial^{p-1}x+\theta\partial,\partial^p)$, telle que la suite exacte

(2.12) $L_\theta : 0 \longrightarrow S \xrightarrow{\Psi} \Lambda_\theta \xrightarrow{\pi} \underline{g} \longrightarrow 0$, $\Psi(x)=(x,0)$, $\pi(x,\partial)=\partial$

est une extension régulière ([Y], p. 444). De plus, L_θ est l'extension régulière triviale si et seulement si $\theta=\delta(\omega)$ pour $\omega \in Z$. L'application

(2.13) $$\sigma : \mathrm{Hom}_S(\underline{g},R)/\mathrm{Im}\delta \longrightarrow E(S,\underline{g}), \quad \sigma(\theta)=L_\theta$$

est alors un isomorphisme ([Y], p. 448).
Nous allons maintenant associer une extension régulière à toute R-algèbre d'Azumaya A contenant S comme sous-algèbre commutative maximale et telle que toute dérivation ∂ de $\underline{g}$ peut être prolongée à une dérivation intérieure de A. Notons D_s la dérivation de A donnée par $D_s x = sx - xs$ pour tout $x \in A$, $s \in A$. Soit

(2.14) $$\Lambda = \{s \in A \mid D_s(S) \subset S\}$$

et notons $\phi(s)$ la restriction de D_s à S. Le S-module Λ est une p-algèbre de Lie par le crochet $[s_1,s_2]=s_1s_2-s_2s_1$ et la p-application $s \longrightarrow s^p$. L'homomorphisme S-linéaire de p-algèbres de Lie $s \longrightarrow \phi(s)$ applique Λ surjectivement sur $\underline{g}$ et il est clair que le noyau de ϕ coïncide avec S. Nous avons ainsi une suite exacte de p-algèbres de Lie

(2.15) $$L : 0 \longrightarrow S \xrightarrow{\iota} \Lambda \xrightarrow{\phi} \underline{g} \longrightarrow 0.$$

D'après ([Ho], lemme 11), L est une extension régulière.

§ 3. Une suite exacte pour le groupe ${}_p\mathrm{Br}(R)$.

Dans tout ce paragraphe, $R \supset R^p$ est une extension d'anneaux commutatifs de caractéristique p ayant une p-base $(X_i)_{1 \leq i \leq n}$, c'est-à-dire telle que $R=R^p[X_1,\ldots,X_n]=\bigoplus_{0 \leq i_j < p} R^p X_1^{i_1}\ldots X_n^{i_n}$. Nous notons $S_i=R[X_i^{1/p}]=R[u_i]$ où u_i est l'image de T dans $R[T]/(T^p-X_i)$ et $S=R[u_1,\ldots,u_n]$. Nous supposons que $S \supset R$ est une extension radicielle finie de hauteur un et que $\mathrm{Br}(S/R) \simeq H^2(S/R,G_m)$ (par exemple si $\mathrm{Pic}(S)=\mathrm{Pic}(S\otimes S)=0$). D'après le résultat de Yuan, on a alors $\mathrm{Br}(S/R) \simeq E(S,\underline{g})$. Nous savons également de (1.6) que

(3.1) $$\mathrm{Br}(S_i/R)=\{[(X_i,f)] \mid f \in R^+/E_{X_i}\}$$

Puisque $S^p=R$, on a de (1.4) et (1.9)

(3.2) $${}_p\mathrm{Br}(R)=\mathrm{Br}(S/R)=\{[\bigotimes_{i=1}^{n}(X_i,f_i)] \mid f_i \in R\}$$

Le critère suivant donne une condition nécessaire et suffisante pour décider si une p-algèbre $\bigotimes_{i=1}^{n}(X_i,f_i)$, $f_i \in R$, est triviale dans ${}_p\mathrm{Br}(R)$.

Proposition 3.3. L'algèbre d'Azumaya $\overset{n}{\underset{i=1}{\otimes}} (X_i, f_i)$ est triviale dans $_p\mathrm{Br}(R)$ si et seulement si $f_i = (\delta\omega)(\partial_i)$ où ∂_i est la R-dérivation de S donnée par $\partial_i u_j = \delta_{ij}$, $\omega \in Z$ (cf. (2.8)) et $\delta : Z \longrightarrow \mathrm{Hom}_S(\underline{g}, R)$ est l'homomorphisme donné par (2.11).

Démonstration. Nous savons que l'algèbre cyclique (X_i, f_i) est engendrée par des générateurs u_i et y_i tels que $u_i^p = X_i$, $y_i^p = f_i$ et $y_i u_i = u_i y_i + 1$. Soit A la R-algèbre libre engendrée par les u_i, y_i, $1 \leq i \leq n$, tels que $u_i^p = X_i$, $y_i^p = f_i$, $y_i y_j = y_j y_i$ et $y_i s = s y_i + \partial_i s$ pour tout $s \in S = R[u_1, \ldots, u_n]$. Les monômes

$$y_1^{i_1} \cdots y_n^{i_n}, \quad 0 \leq i_j < p,$$

forment une base du S-module A. De plus, l'homomorphisme de R-algèbres $\overset{n}{\underset{i=1}{\otimes}} (X_i, f_i) \longrightarrow A$ donné par $\overset{n}{\underset{i=1}{\otimes}} a_i \longrightarrow \prod_{i=1}^{n} a_i$ est un isomorphisme de R-algèbres (utiliser le théorème du commutant). Puisque A contient S comme sous-algèbre commutative maximale et que A est un S-module libre, il suit que toute dérivation ∂ de $\underline{g}$ se prolonge à une dérivation intérieure de A ([Y], p. 430). Notons

$$L : 0 \longrightarrow S \xrightarrow{\ \iota\ } \Lambda \xrightarrow{\ \phi\ } \underline{g} \longrightarrow 0$$

l'extension régulière de p-algèbres de Lie correspondant à A (cf. (2.15)). D'après (2.14), on a $\Lambda = \{x \in A \mid xs - sx \in S \text{ pour tout } s \in S\}$. Ecrivons $x \in \Lambda$ comme polynôme en y_i avec coefficients dans S, le degré en y_i étant au plus p-1. Nous prétendons que si $xu_i - u_i x \in S$, alors le degré de x en y_i est au plus 1. En effet, écrivons $x = x_0 + x_1 y_i + \ldots + x_m y_i^m$ où x_j ne contient pas y_i et $m < p$. Puisque $y_i u_i = u_i y_i + 1$, $y_i^j u_i = u_i y_i^j + j y_i^{j-1}$, nous avons

$$xu_i - u_i x = \sum_{j=0}^{m} (x_j y_i^j u_i - u_i x_j y_i^j) = \sum_{j=0}^{m} x_j (y_i^j u_i - u_i y_i^j) = \sum_{j=0}^{m} j x_j y_i^{j-1}.$$

Or ceci est dans S seulement si $x_m = 0$ lorsque $m > 1$. Il suit facilement que $\Lambda = \{x \in A \mid x = s_o + \sum_{i=1}^{n} s_i y_i,\ s_i \in S\}$. Montrons maintenant que l'extension régulière L est équivalente à

$$L_\theta : 0 \longrightarrow S \xrightarrow{\ \Psi\ } \Lambda_\theta \xrightarrow{\ \pi\ } \underline{g} \longrightarrow 0$$

(cf. (2.12)) où $\theta \in \mathrm{Hom}_S(\underline{g}, R)$ est tel que $\theta(\partial_i) = f_i$. Il suffit de trouver un homomorphisme S-linéaire de p-algèbres de Lie $\xi : \Lambda \longrightarrow \Lambda_\theta$ tel que le diagramme (2.6) soit commutatif. Pour $x = s_o + \sum_{i=1}^{n} s_i y_i$, nous posons $\xi(x) = (s_o, \phi(x))$ où $\phi(x) = \sum_{i=1}^{n} s_i \partial_i$. On vérifie immédiatement que $\xi\iota = \Psi$, $\pi\xi = \phi$, $\xi(sx) = s\xi(x)$ pour $s \in S$ et $\xi[x,y] = [\xi x, \xi y]$. La condition

$\xi(x^p)=\xi(x)^p$ est équivalente aux conditions $\xi(y_i^p)=\xi(y_i)^p$, $1 \leq i \leq n$. Puisque $\theta(\partial_i)=f_i$, on a $\xi(y_i)^p=(0,\partial_i)^p=(\theta(\partial_i),\partial_i^p)=(f_i,0)=\zeta(f_i)=\xi(y_i^p)$, comme désiré. De plus, l'extension régulière L_θ est triviale si et seulement si $\theta=\delta\omega$ pour $\omega \in Z$. Par conséquent, puisque ${}_p\mathrm{Br}(R)\simeq E(S,\underline{g})$, l'algèbre $A\simeq \bigotimes_{i=1}^{n}(X_i,f_i)$ est triviale dans ${}_p\mathrm{Br}(R)$ si et seulement si $(\delta\omega)(\partial_i)=f_i$ pour $1 \leq i \leq n$.

Théorème 3.4. Supposons que R, S_i et S soient comme au début du paragraphe. Alors la suite de groupes abéliens suivante est exacte :

$$0 \longrightarrow \mathrm{Ker}\,\Delta \longrightarrow R^+ \xrightarrow{\ \Delta\ } \bigoplus_{i=1}^{n} \mathrm{Br}(S_i/R) \xrightarrow{\ \varepsilon\ } {}_p\mathrm{Br}(R) \longrightarrow 0$$

où $\Delta(f)=\bigoplus_{i=1}^{n}[(X_i,\frac{\partial f}{\partial X_i})]$, $\varepsilon \bigoplus_{i=1}^{n}[(X_i,f_i)]=[\bigotimes_{i=1}^{n}(X_i,f_i)]$.

Démonstration. D'après (3.2), ε est surjectif. Soit $A=\bigoplus_{i=1}^{n}[(X_i,f_i)]$ dans $\mathrm{Ker}\,\varepsilon$ et montrons que $A=\Delta(f)$ pour $f \in R$. D'après la proposition (3.3), $\varepsilon(A)=0$ dans ${}_p\mathrm{Br}(R)$ si et seulement si $f_i=(\delta\omega)(\partial_i)$ pour une forme différentielle $\omega=ds+\Sigma\, r_j s_j^{p-1}ds_j$, $s, s_j \in S$, $r_j \in R$. Comme $s \in S$ est de la forme

$$s=\sum_{0\leq e_j<p} f_{e_1\ldots e_n} u_1^{e_1}\ldots u_n^{e_n},\quad f_{e_1\ldots e_n} \in R,$$

et en vertu de la linéarité, il suffit de calculer $f_i=(\delta\omega)(\partial_i)$ pour $\omega=ds$ et $\omega=r_o s^{p-1}ds$ où $s=ru_1^{e_1}\ldots u_n^{e_n}$, $r_o, r \in R$, $0 \leq e_j < p$. D'après (2.10) et (2.11), on a d'une part

$$(\delta ds)(\partial_i)=-(\partial_i r u_1^{e_1}\ldots u_n^{e_n})^p=-e_i r^p X_1^{e_1}\ldots X_i^{e_i-1}\ldots X_n^{e_n}=\frac{\partial}{\partial X_i}(-r^p X_1^{e_1}\ldots X_n^{e_n})=$$
$$=\frac{\partial}{\partial X_i}(-s^p).$$

D'autre part, on a

$$(\delta r_o s^{p-1}ds)(\partial_i)=r_o(\partial_i r u_1^{e_1}\ldots u_n^{e_n})^p-(r_o(ru_1^{e_1}\ldots u_n^{e_n})^{p-1}\partial_i r u_1^{e_1}\ldots u_n^{e_n})^p=$$
$$=\begin{cases} r_o r^p X_1^{e_1}\ldots X_i^{e_i-1}\ldots X_n^{e_n}-(r_o r^p)^p X_1^{e_1 p}\ldots X_i^{e_i p-1}\ldots X_n^{e_n p}, & \text{si } e_i\neq 0\\ 0 & \text{si } e_i=0.\end{cases}$$

D'après (3.1), on a $(\delta r_o s^{p-1}ds)(\partial_i) \in E_{X_i}$. Soit maintenant $f_i=(\delta\omega)(\partial_i)$ pour $\omega=ds+\Sigma\, r_j s_j^{p-1}ds_j$, $s, s_j \in S$, $r_j \in R$. Alors, on a $f_i \equiv \frac{\partial}{\partial X_i}(-s^p)$ mod E_{X_i}, c'est-à-dire $A=\Delta(f)$ pour $f=-s^p \in R$. On a ainsi montré que $\mathrm{Ker}\,\varepsilon$ est contenu dans $\mathrm{Im}\,\Delta$. Comme $S^p=R$, l'autre inclusion est immédiatement vérifiée. Il suit que $\mathrm{Im}\,\Delta=\mathrm{Ker}\,\varepsilon$.

Exemples 3.5. Soit K un corps parfait de caractéristique $p > 0$. Alors l'anneau de polynômes $R=K[X_1,\dots,X_n]$ et le corps des fonctions rationnelles $R=K(X_1,\dots,X_n)$ satisfont aux hypothèses du théorème 3.4.

§ 4. Plongement de $R=K[X_1,\dots,X_n]$ dans ${}_p\mathrm{Br}(R)$.

Pour $k \in \{1,\dots,n\}$, posons $X=X_k$ et définissons le sous-anneau R' de R par $R'[X]=R$. Soit d la $R'[X]$-dérivation de $R'[X^{1/p}]$ telle que $d(X^{1/p})=1$ et soit le groupe abélien additif $E_X=\{d^{p-1}(c)+c^p,\ c \in R'[X^{1/p}]\} \subset R$. On vérifie que

(4.1) $E_X=\{f_i^pX^i$ pour $i \neq np-1$, $f_i \in R'$, $-gX^{n-1}+g^qX^{nq-1}$, $n=1,2,\dots$, $q=p^m$, $m=1,2,\dots$, $g \in R'\}$ ([KOS], p. 38).

(4.2) $(X,a)=0$ si et seulement si $a \in E_X$ (suit de (1.6) : $\mathrm{Pic}(R'[X^{1/p}])$ est isomorphe à $\mathrm{Pic}(R)$ qui est nul).

Lemme 4.3. On a $E_X \cap \sum_{i\neq np-1} R'X^i=\{f_i^pX^i$ pour $i\neq np-1$, $f_i \in R'\}$.

Démonstration. Il est trivial que $E_X \cap \sum_{i\neq np-1}R'X^i \supset \{\dots\}$. Soit $f \in E_X$, alors on a

$$f=\sum_{i\neq np-1} f_i^pX^i + \sum_n \sum_{k=1}^{s}(-g_{n,k}X^n+g_{n,k}^{p^k}X^{(n+1)p^k-1})$$

où les sommes sur i et n sont finies et f_i, $g_{n,k} \in R'$. En ordonnant par puissances de X croissantes, on voit que le coefficient de X^i pour $i\neq np-1$ est $f_i^p - \sum_{k=1}^{s} g_{i,k}$ tandis que le coefficient de $X^{(i+1)p^m-1}$ où $m \in \{1,\dots,s\}$ est :

$$T_{i,m}=g_{i,m}^{p^m} + \sum_{\ell=1}^{m-1} g^{p^{m-\ell}}_{(i+1)p^\ell-1,m-\ell} - \sum_{k=1}^{s-m} g_{(i+1)p^m-1,k}$$

où certains $g_{k,\ell}$ peuvent être nuls! Puisque $f \in \sum_{i\neq np-1} R'X^i$, on a $T_{i,m}=0$ pour $m \in \{1,\dots,s\}$. Calculons

$$0=\sum_{m=1}^{s} T_{i,m}^{p^{s-m}} = \sum_{m=1}^{s}(g_{i,m}^{p^s} + \sum_{\ell=1}^{m-1} g^{p^{s-\ell}}_{(i+1)p^\ell-1,m-\ell} - \sum_{k=1}^{s-m} g^{p^{s-m}}_{(i+1)p^m-1,k})$$

$$= \sum_{m=1}^{s} g_{i,m}^{p^s} + \sum_{m=2}^{s}\sum_{\ell=1}^{m-1} g^{p^{s-\ell}}_{(i+1)p^\ell-1,m-\ell} - \sum_{m=1}^{s-1}\sum_{k=1}^{s-m} g^{p^{s-m}}_{(i+1)p^m-1,k}.$$

En changeant l'ordre de sommation, la somme intermédiaire est égale à

$$\sum_{\ell=1}^{s-1}\sum_{m=\ell+1}^{s} g^{p^{s-\ell}}_{(i+1)p^\ell-1,m-\ell} = \sum_{\ell=1}^{s-1}\sum_{k=1}^{s-\ell} g^{p^{s-\ell}}_{(i+1)p^\ell-1,k}.$$

Il suit que $0=\sum_{m=1}^{s} g_{i,m}^{p^s}=(\sum_{m=1}^{s} g_{i,m})^{p^s}$, d'où $\sum_{m=1}^{s} g_{i,m}=0$. Ainsi pour tout $i\neq np-1$, le coefficient de X^i est f_i^p, tandis que le coefficient de X^{np-1} est nul. Le lemme est démontré.

Si K est un corps quelconque de caractéristique $p > 0$, il est possible de plonger le groupe additif de $R=K[X_1,\dots,X_n]$ dans ${}_p\mathrm{Br}(R)$:

Théorème 4.4. Soit $R=K[X_1,\dots,X_n]$ l'anneau de polynômes à $n > 1$ variables sur un corps K de caractéristique $p > 0$. Soit $\Psi : R \longrightarrow {}_p\mathrm{Br}(R)$ l'application définie par $\Psi(f)=[(X_1,f^pX_2)]$ pour tout $f \in R$. Alors Ψ est un homomorphisme de groupes et est injectif.

Démonstration. Puisque d'après (1.3), $\Psi(f+g)=[(X_1,(f+g)^pX_2)]=$ $=[(X_1,f^pX_2+g^pX_2)]=[(X_1,f^pX_2)]+[(X_1,g^pX_2)]=\Psi(f)+\Psi(g)$, l'application Ψ est un homomorphisme de groupes. De plus, si $\Psi(f)=[(X_1,f^pX_2)]=0$, alors d'après (4.2), il faut que $f^pX_2 \in E_{X_1}$. Or $f^pX_2 \in E_{X_1} \cap \sum_{i\neq np-1} R'X^i=$ $=\{f_i^pX_1^i$ pour $i\neq np-1$, $f_i \in R'\}$ (4.3) où $R'=K[X_1,\dots,X_n]$. Par conséquent, on a $f^pX_2=0$, d'où $f=0$.

§ 5. Plongement équivariant de $R=K[X_1,\dots,X_n]$ dans ${}_p\mathrm{Br}(R)$.

Dans ce qui suit, $R=K[X_1,\dots,X_n]$ est l'anneau de polynômes à $n > 1$ variables sur un corps parfait K de caractéristique p, G est un sous-groupe de $GL(n,K)$. Tous les espaces vectoriels sont pris sur K.

Considérons l'action de G sur R donnée par : si $\sigma^{-1}=(a_{ij}) \in G$, $p(X_1,\dots,X_n) \in R$, alors

(5.1) $\sigma.p(X_1,\dots,X_n)=p(\sigma^{-1}(X_1),\dots,\sigma^{-1}(X_n))=p(\sum_j a_{1j}X_j,\dots,\sum_j a_{nj}X_j)$.

Cette action induit une représentation de G sur le groupe de Brauer $\mathrm{Br}(R)$: si A est une R-algèbre d'Azumaya et si A^σ est l'algèbre A munie de la structure de R-module "tordue" $r.a=\sigma(r)a$ ($r \in R$, $a \in A$, $\sigma \in G$), on vérifie que l'application $\rho : G \longrightarrow \mathrm{Aut}\,\mathrm{Br}(R)$ telle que $\rho(\sigma)[A]=[A^\sigma]$ est un homomorphisme de groupes. Nous nous restreignons ici au sous-groupe ${}_p\mathrm{Br}(R)$ de $\mathrm{Br}(R)$ engendré par les classes d'algèbres cycliques de la forme (a,b) où $a, b \in R$ (1.5). Dans la suite, si aucune confusion n'est possible, nous utiliserons la même notation pour une algèbre et sa classe dans ${}_p\mathrm{Br}(R)$. La structure de groupe de ${}_p\mathrm{Br}(R)$ sera notée additivement.

Munissons ${}_p\mathrm{Br}(R)$ d'une structure d'espace vectoriel sur K par

(5.2) $k.(a,b)=(ka,b)=(a,kb)$ pour $k \in K$.

Cette définition a un sens puisque $(ka,b) \simeq (a,kb)$ comme R-algèbres. En effet, soient v et w des générateurs de (ka,b) vérifiant $v^p = ka$, $w^p = b$ et $wv = vw+1$. Puisque K est parfait, il existe α dans K tel que $\alpha^p = k$. L'isomorphisme de R-algèbres $(ka,b) \simeq (a,kb)$ est alors donné par $v \longrightarrow (1/\alpha)v$ et $w \longrightarrow \alpha w$. Notons que l'application $\rho : G \longrightarrow \mathrm{Aut}\ {}_p\mathrm{Br}(R)$ définie ci-dessus munit ${}_p\mathrm{Br}(R)$ d'une structure naturelle de KG-module par $\sigma.(a,b) = (\sigma(a),\sigma(b))$ pour tout $\sigma \in G$. Remarquons que $(a,b)^\sigma = (\sigma(a),\sigma(b))$. En effet, l'algèbre $(a,b)^\sigma$ possède des générateurs v et w tels que $v^p = a.1 = \sigma(a)$ et $w^p = b.1 = \sigma(b)$. Nous montrons que le F_qG-module R se plonge dans le F_qG-module ${}_p\mathrm{Br}(R)$ lorsque G est le groupe unipotent $U_n(q)$ ou le groupe $SL(2,q)$.

<u>Théorème 5.3</u>. Soit $R = F_q[X_1,\ldots,X_n]$ l'anneau de polynômes à $n > 1$ variables sur le corps fini F_q de caractéristique p. Soit $\Psi : R \longrightarrow {}_p\mathrm{Br}(R)$ l'application définie par

(5.4) si $p \neq 2$, $\Psi(f) = (X_n, f^q X_{n-1})$, $f \in R$

(5.5) si $p = 2$, $\Psi(f) = (X_n, f^{2q} X_{n-1})$, $f \in R$.

Supposons que

(5.6) $G = U_n(q)$ est le groupe unipotent formé des $n \times n$-matrices triangulaires supérieures avec coefficients dans F_q

(5.7) $G = SL(2,q)$ lorsque $n = 2$.

Si R et ${}_p\mathrm{Br}(R)$ sont munis des structures naturelles de F_qG-modules, alors l'application Ψ est un homomorphisme de F_qG-modules et est injectif.

<u>Démonstration</u>. Considérons d'abord le cas (5.6) pour $p \neq 2$. Il est clair que Ψ est un homomorphisme d'espace vectoriel. Pour montrer que Ψ est un homomorphisme de F_qG-modules, il suffit de vérifier que Ψ est compatible avec l'action de générateurs de $U_n(q)$. Un élément $\sigma \in U_n(q)$ agit sur X_{n-1} par $\sigma(X_{n-1}) = X_{n-1} + \lambda X_n$ pour $\lambda \in F_q$ et sur X_n par l'identité. On obtient $\sigma.\Psi(f) = (X_n, \sigma(f)^q (X_{n-1} + \lambda X_n)) = \Psi(\sigma(f)) + \lambda(X_n, \sigma(f)^q X_n) = \Psi(\sigma(f))$ dans ${}_p\mathrm{Br}(R)$. Nous avons utilisé le fait que $(X_n, g^q X_n) = 0$ dans ${}_p\mathrm{Br}(R)$ pour tout $g \in R$. En effet, avec les mêmes notations que dans (1.3), on a $g^q X_n = c^p + d_w^{p-1}(c)$ pour $c = g^{q/p} v \in R[v]$. L'injectivité de Ψ est claire (cf. théorème 4.4). On voit de même que pour $p = 2$, l'application $\Psi(f) = (X_n, f^{2q} X_{n-1})$, $f \in R$, est un monomorphisme injectif de $F_qU_n(q)$-modules. Ici $(X_n, g^{2q} X_n) = 0$ dans ${}_2\mathrm{Br}(R)$ car $g^{2q} X_n = c^2 + d_w(c)$ pour $c = g^{q/2} + g^q v \in R[v]$. Considérons maintenant le cas (5.7). On sait ([Di], p. 36) que $SL(2,q)$ est engendré par les $\sigma(\lambda)$, $\tau(\lambda)$ définis comme suit : $\sigma(\lambda)(X_1) = X_1 + \lambda X_2$, $\sigma(\lambda)(X_2) = X_2$, $\tau(\lambda)(X_1) = X_1$, $\tau(\lambda)(X_2) = \lambda X_1 + X_2$ pour tout $\lambda \in F_q$. On vérifie comme ci-dessus que Ψ est compatible avec l'action de ces générateurs.

Soit maintenant $R=F_p[X,Y]$. Etant donné que R se plonge de facon équivariante dans ${}_p\mathrm{Br}(R)$, les représentations de $SL(2,p)$ dans R fournissent des représentations de $SL(2,p)$ dans ${}_p\mathrm{Br}(R)$. On obtient :

<u>Corollaire 5.8</u>. Soit $G=SL(2,p)$ et $R=F_p[X,Y]$. Alors chaque type de F_pG-module de dimension finie apparaît une infinité de fois comme sous-groupe G-invariant de ${}_p\mathrm{Br}(R)$.

<u>Démonstration</u>. C'est vrai pour le F_pG-module R ([G]).

§ 6. <u>Composante p-primaire du groupe de Brauer d'un anneau de polynômes sur un corps parfait de caractéristique p</u>.

Dans tout ce paragraphe, $R=K[X_1,\ldots,X_n]$ est l'anneau de polynômes à n variables sur un corps parfait K de caractéristique p et G est un sous-groupe de $GL(n,K)$ agissant de façon naturelle sur R. D'après le § 5, le groupe ${}_p\mathrm{Br}(R)$ est un KG-module de façon naturelle. Munissons maintenant ${}_p\mathrm{Br}(R)$ d'une structure de R-module. Soient (a,b) et (c,d) des classes d'algèbres cycliques dans ${}_p\mathrm{Br}(R)$ et soit $\lambda \in R$. Nous définissons

$$(6.1)\qquad \begin{aligned} &\lambda.(a,b)=(\lambda^p a,b)=(a,\lambda^p b)\\ &\lambda.\{(a,b)+(c,d)\}=\lambda.(a,b)+\lambda.(c,d)\end{aligned}$$

Cette définition a un sens puisque $(\lambda^p a,b)$ est isomorphe à $(a,\lambda^p b)$ comme S-algèbres, S le corps des fractions de R, et puisque l'homomorphisme $\mathrm{Br}(R) \longrightarrow \mathrm{Br}(S)$ est injectif. Il est aisé de vérifier que le groupe ${}_p\mathrm{Br}(R)$ est ainsi un R-module. Soit Δ le R-module libre de base $\{\sigma\}_{\sigma\in G}$ et munissons Δ de la multiplication $(r\sigma)(s\tau)=r\sigma(s)\sigma\tau$, $r,s\in R$, $\sigma,\tau\in G$. Alors Δ est un anneau de groupe "tordu" ou un produit croisé trivial de R avec G. De plus, le groupe ${}_p\mathrm{Br}(R)$ est un Δ-module. En effet, on a

$$(r\sigma)(s\tau).(a,b)=r\sigma(s^p\tau(a),\tau(b))=(r^p\sigma(s)^p\sigma\tau(a),\sigma\tau(b))=(r\sigma(s)\sigma\tau).(a,b).$$

Le plongement du groupe additif de R dans ${}_p\mathrm{Br}(R)$ donné par le théorème 4.4 n'est pas le seul possible. Nous allons construire un isomorphisme de ${}_p\mathrm{Br}(R)$ avec $(n-1)(p^n-1)$ copies du groupe additif de R. Pour $n>1$, introduisons d'abord l'ensemble I suivant formé de multi-indices. Un multi-indice $\alpha\in I$ est une suite d'indices

$$\alpha=jr_{\pi(1)}r_{\pi(2)}\cdots r_{\pi(i)}$$

variant comme suit. L'indice i parcourt l'ensemble $\{2,\dots,n\}$. Pour i fixé, $j \in \{\pi(1),\dots,\pi(i)\}$ où $\{\pi(1),\dots,\pi(i)\}$ parcourt l'ensemble des suites de longueur i strictement croissantes dans $\{1,\dots,n\}$. Pour i, j fixés, les $r_{\pi(k)}$ varient comme suit : si $j=\pi(1)$, $r_{\pi(1)}=p-1$, $r_{\pi(2)},\dots,r_{\pi(i)} \in \{1,\dots,p-1\}$ et si $j \in \{\pi(2),\dots,\pi(i)\}$, $r_j \in \{0,\dots,p-1\}$, $r_{\pi(1)},\dots,\hat{r}_j,\dots,r_{\pi(i)} \in \{1,\dots,p-1\}$.

<u>Lemme 6.2</u>. On a $\mathrm{card}(I)=(n-1)(p^n-1)$.

<u>Démonstration</u>. Par définition de l'ensemble I, il est immédiat que

$$\mathrm{card}(I)=\sum_{i=2}^{n}\binom{n}{i}\{(p-1)^{i-1}+(i-1)p(p-1)^{i-1}\}=\sum_{i=2}^{n}\binom{n}{i}(p-1)^{i-1}((i-1)p+1).$$

Notons cette dernière somme S_n et montrons par induction sur n que $S_n=(n-1)(p^n-1)$. Pour $n=2$, c'est vrai. Soit $n > 2$. Comme $\binom{n+1}{i}=\binom{n}{i}+\binom{n}{i-1}$, il vient

$$S_{n+1}=(p-1)^n(np+1)+S_n+\sum_{i=2}^{n}\binom{n}{i-1}(p-1)^{i-1}((i-1)p+1).$$

Calculons

$$\sum_{i=2}^{n}\binom{n}{i-1}(p-1)^{i-1}((i-1)p+1)=\sum_{j=1}^{n-1}\binom{n}{j}(p-1)^{j}(jp+1)$$

$$=(p-1)\sum_{j=1}^{n-1}\binom{n}{j}(p-1)^{j-1}((j-1)p+1)\ \pm\ p\sum_{j=1}^{n-1}\binom{n}{j}(p-1)^{j}$$

$$=(p-1)S_n+(p-1)n-(p-1)^n((n-1)p+1)+p\sum_{j=0}^{n}\binom{n}{j}(p-1)^{j}-p(p-1)^n-p.$$

Un calcul évident montre alors que $S_{n+1}=n(p^{n+1}-1)$.

<u>Théorème 6.3</u>. Pour tout entier naturel n, soit $R=K[X_1,\dots,X_n]$ l'anneau de polynômes à n variables sur un corps parfait de caractéristique p. Pour tout multi-indice $\alpha=jr_{\pi(1)}\dots r_{\pi(i)} \in I$, soit e_α la classe de l'algèbre cyclique

$$(X_j, X_{\pi(1)}^{r_{\pi(1)}}\dots X_{\pi(i)}^{r_{\pi(i)}})$$

et soit R_α une copie du groupe additif de R. Alors le groupe ${}_p\mathrm{Br}(R)$ est un R-module libre de rang $(n-1)(p^n-1)$ et de base $(e_\alpha)_{\alpha \in I}$. Plus précisément, soit $\Psi : \bigoplus_{\alpha \in I} R_\alpha \longrightarrow {}_p\mathrm{Br}(R)$ l'application définie par $\Psi(\sum_{\alpha \in I} f_\alpha)=\sum_{\alpha \in I} f_\alpha e_\alpha$, $f_\alpha \in R$. Alors Ψ est un isomorphisme du groupe ${}_p\mathrm{Br}(R)$ avec $(n-1)(p^n-1)$ copies du groupe additif de R.

<u>Exemple 6.4</u>. Soit K un corps parfait de caractéristique 2 et $R=K[X,Y]$. L'application $\Psi : R \oplus R \oplus R \longrightarrow {}_2\mathrm{Br}(R)$ donnée par

$\Psi(f \oplus g \oplus h)=[(X,f^2XY) \otimes (Y,g^2X) \otimes (Y,h^2XY)]$, est un isomorphisme de groupes abéliens.

Démonstration. Soit d'abord $R=K[X]$ l'anneau de polynômes à une variable. Alors toute algèbre cyclique (X,f), $f \in R$, est triviale. En effet, raisonnons par induction sur le degré du polynôme f. Si $\deg(f)=0$, alors $f \in K$ est une puissance de p, car K est parfait, d'où $(X,f)=0$. Soit maintenant $f \in R$ de degré strictement positif et supposons que $(X,g)=0$ pour tout $g \in R$ tel que $\deg(g) < \deg(f)$. Puisque f est de la forme $f=\sum_{0 \leq i < p} f_i^p X^i$, $f_i \in R$, et puisque $(X,f_i^pX^i)=\delta_{i,p-1}(X,f_i)$, on a dans ${}_p\mathrm{Br}(R)$: $(X,f)=(X,f_{p-1}^pX^{p-1})=(X,f_{p-1})=0$ car $\deg(f_{p-1}) < \deg(f)$. Supposons maintenant que $n > 1$. Pour tout $i \in \{2,\ldots,n\}$, soit G_i le sous-groupe de ${}_p\mathrm{Br}(R)$ engendré par les classes de la forme

$$(X_{\alpha_1}, f^pX_{\alpha_1}^{r_1}X_{\alpha_2}^{r_2}\ldots X_{\alpha_i}^{r_i}),$$

$f \in R$, $r_j \in \{1-\delta_{j1},2,\ldots,p-1\}$, avec exactement i variables distinctes X_{α_j} non puissances de p. Nous allons donner un ensemble minimal de générateurs du groupe G_i. On sait qu'il y a $\binom{n}{i}$ façons de choisir i variables distinctes parmi un ensemble de n variables distinctes. Un choix possible $X_{\pi(1)},X_{\pi(2)},\ldots,X_{\pi(i)}$ est donné par une suite $\{\pi(1),\ldots,\pi(i)\}$ de longueur i strictement croissante dans $\{1,\ldots,n\}$. Pour fixer les idées, restreignons-nous au choix $X_1,X_2,\ldots,X_i$ et donnons des générateurs de la forme

$$(X_j, f_j^pX_1^{r_{j1}}X_2^{r_{j2}}\ldots X_i^{r_{ji}}),$$

$f_j \in R$, $j \in \{1,\ldots,i\}$, $r_{jk} \in \{1-\delta_{jk},2,\ldots,p-1\}$, engendrant le sous-groupe H_i de G_i. Un élément quelconque A de H_i s'écrit :

$$A=\sum_{j=1}^{i}\sum_{\substack{r_k=1\\k\neq j}}^{p-1}(X_j, f^p_{jr_1\ldots p-1\ldots r_i}X_1^{r_1}\ldots X_j^{p-1}\ldots X_i^{r_i}) + \sum_{j=1}^{i}(X_j,\frac{\partial f_j}{\partial X_j})$$

où les $f_{jr_1\ldots p-1\ldots r_i} \in R$, $f_j \in R^p[X_1,\ldots,X_i]$. D'après le théorème 3.4, on a :

$$(X_1,\frac{\partial f_1}{\partial X_1})=-\sum_{j=2}^{i}(X_j,\frac{\partial f_1}{\partial X_j}).$$

Il vient, en posant $\overline{f}_j=f_j-f_1$:

$$A=\sum_{j=1}^{i}\sum_{\substack{r_k=1\\k\neq j}}^{p-1}(X_j, f^p_{jr_1\ldots p-1\ldots r_i}X_1^{r_1}\ldots X_j^{p-1}\ldots X_i^{r_i}) + \sum_{j=2}^{i}(X_j,\frac{\partial \overline{f}_j}{\partial X_j}).$$

Vérifions que cette expression pour A est unique. Il faut montrer que si $A=0$, alors tous les $f_{jr_1\ldots p-1\ldots r_i}$ et $\frac{\partial \overline{f}_j}{\partial X_j}$ sont nuls. Or d'après le théorème 3.4, on a $A=0$ si et seulement si

$$\sum_{r_k=1}^{p-1} f^p_{1p-1r_2\ldots r_i} X_1^{p-1}X_2^{r_2}\ldots X_i^{r_i} \equiv \frac{\partial f}{\partial X_1} \mod E_{X_1}$$

$$\sum_{r_k=1}^{p-1} f^p_{jr_1\ldots p-1\ldots r_i} X_1^{r_1}\ldots X_j^{p-1}\ldots X_i^{r_i} + \frac{\partial \bar{f}_j}{\partial X_j} \equiv \frac{\partial f}{\partial X_j} \mod E_{X_j}, \quad 2\leq j\leq n,$$

pour un certain $f \in R$. Ceci n'est possible que si les $f_{jr_1\ldots p-1\ldots r_i}=0$ et $\frac{\partial f}{\partial X_1} \in E_{X_1}$. Il suit que

$$f \equiv \sum_{r=1}^{p-1} g_r^p X_1^r \mod R^p[X_2,\ldots,X_n], \quad g_r \in R.$$

Puisque $\bar{f}_j \in R^p[X_1,\ldots,X_i]$ et $\frac{\partial \bar{f}_j}{\partial X_j} \equiv \frac{\partial f}{\partial X_j}$ mod E_{X_j}, il faut que $\frac{\partial \bar{f}_j}{\partial X_j} \in R^p[X_2,\ldots,X_i]$. Mais alors les expressions $(X_j,(\partial \bar{f}_j/\partial X_j))$ ne contiennent que $i-1$ variables distinctes non puissances de p. Par définition de H_i, il faut que $\partial \bar{f}_j/\partial X_j=0$, c'est-à-dire l'expression pour A ci-dessus est unique. Par conséquent, pour engendrer tout le sous-groupe H_i, il faudra prendre les $i(p-1)^{i-1}$ générateurs de la forme

$$(X_j, f^p_{jr_1\ldots r_i} X_1^{r_1}\ldots X_i^{r_i}), \quad f_{jr_1\ldots r_i} \in R, \quad j \in \{1,\ldots,i\}, \quad r_j=p-1,$$
$$r_k \in \{1,\ldots,p-1\} \quad \text{lorsque} \quad k\neq j,$$

ainsi que les $(i-1)(p-1)^i$ générateurs de la forme

$$(X_j, f^p_{jr_1\ldots r_i} X_1^{r_1}\ldots X_i^{r_i}), \quad f_{jr_1\ldots r_i} \in R, \quad j \in \{2,\ldots,i\}, \quad r_j \in \{0,\ldots,p-2\},$$
$$r_k \in \{1,\ldots,p-1\} \quad \text{lorsque} \quad k\neq j.$$

Pour engendrer tout G_i, il suffira de remplacer successivement la suite $\{1,2,\ldots,i\}$ ci-dessus par les autres suites $\{\pi(1),\ldots,\pi(i)\}$ de longueur i strictement croissantes dans $\{1,2,\ldots,n\}$. Par construction, ces générateurs sont linéairement indépendants dans ${}_p\mathrm{Br}(R)$. Soit I_i l'ensemble parcouru par le multi-indice $jr_{\pi(1)}\ldots r_{\pi(i)}$ décrit ci-dessus. L'application $\Psi_i : \bigoplus_{I_i} R_{jr_{\pi(1)}\ldots r_{\pi(i)}} \longrightarrow G_i$ définie par

$$\Psi_i\Big(\sum_{I_i} f_{jr_{\pi(1)}\ldots r_{\pi(i)}}\Big) = \sum_{I_i} (X_j, f^p_{jr_{\pi(1)}\ldots r_{\pi(i)}} X_{\pi(1)}^{r_{\pi(1)}}\ldots X_{\pi(i)}^{r_{\pi(i)}})$$

est un isomorphisme du groupe G_i avec $\binom{n}{i}(p-1)^{i-1}((i-1)p+1)$ copies de R^+. Une application du théorème 3.4 (analogue à celle déjà faite ci-dessus) montre que la somme $\sum_{i=2}^{i} G_i$ est directe. Finalement, il reste à montrer que le groupe ${}_p\mathrm{Br}(R)$ est engendré par les sous-groupes G_i. D'après (3.2) et le fait que $R=\bigoplus_{0\leq r_j<p} R^pX_1^{r_1}\ldots X_n^{r_n}$, il suffit de montrer qu'une classe d'algèbres de la forme $(X_j,f^pX_j^r)$, $f \in R$, $r \in \{0,\ldots,p-1\}$, appartient à $\sum_{i=2}^{n} G_i$. Mais ceci est vrai en vertu des relations $(X_j,f^pX_j^r)=\delta_{r,p-1}(X_j,f)$ et d'un raisonnement par induction sur le degré

du polynôme f (voir le cas $n=1$). Il suit, en utilisant le lemme 6.2 (démonstration), que

$$ {}_p\mathrm{Br}(R) = \bigoplus_{i=2}^{n} G_i \simeq \bigoplus_{i=2}^{n} \binom{n}{i}(p-1)^{i-1}((i-1)p+1)R^{+} = (n-1)(p^n-1)R^{+}. $$

En outre, l'isomorphisme Ψ est induit par les isomorphismes Ψ_i.

cqfd

Soit maintenant R^{1/p^m} le groupe abélien additif engendré par les racines p^m-èmes des éléments de R avec les relations $p^{m+1}.f=0$ pour tout $f \in R^{1/p^m}$. Pour $m \leq k$, définissons les homomorphismes

$$ \pi_m^k : R^{1/p^m} \longrightarrow R^{1/p^k}, \quad \pi_m^k(f) = p^{k-m}.f, $$

alors $(R^{1/p^m}, \pi_m^k)$ est un système inductif filtrant de groupes abéliens et nous notons la limite directe de ce système par

$$ R^{1/p^\infty} := \varinjlim_{m \in \mathbb{N}} R^{1/p^m}. $$

<u>Corollaire 6.5</u>. Soit $R=K[X_1,\ldots,X_n]$, K parfait de caractéristique p. Alors on a $\mathrm{Br}(R)_p \simeq (n-1)(p^n-1)R^{1/p^\infty}$. De plus, si $K=F_q$ est un corps fini, on a $\mathrm{Br}(F_q[X_1,\ldots,X_n]) \simeq (n-1)(p^n-1)F_q[X_1,\ldots,X_n]^{1/p^\infty}$.

<u>Démonstration</u>. Nous identifions $R^{1/p^m} = K[X_1^{1/p^m},\ldots,X_n^{1/p^m}] = R[X_1^{1/p^m},\ldots,X_n^{1/p^m}]$. D'après le théorème de ([KO], p. 146), l'homomorphisme

$$ \mathrm{Br}(R) \longrightarrow \mathrm{Br}(R^{1/p^m}) $$

est surjectif. Le noyau de cet homomorphisme est $\mathrm{Br}(R^{1/p^m}/R) = {}_{p^m}\mathrm{Br}(R)$ (1.4). Il suit que

$$ \mathrm{Br}(R^{1/p^m}) \simeq \mathrm{Br}(R)/{}_{p^m}\mathrm{Br}(R). $$

Ainsi, le groupe ${}_p\mathrm{Br}(R^{1/p^m})$ est isomorphe au sous-groupe de $\mathrm{Br}(R)$ formé des éléments d'ordre exactement p^{m+1}. L'affirmation suit du théorème 6.4 :

$$ \mathrm{Br}(R)_p = \varinjlim {}_p\mathrm{Br}(R^{1/p^m}) = \varinjlim (n-1)(p^n-1)R^{1/p^m} = (n-1)(p^n-1)R^{1/p^\infty}. $$

La seconde affirmation est vraie puisque $\mathrm{Br}(F_q[X_1,\ldots,X_n])$ est de p-torsion ([KOS]).

CHAPITRE II. Théorie des invariants.

Ce chapitre traite de la théorie classique des invariants. Dans le § 1, nous généralisons à la caractéristique p une formule bien connue en caractéristique 0 (cf. [Sp], corollaire 4.1.5, p. 74) : si R est l'anneau de polynômes à n variables sur un corps k, S son corps des fractions, G un sous-groupe fini de $GL(n,k)$ opérant sur R et $f_1, \ldots, f_n$ un système homogène de paramètres de l'anneau des invariants R^G, alors $\prod_{i=1}^{n} \deg f_i = (S^G : k(f_1,\ldots,f_n)) \operatorname{card}(G)$ (remarquer que S^G est le corps des fractions de R^G). A l'aide de cette formule et des calculs de Klein ([Kl]) et de Dickson ([D]$_2$), nous construisons au § 2 l'anneau des invariants $k[X,Y]^G$ pour tout sous-groupe fini G de $SL(2,k)$, k un corps de caractéristique $p > 2$. Le résultat obtenu est semblable à celui de Klein pour les sous-groupes finis de $SL(2,\mathbb{C})$: $k[X,Y]^G$ est ou un anneau de polynômes, ou l'anneau de coordonnées d'une hypersurface normale avec l'origine comme point singulier isolé. Nous en tirons alors des conditions nécessaires et suffisantes pour que $k[X,Y]^G$ soit un anneau de polynômes. Finalement nous avons ajouté au § 3 une démonstration élémentaire par la combinatoire du "théorème de Dickson" qui affirme que $F_q[X,Y]^{SL(2,q)}$ est un anneau de polynômes.

§ 1. Une formule de la théorie des invariants.

Dans tout ce paragraphe, $R=k[X_1,\ldots,X_n]$ est l'anneau de polynômes à n variables sur un corps k, S son corps des fractions et G est un sous-groupe fini de $GL(n,k)$. Le groupe linéaire opère comme suit sur R : si $\sigma^{-1}=(a_{ij})$ appartient à $GL(n,k)$, $f=f(X_1,\ldots,X_n)$ appartient à R, alors

$$\sigma.f=f(\sigma^{-1}(X_1),\ldots,\sigma^{-1}(X_n)) \quad \text{où} \quad \sigma^{-1}(X_i)=\sum_{j=1}^{n} a_{ij}X_j \quad (1 \leq i \leq n).$$

Ecrivons $R=\bigoplus_{d\geq 0} R_d$ où R_d est le k-espace vectoriel des polynômes homogènes de degré d. Pour l'anneau des invariants, on a $R^G=\bigoplus_{d\geq 0} R_d^G$ où $R_d^G=R^G \cap R_d$. Considérons la série de Poincaré :

$$P_G(T)=\sum_{d\geq 0} (\dim_k R_d^G) T^d.$$

Le résultat suivant est bien connu :

Théorème 1.1. Supposons que $f_1,\ldots,f_n$ soit un système homogène de paramètres de l'anneau des invariants R^G. Alors

$$P_G(T)=F(T) \prod_{i=1}^{n} (1 - T^{\deg f_i})^{-1}$$

où $F(T) \in \mathbb{Z}[T]$ avec $F(1)=(S^G:L)$, $L=k(f_1,\ldots,f_n)$.
(cf. [Sp], proposition 2.5.6)

Théorème 1.2. Soit G un sous-groupe fini de $GL(n,k)$ et soit f_1, $\ldots,f_n$ un système homogène de paramètres de R^G. Posons $L=k(f_1,\ldots,f_n)$. Alors la formule suivante est valable :

$$\prod_{i=1}^{n} \deg f_i=(S^G:L)\,\mathrm{card}(G) \quad (*).$$

Démonstration. Puisque $f_1,\ldots,f_n$ est un système homogène de paramètres de R^G, l'anneau R^G est entier sur $k[f_1,\ldots,f_n]$ ([ZS], p.198). D'autre part, il est bien connu que R est entier sur R^G. Ainsi R est entier sur $k[f_1,\ldots,f_n]$. Faisons opérer le groupe trivial $H=\{1\}$ sur R. On sait que la série de Poincaré de l'anneau de polynômes R est $P_H(T)=(1-T)^{-n}$, d'où $(1-T)^n P_H(T)=1$. En appliquant maintenant le théorème 1.1 au groupe trivial H, nous obtenons

$$1=(1-T)^n P_H(T)(T=1)=(S:L)\prod_{i=1}^{n}\left(\frac{1-T}{1-T^{\deg f_i}}\right)(T=1)=(S^G:L)\,\mathrm{card}(G)\prod_{i=1}^{n}\deg f_i^{-1},$$

d'où la formule (*).

§ 2. Anneaux d'invariants des sous-groupes finis de SL(2,k) en caractéristique p.

Théorème 2.1. Soit $R=k[X,Y]$ l'anneau de polynômes à deux variables sur un corps k de caractéristique $p > 2$, et soit G un sous-groupe fini de $SL(2,k)$. Alors il existe un système homogène de paramètres u, v de R^G, un polynôme homogène $w \in R^G$ et un polynôme $f(X,Y) \in R$ tels que

$$R^G=k[u,v,w]\simeq k[X,Y,Z]/(Z^t-f(X,Y))$$

avec la relation $w^t=f(u,v)$ où $t=\deg u.\deg v.\mathrm{card}(G)^{-1}$. Plus précisément, deux cas sont possibles :

a) si $w=0$, $f=0$, alors R^G est un anneau de polynômes et alors $\deg u.\deg v=\mathrm{card}(G)$,

b) si $w\neq 0$, alors la variété affine correspondant à l'anneau de coordonnées R^G est une hypersurface normale avec l'origine comme point singulier isolé.

Théorème 2.2. Les hypothèses sont les mêmes que pour le théorème 2.1. Les affirmations suivantes sont équivalentes :

(1) G est engendré par des pseudo-réflexions

(2) R est un R^G-module libre gradué de rang card(G)

(3) R^G est une k-algèbre graduée de polynômes

(4) a) G est isomorphe à $SL(2,p^s)$ pour un entier s, ou

b) G est isomorphe à un groupe de matrices triangulaires de la forme $\begin{pmatrix} 1 & x \\ 0 & 1 \end{pmatrix}$, $x \in k$, ou

c) G est isomorphe au groupe de l'icosaèdre d'ordre 120 dans $SL(2,3^r)$ pour 3^r de la forme $10h\pm 1$.

<u>Remarques 2.3.</u> (i) Pour un sous-groupe fini G de SL(2,k), k un corps de caractéristique $p > 2$, le théorème 2.1 nous dit que R^G est isomorphe à $k[X,Y,Z]/(Z^t-f(X,Y))$. Dans le langage géométrique, ceci signifie que l'hypersurface définie dans l'espace affine $\mathbb{A}^3_k$ par l'équation $Z^t=f(X,Y)$ est la variété quotient $\mathbb{A}^2_k/G$. On sait en effet que cette variété correspond à l'anneau de coordonnées R^G ([DC] ou [M], p. 66).

(ii) Le théorème 2.2 n'est plus valable sous cette forme pour des groupes linéaires finis plongés dans SL(n,k) pour $n > 2$. Le contre-exemple suivant est dû à Bertin-Dumas ($[BD]_2$, p. 655). Soit G le sous-groupe de SL(3,3) engendré par la matrice

$$\sigma = \begin{pmatrix} 1 & 0 & 0 \\ 1 & 1 & 0 \\ 0 & 1 & 1 \end{pmatrix}.$$

Alors G est un groupe de réflexion, mais l'anneau des invariants $k[X_1,X_2,X_3]^G$ n'est pas un anneau de polynômes. Une généralisation adéquate du théorème 2.2 est à chercher dans la direction du résultat suivant de Dress ([Dr]) : Soit $G \subset GL(n,k)$ un groupe fini engendré par des pseudo-réflexions agissant de façon naturelle sur $R=k[X_1,\dots,X_n]$. Alors l'anneau des invariants R^G est un anneau factoriel.

Divisons l'ensemble des sous-groupes finis de SL(2,k) en deux classes : la classe C_1 des sous-groupes finis de matrices triangulaires de SL(2,k) et la classe C_2 des autres sous-groupes finis de SL(2,k). Si le groupe G appartient à la classe C_2, alors il se plonge dans SL(2,q) pour un q approprié. En effet, d'après le théorème 2.4, il existe $g \in GL(2,k)$ tel que $gGg^{-1} \subset SL(2,q)$ pour un q approprié. Mais alors l'anneau des invariants R^G est égal à $g^{-1}.R^{gGg^{-1}}$ puisque $f \in R^G \iff \sigma(f)=f$ pour tout $\sigma \in G \iff \sigma g^{-1}g(f)=f$ pour tout $\sigma \in G$ $\iff g\sigma g^{-1}(g(f))=g(f)$ pour tout $\sigma \in G \iff g(f) \in R^{gGg^{-1}} \iff f \in g^{-1}.R^{gGg^{-1}}$. Ainsi, pour démontrer les théorèmes 2.1 et 2.2, on peut se restreindre à la classe C_1 et à la classe des groupes appartenant à C_2 et contenus dans les différents SL(2,q).

<u>Théorème 2.4</u>. Soient k un corps infini de caractéristique $p > 0$ et G un sous-groupe fini de $SL(2,k)$. Supposons que G ne soit pas un groupe de matrices triangulaires. Alors il existe $g \in GL(2,k)$ tel que gGg^{-1} est contenu dans $SL(2,q)$ pour un q approprié.

<u>Démonstration</u>. Rappelons quelques résultats élémentaires relatifs aux groupes classiques :

(2.5) $SL(2,k)$ est le sous-groupe de $GL(2,k)$ engendré par les transvections de $GL(2,k)$ ([Di], p. 36).

(2.6) Deux transvections quelconques sont toujours conjuguées dans $GL(2,k)$ ([Di], p. 4).

(2.7) Supposons que $GL(2,k)$ soit le groupe des transformations linéaires du k-espace vectoriel V. Une transvection u de V s'écrit $u(x)=x+a\rho(x)$, $x \in V$, $\rho \in V^*=Hom_k(V,k)$, $a \in \rho^{-1}(0)$. D'après ([B]$_2$, p. 66), u est une pseudo-réflexion et a la forme générale

(2.8) $u=\begin{pmatrix} 1+x & y \\ z & 1-x \end{pmatrix}$, $x^2=-yz$, $(y,z)\neq(0,0)$, $u^p=1$.

D'après (2.5), G est engendré par des transvections de $GL(2,k)$. Selon (2.6), on peut supposer sans restriction de généralité que $\begin{pmatrix} 1 & 1 \\ 0 & 1 \end{pmatrix}$ est un générateur de G (sinon on conjugue une transvection de la forme $\begin{pmatrix} 1+x & y \\ z & 1-x \end{pmatrix}$, $y\neq 0$, avec l'élément $g=\begin{pmatrix} 1 & 0 \\ x & y \end{pmatrix}$!). Soit T l'ensemble des transvections de $GL(2,k)$ engendrant G et soit S le sous-ensemble non vide de T formé des transvections triangulaires supérieures $\begin{pmatrix} 1 & x \\ 0 & 1 \end{pmatrix}$, $x \in k$. Considérons le sous-groupe $H=\langle S, \sigma\rangle$ de G engendré par S et $\sigma \in T \setminus S$. On a d'après (2.8) : $\sigma=\begin{pmatrix} 1+x & y \\ z & 1-x \end{pmatrix}$, $x^2=-yz$, $z\neq 0$. Posons $\tau=\begin{pmatrix} 1 & 1 \\ 0 & 1 \end{pmatrix}$ $\in H$. La transformation $s=\begin{pmatrix} 1 & -x/z \\ 0 & 1 \end{pmatrix}$ est telle que $s\tau s^{-1}=\tau$ et $s\sigma s^{-1}=\bar{\sigma}=\begin{pmatrix} 1 & 0 \\ z & 1 \end{pmatrix}$. On obtient $\tau\bar{\sigma}=\begin{pmatrix} 1+z & 1 \\ z & 1 \end{pmatrix} \in sHs^{-1}$. La relation $\overline{\tau\sigma}^{card(H)}=1$ donne une équation algébrique (non triviale!) de degré $card(H)$ en z avec coefficients dans F_p. Par conséquent, z appartient à un corps fini. Soit maintenant $t=\begin{pmatrix} 1 & 0 \\ -x/z & 1 \end{pmatrix}$. On obtient $t\bar{\sigma}t^{-1}=\bar{\sigma}$ et $t\tau t^{-1}=\bar{\tau}=\begin{pmatrix} 1+(x/z) & 1 \\ -(x/z)^2 & 1-(x/z) \end{pmatrix}$. De même que ci-dessus, la relation $\overline{\tau\sigma}^{card(H)}=1$ fournit une équation algébrique pour x/z avec coefficients dans $F_p(z)$. Il suit que $x=z.(x/z)$, $y=-x^2/z$ appartiennent à un corps fini. Par conséquent σ a ses coefficients dans un corps fini. Soit maintenant $\nu=\begin{pmatrix} 1 & \alpha \\ 0 & 1 \end{pmatrix}$ un élément de S. On obtient $s\nu s^{-1}=\nu$ et $\nu\bar{\sigma}=\begin{pmatrix} 1+z & 1 \\ z & 1 \end{pmatrix} \in sHs^{-1}$. A nouveau l'équation $\nu\bar{\sigma}^{card(H)}=1$ montre que α doit appartenir à un corps fini. Par conséquent H est contenu dans $SL(2,q)$ pour un q approprié. On procède maintenant par récurrence. On montre que $H'=\langle H, \sigma' \mid \sigma' \in T\setminus\{S,\sigma\}\rangle$ est contenu dans $SL(2,q')$, etc.... Puisque T est un ensemble fini, le résultat est vrai.

Nous avons encore besoin des résultats suivants :

Théorème 2.9. Soit l'anneau $A=k[u_1,\dots,u_n]$, une k-algèbre de type fini sur un corps k. Si A est un anneau intègre, de Cohen-Macaulay et si la variété ayant A pour anneau affine est non singulière en codimension 1, alors A est un anneau intégralement clos ($[BD]_2$, th. 3).

Théorème 2.10. Soit $R=k[X_1,\dots,X_n]$ l'algèbre de polynômes à n variables sur un corps k et soit $A=k[f_1,\dots,f_n]$ la sous-algèbre de R engendrée par n éléments homogènes de degrés positifs de R. Alors les propriétés suivantes sont équivalentes :

(1) $(f_1,\dots,f_n)$ est une suite A-régulière

(2) R est entier sur A

(3) L'idéal de R engendré par $(f_1,\dots,f_n)$ est de codimension finie dans R

(4) Pour toute extension $\bar{k}$ de k, le système d'équations $f_i(x_1,\dots,x_n)= =0$ $(1 \leq i \leq n,\ x_i \in \bar{k})$ n'a que la solution triviale $(0,\dots,0)$.

Si ces propriétés sont vérifiées, les f_i sont algébriquement indépendants sur k et R est un A-module libre de rang $\prod_{i=1}^{n} \deg f_i$. En particulier, $f_1,\dots,f_n$ est un système homogène de paramètres de l'anneau R ($[B]_2$, exercice 5, p. 137).

Démonstration du théorème 2.1. Considérons d'abord le cas d'un sous-groupe G de SL(2,q). Nous argumentons selon la méthode générale exposée par Bertin-Dumas (cf. $[BD]_1$). La détermination explicite des polynômes invariants u, v, w de R^G ainsi que la relation $w^t=f(u,v)$ se fera plus tard. Soit le diagramme suivant :

$$\begin{array}{ccccc} A=k[u,v] & \subset & A'=k[u,v,w] & \subset & R^G \\ \cap & & \cap & & \cap \\ L & \subset & L' & \subset & S^G \end{array}$$

où les inclusions verticales vont partout dans les corps des fractions des anneaux donnés. L'élément $w \in L'$ étant entier sur A de degré $t=(S^G:L)$ (théorème 1.2), il suit que $L'=S^G$. De plus, l'anneau R^G est entier sur A' puisque R^G est entier sur A. Pour montrer que $A'=R^G$, il suffit de vérifier que A' est intégralement clos. Ceci se déduit du théorème 2.9. En effet, comme A' est une hypersurface, c'est un anneau de Cohen-Macaulay. Considérons maintenant l'ensemble des points singuliers de la variété affine $V \subset \mathbb{A}_k^3$ d'anneau de coordonnées A'. On a

$$\text{Sing } V=\begin{cases}\emptyset \text{ si } (S^G:L)=1\\ \{(u,v,w)\ \varepsilon\ \mathbb{A}_k^3 \mid (-\frac{\partial f}{\partial u},-\frac{\partial f}{\partial v},(S^G:L)w^{(S^G:L)-1})=(0,0,0)\} \text{ sinon}\end{cases}$$

$$=\begin{cases}\emptyset \text{ si } (S^G:L)=1\\ \{(u,v,w) \mid \frac{\partial f}{\partial u}=\frac{\partial f}{\partial v}=0\} \text{ si } p \mid (S^G:L)\\ \{(u,v,w) \mid w=\frac{\partial f}{\partial u}=\frac{\partial f}{\partial v}=0\} \text{ si } p \nmid (S^G:L)\end{cases}$$

Nous verrons plus tard que le nombre premier impair p ne divise jamais $(S^G:L)$. De plus, le système d'équations $\frac{\partial f}{\partial u}=\frac{\partial f}{\partial v}=0$ ne possède que la solution triviale $u=v=0$. Il suit que V est lisse (sans points singuliers) si $(S^G:L)=1$. Sinon V a un point singulier isolé à l'origine. Ceci montre que V est non singulière en codimension 1. Par conséquent $A'=R^G$ est un anneau de Cohen-Macaulay intégralement clos. Donnons maintenant une description des symboles u, v, w, $f(u,v)$ et $t=(S^G:L)$. Le lecteur vérifiera facilement que u et v forment un système homogène de paramètres en appliquant le critère (4) du théorème 2.10. La liste complète des sous-groupes de $SL(2,q)$ est donnée par Dickson ($[D]_2$, p. 185).

(2.11) Groupe cyclique C_d d'ordre un diviseur d de $q-1$. Par conjugaison, on peut se restreindre à $C_d=\langle\begin{pmatrix}\delta & 0\\ 0 & \delta^{-1}\end{pmatrix}\rangle$ où δ est une racine primitive de T^d-1. On a $u=X^d$, $v=Y^d$, $w=XY$, $f(u,v)=uv$, $(S^{C_d}:L)=d$, p ne divise pas d.

(2.12) Groupe cyclique C_d d'ordre un diviseur $d>2$ de $q+1$. On a $C_d=\langle S=\begin{pmatrix}\ell & -1\\ 1 & 0\end{pmatrix}\rangle$ où le polynôme caractéristique $T^2-\ell T+1$ est irréductible. Comme S a la forme canonique $\begin{pmatrix}\rho & 0\\ 0 & \rho^{-1}\end{pmatrix}$ ($[D]_2$), ρ est une racine primitive de T^d-1. Posons $\lambda=X-\rho Y$, $\mu=X-\rho^{-1}Y$. On a $S\lambda=\rho^{-1}\lambda$, $S\mu=\rho\mu$. Par conséquent $u=\lambda\mu=X^2-\ell XY+Y^2$, $v=\frac{1}{2}(\lambda^d+\mu^d)$, $w=(\rho^{-1}-\rho)^{-1}(\lambda^d-\mu^d)$, $f(u,v)=4(\ell^2-4)^{-1}(v^2-u^d)$, $(S^{C_d}:L)=2$.

(2.13) Groupe dicyclique G_{4m} d'ordre $4m$ pour tout m tel que $2m$ divise $q-1$. On a $G_{4m}=\langle S=\begin{pmatrix}0 & 1\\ -1 & 0\end{pmatrix}$, $T=\begin{pmatrix}\varepsilon & 0\\ 0 & \varepsilon^{-1}\end{pmatrix}$, ε une racine primitive de $Z^{2m}-1\rangle$ avec les relations $S^4=E$, $T^{2m}=E$, $T^m=S^2$, $S^{-1}TS=T^{-1}$. On obtient $u=X^2Y^2$, $v=X^{2m}+Y^{2m}$, $w=XY(X^{2m}-Y^{2m})$, $f(u,v)=u(v^2-4u^m)$, $(S^{G_{4m}}:L)=2$.

(2.14) Groupe dicyclique G_{4m} d'ordre $4m$ pour tout m tel que $2m$ divise $q+1$. On a $G_{4m}=\langle S=\begin{pmatrix}\ell & -1\\ 1 & 0\end{pmatrix}$, $Z^2-\ell Z+1$ irréductible, $T=\begin{pmatrix}\alpha & \beta\\ \beta+\alpha\ell & \alpha\end{pmatrix}$, $\alpha^2+\beta^2+\alpha\beta\ell=-1\rangle$ avec les mêmes relations qu'en (2.13). Soit ρ une racine primitive de $Z^{2m}-1$ et introduisons les variables imaginaires conjuguées $\lambda=X-\rho Y$, $\mu=X-\rho^{-1}Y$. Dans cette nouvelle base, $S=\begin{pmatrix}\rho & 0\\ 0 & \rho^{-1}\end{pmatrix}$,

$T=\begin{pmatrix}0 & -r\\ r^{-1} & 0\end{pmatrix}$ où $r=\rho(\beta+\rho\alpha)$, $-r^{-1}=\rho^{-1}(\beta+\rho^{-1}\alpha)$. On a $S\lambda=\rho^{-1}\lambda$, $S\mu=\rho\mu$, $T\lambda=-r\mu$, $T\mu=r^{-1}\lambda$. Par conséquent $u=(\lambda\mu)^2$, $v=\lambda^{2m}+(r\mu)^{2m}$, $w=\lambda\mu(\lambda^{2m}-(r\mu)^{2m})$, $f(u,v)=u(v^2-4(r^2u)^m)$, $(S^{G_{4m}}:L)=2$.

(2.15) Sous-groupe unipotent G_{p^m} de $SL(2,p^r)$ pour $m=1,\ldots,r$. Nous savons que k est une extension galoisienne de F_p de degré r et l'automorphisme de Frobenius $\phi(x)=x^p$ engendre le groupe de Galois de k sur F_p. D'après le théorème de la base normale ([La], p. 229), il existe un élément $\rho \in k$ tel que les éléments $\phi^i(\rho)=\rho^{p^i}$, $i=0,\ldots,r-1$, forment une base de k sur F_p. Pour $m=1,\ldots,r$, définissons

$$G_{p^m}=<\begin{pmatrix}1 & \rho^{p^i}\\ 0 & 1\end{pmatrix},\quad i=0,\ldots,m-1>.$$

Ce groupe est p-élémentaire : il est isomorphe à $\bigoplus_{i=0}^{m-1} C_p(\rho^{p^i})$. $R^{G_{p^m}}$ est l'anneau de polynômes $F_q[u,v]$ où $u=Y$ et

$$v=\prod_{i_1,\ldots,i_m \in F_p} (X+(i_1\rho+\ldots+i_m\rho^{p^{m-1}})Y).$$

(voir aussi $[BD]_1$).

(2.16) Sous-groupe $G_{p^m d}$ pour tout diviseur d de $q-1$ et pour tout $m=1,\ldots,r$. Nous conservons les notations de (2.15). Soit $I=I(X,Y)$ l'invariant de $R^{G_{p^m}}$ de degré p^m construit en (2.15). Nous prenons

$G_{p^m d}=<\begin{pmatrix}\delta^j & 0\\ 0 & \delta^{-j}\end{pmatrix}, \begin{pmatrix}1 & \rho^{p^i}\\ 0 & 1\end{pmatrix}$, $i=0,\ldots,m-1$, $j=0,\ldots,d-1$, δ une racine primitive de $T^d-1>$.

Nous obtenons $u=Y^d$, $v=I^d$, $w=YI$, $f(u,v)=uv$, $(S^{G_{p^m d}}:L)=d$, $p \nmid d$.

(2.17) Sous-groupe $SL(2,s)$ pour tout $s=p^i$, i un diviseur de r. L'anneau des invariants $R^{SL(2,s)}$ est l'anneau de polynômes $F_q[f_s,g_s]$ où $f_s=X^sY-XY^s$, $g_s=\sum_{j=0}^{s} X^{(s-j)(s-1)}Y^{j(s-1)}$ (cf. $[D]_1$ et $[BD]_1$).

(2.18) Sous-groupe $G_{2s(s^2-1)}=<SL(2,s), \begin{pmatrix}\varepsilon & 0\\ 0 & \varepsilon^{-1}\end{pmatrix}>$ pour $s=p^i$, i un diviseur de r tel que r/i soit pair, $\varepsilon=\rho^{1/2}$, ρ une racine primitive de $T^{q-1}-1$. Nous avons $u=f_s^2$, $v=g_s^2$, $w=f_sg_s$, $f(u,v)=uv$, $(S^{G_{2s(s^2-1)}}:L)=2$.

(2.19) <u>Groupe du tétraèdre G_{24} pour $p > 3$</u>. Il est à remarquer que pour $p=3$, $G_{24}\simeq SL(2,3)$. Le groupe abstrait G_{24} est engendré par 3 générateurs A, B, C vérifiant les relations $A^4=E$, $B^2=A^2$, $B^{-1}AB=A^{-1}$, $C^3=E$, $C^{-1}AC=B$, $C^{-1}BC=AB$.

<u>1er cas</u> : $q=4\ell+1$.

L'entier imaginaire $i \in F_q$, car -1 est résidu quadratique. On a $G_{24}=\langle A=\begin{pmatrix} i & 0\\ 0 & -i\end{pmatrix},\ B=\begin{pmatrix} 0 & -i\\ -i & 0\end{pmatrix},\ C=\frac{1}{2}\begin{pmatrix} -(1+i) & 1-i\\ -(1+i) & i-1\end{pmatrix}\rangle$. Klein ([Kl], p. 51) donne les invariants relatifs avec coefficients dans $F_q(\sqrt{3})$ suivants :

$$(2.20)\qquad \Phi=X^4+2i\sqrt{3}X^2Y^2+Y^4,\quad \Psi=X^4-2i\sqrt{3}X^2Y^2+Y^4 \quad \text{et} \quad t=XY(X^4-Y^4).$$

Comme invariants absolus, nous prenons

$$u=\frac{\Phi^3+\Psi^3}{2}=X^{12}-33(X^8Y^4+X^4Y^8)+Y^{12},\quad v=t \quad \text{et} \quad w=\Phi\Psi=X^8+14X^4Y^4+Y^8.$$

Nous avons la relation $w^3=u^2+108v^4$ qui se déduit de la relation $12i\sqrt{3}t^2=\Phi^3-\Psi^3$. On a $(S^{G_{24}}:L)=3$.

<u>2ème cas</u> : $q=4\ell-1$.

On a $G_{24}=\langle A=\begin{pmatrix} 0 & 1\\ -1 & 0\end{pmatrix},\ B=\begin{pmatrix} \alpha & \beta\\ \beta & -\alpha\end{pmatrix},\ C=\frac{1}{2}\begin{pmatrix} \alpha-\beta-1 & \alpha+\beta+1\\ \alpha+\beta-1 & -\alpha+\beta-1\end{pmatrix},\ \alpha^2+\beta^2=-1\rangle$ ($[D]_2$, p. 189 et p. 192). Posons $z=(\alpha^2-\beta^2)(\alpha\beta)^{-1}$ et $t=\alpha\beta$. Nous avons l'invariant absolu ($[D]_2$, p. 192) :

$$(2.29)\qquad f=X^6 + 2z(X^5Y-XY^5) - 5(X^4Y^2+X^2Y^4) + Y^6.$$

La forme de Hesse de f donne l'invariant

$$g=\frac{-1}{100(3+z^2)}\left(\frac{\partial^2 f}{\partial X^2}\cdot\frac{\partial^2 f}{\partial Y^2} - \left(\frac{\partial^2 f}{\partial X\partial Y}\right)^2\right)=\frac{1}{(3+z^2)}((3+z^2)(X^8+Y^8)-4z(X^7Y-XY^7)+$$
$$+28(X^6Y^2+X^2Y^6)+28z(X^5Y^3-X^3Y^5)+14(z^2-1)X^4Y^4).$$

Décomposons l'invariant g en un produit de deux facteurs de degré 4 invariants pour A : $g=\Phi\Psi$ où $\Phi=X^4+Y^4+a(X^3Y-XY^3)+bX^2Y^2$ et $\Psi=X^4+Y^4+\alpha(X^3Y-XY^3)+\beta X^2Y^2$. On a $B\Phi=\Phi \iff b=2+az$ et $B\Psi=\Psi \iff \beta=2+\alpha z$. Calculons

$$\Phi\Psi=(X^8+Y^8)+(a+\alpha)(X^7Y-XY^7)+(a\alpha+(a+\alpha)z+4)(X^6Y^2+X^2Y^6)+$$
$$+(2a\alpha z+a+\alpha)(X^5Y^3-X^3Y^5)+(6+a\alpha(z^2-2)+2(a+\alpha)z)X^4Y^4.$$

Il suit par comparaison des coefficients que

$$(*)\qquad a^2(3+z^2)+4az+16=0.$$

On en tire $a=\frac{-2z+(2/t)\sqrt{-3}}{3+z^2}$, $\alpha=\frac{-2z-(2/t)\sqrt{-3}}{3+z^2}$. Nous avons $A(C\Phi)=CB\Phi=C\Phi$, $B(C\Phi)=CAB\Phi=C\Phi$. Les formes Φ, Ψ étant définies de façon unique par les propriétés $g=\Phi\Psi$, Φ et Ψ invariants pour A et B, il suit que

$C\Phi=\omega\Phi$ ou $\omega\Psi$ (en fait $C\Phi=\omega\Phi$!) avec ω une racine de T^3-1. Il suit que

$$u=\frac{\Phi^3+\Psi^3}{2}=\frac{1}{2}((\Phi+\Psi)^3-3\Phi\Psi(\Phi+\Psi))$$

est un invariant absolu de R. On calcule

$$H(\Phi)=\frac{\partial^2\Phi}{\partial X^2}.\frac{\partial^2\Psi}{\partial Y^2} - (\frac{\partial^2\Phi}{\partial X\partial Y})^2=3(8b-3a^2)\Psi$$

$$J(\Phi,\Psi)=\frac{\partial\Phi}{\partial X}.\frac{\partial\Psi}{\partial Y} - \frac{\partial\Phi}{\partial Y}.\frac{\partial\Psi}{\partial X}=4(\alpha-a)f.$$

Posons $x=\frac{1}{144}H(\Phi)$, $y=\frac{1}{8}J(\Phi,\Psi)$. La forme biquadratique binaire Φ s'écrit comme suit :

$$\Phi=a_0X^4 + 4a_1X^3Y + 6a_2X^2Y^2 + 4a_3XY^3 + a_4Y^4 \quad \text{pour des } a_i \in F_q.$$

Nous avons alors la relation suivante entre les formes Φ, x et y ([Kl], p. 53) :

$$4x^3 - g_2\Phi^2x + g_3\Phi^3 + y^2=0$$

où $g_2=a_0a_4-4a_1a_3+3a_2^2$, $g_3=\begin{vmatrix} a_0 & a_1 & a_2 \\ a_1 & a_2 & a_3 \\ a_2 & a_3 & a_4 \end{vmatrix}$

Mais $g_2=1+\frac{a^2}{4}+\frac{b^2}{12}=\frac{1}{12}(12+3a^2+b^2)=0$ d'après (*). Ceci mène à une relation de la forme

$$(**) \quad f^2=r\Phi^3 + s\Psi^3, \quad r,s \in F_q(\sqrt{-3}).$$

Les invariants absolus de R sont $u=\frac{\Phi^3+\Psi^3}{2}$, $v=f$ et $w=\Phi\Psi=g$. Un calcul aisé (utiliser (**)) donne la relation avec coefficients dans F_q : $w^3=-(r-s)^{-2}(v^4-2(r+s)uv^2+4rsu^2)$. Remplaçant u par $u-(\frac{r+s}{2rs}.v)^2$, cette relation prend la forme $w^3=r'v^4 + s'u^2$ avec $r', s' \in F_q$.

(2.22) <u>Groupe de l'octaèdre G_{48} pour $p > 3$</u>. Pour $p=3$, G_{48} est du type $G_{2p(p^2-1)}$ traité dans (2.18). On a $G_{48}=\langle S_1, S_2, D\rangle$ avec les relations $S_1^4=D$, $S_2^3=E$, $(S_1S_2)^2=D$, $D^2=E$, $DS_1=S_1D$, $DS_2=S_2D$. Le groupe G_{48} s'obtient du groupe $G_{24}=\langle A, B, C\rangle$ en y adjoignant un opérateur S_1 tel que $S_1^2=A$. On a $S_2=S_1BC$. D'après Dickson, le groupe G_{48} existe pour $q=8h\pm1$.

<u>1er cas</u> : $q=8h+1$.

Dans ce cas, l'entier imaginaire i appartient à F_q. On a

$$G_{48}=\langle S_1=\begin{pmatrix}(1+i)/\sqrt{2} & 0 \\ 0 & (1-i)/\sqrt{2}\end{pmatrix}, S_2=S_1BC, D=\begin{pmatrix}-1 & 0 \\ 0 & -1\end{pmatrix}\rangle$$

(B et C comme dans le 1er cas du tétraèdre).

Les invariants absolus sont $u=t^2$, $v=\Phi\Psi$, $w=(\frac{\Phi^3+\Psi^3}{2})t$ avec la relation $w^2=u(v^3-108u^2)$ ([Kl], p. 54-55).

<u>2ème cas</u> : $q=8h-1$.

On a $G_{48}=\langle S_1=\begin{pmatrix}\omega & \omega\\ -\omega & \omega\end{pmatrix}$ où $\omega^2=\frac{1}{2}$, $\omega \in F_q$ car 2 est un carré, $S_2=S_1BC$, $D=\begin{pmatrix}-1 & 0\\ 0 & -1\end{pmatrix}$, B et C comme dans le 2ème cas du tétraèdre$\rangle$.

Les invariants sont $u=f^2$ (2.21), $v=\Phi\Psi$ (2.20) et $w=\frac{\Phi^3+\Psi^3}{2}.f$ (remarquer que $S_1f=-f$, $S_1(\frac{\Phi^3+\Psi^3}{2})=-(\frac{\Phi^3+\Psi^3}{2})$!). En modifiant w (voir le cas du tétraèdre), nous obtenons une relation de la forme $w^2=u(rv^3+su^2)$ avec $r, s \in F_q$.

(2.23) <u>Groupe de l'icosaèdre G_{120}</u>. Pour la caractéristique $p=5$, le groupe G_{120} est isomorphe à $SL(2,5)$ et alors $R^{G_{120}}$ est un anneau de polynômes (cf. (2.17)). D'après Dickson, lorsque $p\neq 2,5$, le groupe G_{120} existe pour tout $q=10h\pm 1$.

<u>1er cas</u> : $q=10h+1$.

Le corps F_q contient une racine primitive ε de T^5-1. D'après ([Sp], p. 93), on a

$$G_{120}=\langle A=-\begin{pmatrix}\varepsilon^2 & 0\\ 0 & \varepsilon^{-2}\end{pmatrix},\ B=\begin{pmatrix}0 & 1\\ -1 & 0\end{pmatrix},\ C=(\varepsilon^2-\varepsilon^{-2})^{-1}\begin{pmatrix}\varepsilon+\varepsilon^{-1} & 1\\ 1 & -(\varepsilon+\varepsilon^{-1})\end{pmatrix}\rangle$$

avec les relations $BAB^{-1}=A^{-1}$, $BCB^{-1}=-C$, $CAC=ACBA$, $CA^2C=A^{-2}CA^{-2}$, $A^5=B^2=C^2=-E$. Klein ([Kl], p. 56) donne les invariants

$u=XY(X^{10}+11X^5Y^5-Y^{10})$,

$v=-(X^{20}+Y^{20})+228(X^{15}Y^5-X^5Y^{15})-494X^{10}Y^{10}$ et

$w=(X^{30}+Y^{30})+522(X^{25}Y^5-X^5Y^{25})-10005(X^{20}Y^{10}-X^{10}Y^{20})$.

La relation est $w^2=1728u^5-v^3$.

<u>2ème cas</u> : $q=10h-1$.

D'après Dickson ($[D]_2$, p. 184), le groupe G_{120} est isomorphe à

$$SU(q^2)=\langle\begin{pmatrix}\alpha & \beta\\ -\bar\beta & \bar\alpha\end{pmatrix},\ \alpha\bar\alpha+\beta\bar\beta=1,\ \bar\alpha=\alpha^q,\ \bar\beta=\beta^q,\ \alpha,\beta \in F_{q^2}\rangle.$$

Posons $r=\frac{1}{5}(q+1)=2h$, un entier naturel. Soit J une racine primitive de $T^{q+1}-1$ et posons $t=J^r$, $\alpha=\bar t/(t-\bar t)$. Alors on a $\bar\alpha=1+(t-\bar t)^{-2}$ et

$$G_{120}=\langle A=\begin{pmatrix}-1 & 0\\ 0 & -1\end{pmatrix},\ B=\begin{pmatrix}\alpha & \beta\\ -\bar\beta & \bar\alpha\end{pmatrix},\ C=\begin{pmatrix}t & 0\\ 0 & \bar t\end{pmatrix}\rangle$$

avec les relations $A^2=E$, $AB=BA$, $AC=CA$, $B^3=E$, $C^5=E$, $(BC)^2=A$. La caractéristique du corps étant différente de 11 (car $q=11^r$ est de la forme $10h+1$), l'élément

(2.24) $$a=11^{-1}(\alpha^{10}-\bar\beta^{10})(\alpha\bar\beta)^{-5} \in F_q$$

est non nul. Le polynôme homogène $u=XY(X^{10}+11aX^5Y^5-Y^{10})$ est un invariant absolu. En effet, pour des raisons géométriques ([Kl], p. 47 et 55) aussi bien que pour des raisons algébriques ([W], chapitre IX, § 74, p. 280), l'anneau $R^{G_{120}}$ contient un invariant de degré 12, soit le polynôme homogène u'. Comme $Cu'=u'$, u' est de la forme $a_0X^{11}Y+a_1X^6Y^6+a_2XY^{11}$. Comme $\begin{pmatrix}0 & 1\\ -1 & 0\end{pmatrix}u'=u'$, on a $a_2=-a_0$. On choisit ainsi $u'=X^{11}Y+a_1X^6Y^6-XY^{11}$. Mais $Bu'=u'$, par conséquent le coefficient de X^{12} dans Bu' doit être nul : $-\alpha^{11}\bar{\beta}+a_1(\alpha\bar{\beta})^6+\alpha\bar{\beta}^{11}=0$, d'où $a_1=(\alpha^{10}-\bar{\beta}^{10})(\alpha\bar{\beta})^{-5}$, donc $u'=u$. Les autres invariants sont :
$v=H(u)/121=-(X^{20}+Y^{20})+228a(X^{15}Y^5-X^5Y^{15})-(396a^2+98)X^{10}Y^{10}$ et
$w=J(u,v)/20=(X^{30}+Y^{30})+522a(X^{25}Y^5-X^5Y^{25})-(9504a^2+501)(X^{20}Y^{10}+X^{10}Y^{20})$.
Cherchons une relation de la forme $v^3+bw^2=cu^5$ avec $b,c \in F_q$. Posons $\lambda=X^{10}-Y^{10}$, $\mu=X^5Y^5$. Alors on a : $u^5=\mu(\lambda+11a\mu)^5$, $v=-\lambda^2+228a\lambda\mu-(396a^2+100)\mu^2$ et $w^2=(\lambda^2+4\mu^2)(\lambda^2+522\lambda\mu-(9504a^2+500)\mu^2)^2$. Faisons $\mu=0$ dans la relation cherchée. Nous obtenons $b=1$. Pour $\lambda=0$, $\mu=1$, nous obtenons $c=12^3(125+1503a^2-297a^4)(11a)^{-3}$.

<u>Remarque 2.25</u>. Lorsque la caractéristique du corps est 3, les invariants v et w dégénèrent en $v=-(X^{10}+Y^{10})^2$ et $w=(X^{10}+Y^{10})^3$. L'anneau des invariants $R^{G_{120}}$ est alors l'anneau de polynômes $F_q[u,X^{10}+Y^{10}]$.

Il reste à discuter les groupes de matrices triangulaires dans $SL(2,k)$ pour un corps k infini de caractéristique $p > 2$. Deux cas sont possibles :

(2.26) Si $G=<\begin{pmatrix}1 & x_i\\ 0 & 1\end{pmatrix}$, $x_i \in k>$, $(x_i)_{1 \leq i \leq m}$ linéairement indépendants sur F_p, alors l'anneau des invariants

$$R^G=k[Y, \prod_{i_1,\dots,i_m \in F_p}(X+(\sum_{j=1}^{m} i_j x_j)Y)]$$

est un anneau de polynômes.

(2.27) Supposons que $G=<\begin{pmatrix}1 & x_i\\ 0 & 1\end{pmatrix}$, $\begin{pmatrix}\varepsilon_j & 0\\ 0 & \varepsilon_j^{-1}\end{pmatrix}>$, $(x_i)_{1 \leq i \leq m}$ linéairement indépendants sur F_p, $x_i \in k$, $\varepsilon_j^{k_j}=1$ avec $j=1,\dots,n$, $(k_j,k_\ell)=1$ pour $j\neq\ell_n$. Soit I l'invariant d'ordre p^m construit en (2.26), et soit $d=\prod_{j=1} k_j$. On obtient $u=Y^d$, $v=I^d$, $w^d=uv$. L'entier p ne divise pas d, car p ne divise pas k_j pour tout $j=1,\dots,n$. En effet, ε_j appartient à une extension finie de F_p, qui est un corps fini F_q, donc $\varepsilon_j^{q-1}=1$ et $(p,k_j)=(p,q-1)=1$.

Démonstration du théorème 2.2. Bourbaki ($[B]_2$, p. 115) montre que les implications (1) à (3) sont équivalentes lorsque la caractéristique du corps k ne divise pas l'ordre du groupe G. Lorsque la caractéristique du corps divise l'ordre du groupe, (2) et (3) sont équivalents, (1) suit de (2), mais il n'est pas vrai en général que (2) suit de (1). On voit facilement que seul un élément d'ordre p dans $SL(2,k)$ peut être une pseudo-réflexion. En vertu du théorème 2.4, il suffit donc de considérer les sous-groupes finis de matrices triangulaires dans $SL(2,k)$ ainsi que les autres sous-groupes de $SL(2,p^r)$ dont l'ordre est divisible par p :

(2.28) Si G est un groupe fini de matrices triangulaires de la forme $\begin{pmatrix}1 & x\\ 0 & 1\end{pmatrix}$, $x \in k$, alors R^G est un anneau de polynômes (cf. (2.15) et (2.26)).

(2.29) Si G est un groupe fini de matrices triangulaires engendré par des éléments $\begin{pmatrix}1 & x\\ 0 & 1\end{pmatrix}$, $x \in k$, $\begin{pmatrix}\varepsilon & 0\\ 0 & \varepsilon^{-1}\end{pmatrix}$, ε une racine card(G)-ème de l'unité, alors R^G n'est pas un anneau de polynômes et G n'est pas engendré par des pseudo-réflexions (cf. (2.16) et (2.27)).

(2.30) Si $G=SL(2,p^i)$ pour un i, alors R^G est un anneau de polynômes (cf. (2.17)).

(2.31) Si $G=G_{2p^i(p^{2i}-1)}$ (cf. (2.18)), alors R^G n'est pas un anneau de polynômes et G n'est pas engendré par des pseudo-réflexions.

(2.32) Si G est isomorphe au groupe de l'icosaèdre d'ordre 120 dans $SL(2,3^r)$ pour 3^r de la forme $10h\pm1$, alors R^G est un anneau de polynômes (remarque 2.25).

§ 3. Une autre démonstration d'un théorème de Dickson.

Le "théorème de Dickson" ($[D]_1$) affirme que l'anneau des invariants de $F_q[X_1,\dots,X_n]$ par $SL(n,q)$ est engendré par n polynômes homogènes algébriquement indépendants. Une façon moderne de démontrer ce théorème se trouve dans Bourbaki ($[B]_2$, exercice 6, p. 137). Notre démonstration, qui se limite au cas $n=2$, est purement combinatoire.

Montrons d'abord comment se construisent les invariants de $SL(n,q)$. Pour tout n-uple $\underline{e}=(e_1,\dots,e_n)$ d'entiers positifs, on pose

$$L(\underline{e})=\det(X_i^{q^{e_j}}).$$

Soit $\sigma \in SL(n,q)$ tel que $\sigma X_i = \sum_{j=1}^{n} a_{ij}X_j$. Il suit que

$$\sigma X_i^{q^{e_k}} = \sum_{j=1}^{n} a_{ij} X_j^{q^{e_k}}$$

dans F_q pour tout entier positif $k \leqslant n$. Cette égalité signifie que

$$\det(\sigma X_i^{q^{e_k}}) = \sigma(L(\underline{e})) = \det(\sigma).L(\underline{e}) = L(\underline{e}).$$

Ainsi $L(\underline{e})$ est invariant par $SL(n,q)$. Pour $i=1,\ldots,n$, on pose $\underline{e}_i=(0,1,\ldots,i-1,i+1,\ldots,n-1,n)$. Dickson a montré que

$$F_q[X_1,\ldots,X_n]^{SL(n,q)} = F_q[L(\underline{e}_n), \frac{L(\underline{e}_1)}{L(\underline{e}_n)}, \ldots, \frac{L(\underline{e}_{n-1})}{L(\underline{e}_n)}].$$

Pour $n=2$, on obtient $f:=L(\underline{e}_1)/L(\underline{e}_2) = \sum_{i=0}^{q} X^{(q-i)(q-1)} Y^{i(q-1)}$ et $g:=L(\underline{e}_2)=X^qY-XY^q$.

<u>Théorème 3.1</u>. Soit $R=F_q[X,Y]$ et $G=SL(2,q)$. Alors R est un $F_q[f,g]$-module libre de rang card(G).

<u>Corollaire 3.2</u>. On a $F_q[X,Y]^{SL(2,q)} = F_q[f,g]$.

<u>Démonstration</u>. Il est évident qu'une base $(e_i)_{1 \leqslant i \leqslant \text{card}(G)}$ de R sur $F_q[f,g]$ est aussi une base du corps des fractions S de R sur le corps des fractions $F_q(f,g)$. En effet, un élément u/v de S, $u,v \in R$, s'écrit comme quotient d'un élément de R par $N(v) \in F_q[f,g]$ où N est la norme de l'extension algébrique $S/F_q(f,g)$. Mais alors on a $S^G=F_q(f,g)$ puisque l'extension S/S^G est galoisienne de degré card(G). Comme R est libre de rang fini sur $F_q[f,g]$ (théorème 3.1), R est entier sur $F_q[f,g]$ et en particulier R^G est entier sur $F_q[f,g]$:

$$\begin{array}{ccc} F_q[f,g] & \overset{\text{entier}}{\subset} & R^G \\ \cap & & \cap \\ F_q(f,g) & = & S^G \end{array}$$

Mais $F_q[f,g]$ est intégralement clos, donc $R^G=F_q[f,g]$.

<u>Démonstration du théorème 3.1</u>. Nous allons construire une base de R sur $F_q[f,g]$ à l'aide des monômes X^iY^j de degré $d=i+j \leqslant (q-1)(q+1)$. Pour $d=0,\ldots,q-1$, il faut tous les prendre : il y en a $N_1=\frac{1}{2}q(q+1)$. Pour chaque degré $d=q,\ldots,q(q-1)-1$, $q+1$ monômes suffisent. En effet, pour $d=q$, c'est clair. Soit $q < d \leqslant q(q-1)-1$. Il suffit de prendre

$$X^d,\ X^{d-1}Y,\ X^{d-2}Y^2,\ \ldots,\ X^{d-(q-1)}Y^{q-1},\ Y^q$$

puisqu'un terme de la forme $X^{d-n}Y^n$ $(q-1 < n < d)$ s'écrit :

$$X^{d-n}Y^n = X^{d-(n+1-q)}Y^{n+1-q} - X^{d-n-1}Y^{n-q}g. \tag{3.3}$$

Les monômes ainsi décrits sont au nombre de $N_2=(q-2)q(q+1)$. Pour $d=q(q-1)+i$, $0 \leqslant i < q$, les $q-i$ monômes suivants suffisent :

$$X^{q(q-1)}Y^i,\ X^{q(q-1)-1}Y^{i+1},\ \ldots,\ X^{(q-1)^2+i}Y^{q-1}.$$

Ceux-ci sont au nombre de $N_3=\frac{1}{2}q(q+1)$. Montrons par induction sur d que les monômes restant appartiennent à $M_d := \bigoplus_{(i,j)\in A_d} F_q[f,g]X^iY^j$ où A_d est l'ensemble des indices (i,j) des monômes déjà choisis pour $i+j \leqslant d$.

Nous distinguons deux cas :

<u>1er cas</u> : soit tout d'abord $d=q(q-1)$. Il faut montrer que $X^{d-j}Y^j \in M_d$ pour $j=q,\ldots,d$. En posant $d'=d-1$, on obtient à l'aide de (3.3) pour $q \leqslant j < d$: $X^{d-j}Y^j = Y(X^{d'-j+1}Y^{j-1}) = Y(X^{d'-j+q}Y^{j-q} - X^{d'-j}Y^{j-1-q}g)$, d'où

$$X^{d-j}Y^j = X^{d-(j-q+1)} - X^{d-j-1}Y^{j-q}. \tag{3.4}$$

En utilisant la formule (3.4), on montre maintenant par induction sur j que $X^{d-j}Y^j \in M_d$ pour tout $j=q,q+1,\ldots,q(q-1)-1$. En outre, on a :
$Y^d = Y^{q(q-1)} = f - \sum_{k=0}^{q-1} X^{(q-k)(q-1)}Y^{k(q-1)} \in M_d$.

<u>2ème cas</u> : soit maintenant $d=q(q-1)+i$, $1 \leqslant i < q$. Il faut montrer que

$$X^d,\ X^{d-1}Y,\ \ldots, X^{d-i+1}Y^{i-1},\ X^{d-q}Y^q,\ X^{d-q+1}Y^{q+1},\ \ldots, XY^{d-1},\ Y^d$$

appartiennent à M_d. Nous allons utiliser la formule

$$X^{q(q-1)+1} = Xf - \sum_{k=1}^{q-1} kX^{(q-k-1)(q-1)}Y^{k(q-1)-1}g. \tag{3.5}$$

(i) Montrons que $X^{d-k}Y^k \in M_d$ pour $k=0,\ldots,i-1$. On a d'après (3.5) :

$$X^{d-k}Y^k = X^{q(q-1)+i-k}Y^k = X^{i-1-k}Y^k.X^{q(q-1)+1} =$$
$$= X^{i-k}Y^k - \sum_{m=1}^{q-1} mX^{(q-m-1)(q-1)+i-k}Y^{m(q-1)-1+k}g.$$

Le degré d'un monôme en X, Y sous le signe somme étant

$$q^2-2q+i-1 < q^2-2q+q-1 = q(q-1)-1 < d,$$

il suit par hypothèse d'induction que $X^{d-k}Y^k \in M_d$ pour tout $k=0,\ldots,i-1$.

(ii) Montrons que $X^{d-k}Y^k \in M_d$ pour $k=q,q+1,\dots,d$. Pour $k=q,q+1,\dots,d-1$, on a d'après (3.4) : $X^{d-k}Y^k=X^{d-(k-q+1)}Y^{k-q+1}Y^{k-q+1} - X^{d-k-1}Y^{k-q}g \in M_d$. Pour $k=d$, on obtient : $Y^d=Y^{q(q-1)+i}=Y^i f - \sum_{m=0}^{q-1} X^{(q-m)(q-1)}Y^{m(q-1)+i}$. Le degré en Y d'un monôme sous le signe somme est :

$$m(q-1)+i < (q-1)^2+q=q(q-1)+1 \leqslant d.$$

Par conséquent, $Y^d \in M_d$ par ce qui a déjà été montré.

Par construction, l'ensemble $(X^iY^j)_{(i,j)\in A_{(q-1)(q+1)}}$ des monômes choisis est linéairement indépendant sur $F_q[f,g]$. On a

$$\mathrm{card}(A_{(q-1)(q+1)})=N_1+N_2+N_3=q(q+1)(q-1)=\mathrm{card}(G).$$

Il reste à montrer que pour $d=i+j > (q-1)(q+1)$, chaque monôme X^iY^j appartient à

$$M:=\bigoplus_{(i,j)\in A_{(q-1)(q+1)}} F_q[f,g]X^iY^j,$$

ce qui montre $M=R$. Ecrivons $d=q(q-1)+\alpha+1$, $\alpha \geqslant q-1$, et raisonnons par induction sur α.

(i) Pour $k=0,\dots,q-1$, on a d'après (3.5) :
$X^{d-k}Y^k=X^{\alpha-k}Y^k.X^{q(q-1)+1}=X^{\alpha-k+1}Y^k f-\sum_{m=1}^{q-1} mX^{(q-m-1)(q-1)+\alpha-k}Y^{m(q-1)-1-k}g.$

Le degré d'un monôme en X, Y sous le signe somme est :
$q^2-2q+\alpha < q^2-2q+\alpha+1=(q-1)^2+\alpha < q(q-1)+\alpha+1=d$. Le raisonnement par induction sur α montre que $X^{d-k}Y^k \in M$ pour $k=0,\dots,q-1$.

(ii) Pour $k=q,q+1,\dots,d-1$, on a d'après (3.4) :
$X^{d-k}Y^k=X^{d-(k-q+1)}Y^{k-q+1}-X^{d-k-1}Y^{k-q}g$. Le deuxième terme appartient à M par hypothèse d'induction sur α. Le degré en Y du monôme est strictement inférieur à k. En faisant une induction sur k et en utilisant ce qui a été montré en (i), on voit que $X^{d-k}Y^k \in M$ pour tout $k=q,\dots,d-1$. Pour $k=d$, on obtient :

$$Y^d=Y^{q(q-1)+\alpha+1}=Y^{\alpha+1}f-\sum_{m=0}^{q-1} X^{(q-m)(q-1)}Y^{m(q-1)+\alpha+1}.$$

Le degré en Y d'un monôme sous le signe somme étant $m(q-1)+\alpha+1 < q(q-1)+\alpha+1=d$, il suit que $Y^d \in M$ par induction.

cqfd.

CHAPITRE III. Action de groupes sur le groupe de Brauer.

L'objet de ce chapitre est l'étude des algèbres d'Azumaya G-normales, étudiées déjà par Teichmüller, Eilenberg-MacLane, Hochschild-Serre, Childs et autres, ainsi que des algèbres d'Azumaya G-invariantes. Pour un groupe d'automorphismes G de L'anneau commutatif R, une R-algèbre d'Azumaya est appelée G-invariante si sa classe est invariante par l'action de G induite sur $Br(R)$. Nous montrons au § 1 que si tout R-module projectif de type fini est libre, alors les concepts d'algèbre normale et d'algèbre invariante sont les mêmes. Dans le § 2, nous nous intéressons à l'homomorphisme $Br(R^G) \longrightarrow Br(R)^G$ induit par l'inclusion d'anneaux $R^G \longrightarrow R$. Nous considérons d'abord un corps K et un groupe fini d'automorphismes G de K. Nous supposons que G est un groupe parfait. Si L/K est une extension cyclique de groupe C_n telle que l'extension L/K^G est galoisienne, alors le cocycle de Teichmüller d'une algèbre cyclique G-normale (L,C_n,a), $a \in K^*/NL^*$, est trivial. Par exemple, si G est un sous-groupe parfait du groupe projectif linéaire $PGL(2,q)$ agissant sur le corps de fonctions rationnelles à une variable $F_q(X)$, alors l'homomorphisme $Br(F_q(X)^G) \longrightarrow Br(F_q(X))^G$ est surjectif. Pour les anneaux, il est plus approprié de remplacer $Br(R)^G$ par le groupe $H^0(R/R^G,Br)$ en cohomologie d'Amitsur. Si G est un groupe fini d'automorphismes de R et si l'extension R/R^G est fidèlement plate et isotriviale (voir définition au § 2), alors l'homomorphisme de p-groupes abéliens $Br(R^G)_p \longrightarrow H^0(R/R^G,Br)_p$ est surjectif. Nous appliquons ceci aux anneaux de polynômes sur un corps fini et à leurs corps de fractions. Pour tout sous-groupe G de $GL(n,q)$, l'homomorphisme de p-groupes abéliens $Br(F_q(X_1,\ldots,X_n)^G)_p \longrightarrow Br(F_q(X_1,\ldots,X_n))^G_p$ est surjectif. De plus, si p ne divise pas $card(G)$, c'est un isomorphisme. Par contre, si p divise $card(G)$, cet homomorphisme n'est pas injectif. Lorsque $R=F_q[X_1,\ldots,X_n]$ est l'anneau de polynômes à $n > 1$ variables sur le corps fini F_q, nous obtenons un homomorphisme surjectif $Br(R^G) \longrightarrow H^0(R/R^G,Br)$ pour $G=GL(n,q)$, $SL(n,q)$ ou $U_n(q)$. Par contre l'homomorphisme $Br(R^G) \longrightarrow Br(R)^G$ n'est pas surjectif si $n=2$ et si G est un sous-groupe de $SL(2,q)$ dont l'ordre est divisible par p. Le conoyau de cet homomorphisme contient même une copie de R^G.

§ 1. Algèbres d'Azumaya invariantes et algèbres d'Azumaya normales.

Dans tout ce paragraphe, R est un anneau commutatif et $G \subset Aut(R)$ est un groupe fini d'automorphismes de R. Le groupe de Brauer $Br(R)$ est muni de l'action $\rho : G \longrightarrow Aut\,Br(R)$ telle que $\rho(\sigma)[A]=[A^\sigma]$, $\sigma \in G$, $[A] \in Br(R)$ (cf. chapitre I, § 5).

<u>Définitions 1.1</u>. Une R-algèbre d'Azumaya A est appelée <u>G-invariante</u> si sa classe est invariante par l'action de G sur Br(R), c'est-à-dire si $[A]=[A^{\sigma}]$ pour tout $\sigma \in G$. On note $Br(R)^G$ le sous-groupe de Br(R) engendré par les classes d'algèbres G-invariantes. Une R-algèbre d'Azumaya A est appelée <u>G-normale</u> s'il existe une application $w : G \longrightarrow Aut_{R^G}(A)$ qui prolonge l'action de G sur R. (w n'est pas forcément un homomorphisme de groupes!). <u>Une classe</u> dans Br(R) <u>est dite G-normale</u> s'il existe un représentant de la classe qui est G-normal. L'ensemble des classes G-normales dans Br(R) est noté Br(R,G). Il est clair que Br(R,G) est un sous-monoïde de Br(R) et comme A^o est G-normal si A l'est, Br(R,G) est un sous-groupe de Br(R).

Puisque sur un corps K la propriété de cancellation "$A \otimes_K C \simeq B \otimes_K C$ implique $A \simeq B$ pour des K-algèbres centrales simples A, B et C" est valable sans restrictions, on voit aisément qu'une algèbre est G-normale si et seulement si elle est G-invariante. Pour un anneau, la question n'est pas aussi simple. On a par exemple :

<u>Proposition 1.2</u>. Soit R un anneau commutatif tel que tout R-module projectif de type fini est libre. On a alors $Br(R,G)=Br(R)^G$ pour tout sous-groupe fini G de Aut(R).

<u>Exemples 1.3</u>. Un anneau local ou un anneau de polynômes sur un anneau principal (cf. [L]).

<u>Démonstration</u>. On vérifie facilement que Br(R,G) est contenu dans $Br(R)^G$ pour un anneau commutatif quelconque. Soit $[A] \in Br(R)^G$. Il existe des R-modules fidèlement projectifs $P(\sigma)$ et $Q(\sigma)$ tels que

$$A^{\sigma} \otimes End_R(P(\sigma)) \simeq A \otimes End_R(Q(\sigma)) \quad \text{pour tout } \sigma \in G.$$

Les modules A, A^{σ}, $P(\sigma)$ et $Q(\sigma)$ sont en particulier des R-modules projectifs de type fini, donc libres par hypothèse. Il suit que

$$A^{\sigma} \otimes M_{n(\sigma)}(R) \simeq A \otimes M_{n(\sigma)}(R)$$

pour un entier naturel $n(\sigma)$. Soit n le plus petit commun multiple des $n(\sigma)$, $\sigma \in G$. On a alors

$$M_n(A) \simeq M_n(A^{\sigma}) \simeq M_n(A)^{\sigma}.$$

Nous obtenons de cette façon un élément de $Aut(M_n(A))$ qui prolonge σ.

Par conséquent la classe [A] est G-normale.

§ 2. Algèbres invariantes sur les anneaux de polynômes et sur les corps de fonctions rationnelles en caractéristique p.

Pour un corps K, nous savons que si G est cyclique, alors l'homomorphisme $Br(K^G) \longrightarrow Br(K)^G$ est surjectif ([EM], p. 10). Nous donnons d'autres exemples où ceci reste valable.

Proposition 2.1. Soit K un corps et G un groupe fini d'automorphismes de K. Soit L/K une extension galoisienne finie telle que L/K^G soit également galoisienne et soit A la K-algèbre produit croisé $A=(L,G_{L/K},f)$, $f \in H^2(G_{L/K},L^*)$. Alors une condition nécessaire et suffisante pour que A soit G-normale est que tout automorphisme de L sur K^G se prolonge à un automorphisme de A. ([EM], p. 11)

Définition 2.2. Soit G un groupe fini. On dit que G est parfait si G est égal à son groupe de commutateurs (G,G).

Théorème 2.3. Soit K un corps et G un groupe fini d'automorphismes de K, G un groupe parfait. Soit L/K une extension cyclique de degré n telle que L/K^G soit galoisienne. Alors le cocycle de Teichmüller d'une algèbre cyclique G-normale $A=(L,C_n(s),a)$, $a \in K^*/NL^*$, est trivial (N désigne la norme $L^* \longrightarrow K^*$).

Démonstration. Notons H le groupe de Galois de l'extension L/K^G. D'après l'interprétation de Hochschild-Serre ([HS], p. 131), on a :

$$Br(L/K)^G \simeq H^2(C_n(s),L^*)^H \tag{2.4}$$

où H agit sur le groupe de cohomologie par conjugaison. De façon précise, pour $h \in H$ et f un 2-cocycle, on a ([ML], p. 351) : $h.\mathrm{cls}\, f=\mathrm{cls}\, h.f$ où

$$(h.f)(s^i,s^j)=h(f(h^{-1}s^ih,h^{-1}s^jh)) \quad \text{pour } i,j \in \{1,\dots,n\} \tag{2.5}$$

Puisque $C_n(s)$ est un groupe cyclique, on sait que $H^2(C_n(s),L^*) \simeq K^*/NL^*$. Soit f_a un 2-cocycle correspondant à $a \in K^*/NL^*$, par exemple

$$f_a(s^i,s^j)=\begin{cases} 1 & \text{si } 0<i+j<n \\ a & \text{si } i+j \geq n \end{cases} \qquad (1 \leq i,j \leq n)$$

De plus, si $\overline{f_a} \in \operatorname{cls} f_a$ est un 2-cocycle cohomologue, alors $\overline{f_a}(s^i,s^j)=f_a(s^i,s^j)N(s^i,s^j)$ où

$$N(s^i,s^j)=\begin{cases}1 & \text{si } 0 < i+j < n\\ N & \text{si } i+j \geq n\end{cases} \qquad (1 \leq i,j \leq n,\ \ N \in NL^*)$$

Il suit de (2.4) et (2.5) que l'algèbre cyclique A est G-invariante si et seulement si $\operatorname{cls} h.f_a = \operatorname{cls} f_a$ pour tout $h \in H$, c'est-à-dire si

(2.6) $h(f_a(h^{-1}s^ih,h^{-1}s^jh))=f_a(s^i,s^j)N_h(s^i,s^j)$ pour tout $h \in H$, $i,j \in \{1,\dots,n\}$, N_h un cobord.

Supposons d'abord que $h^{-1}sh=s^k$ pour un $h \in H$ et un k différent de un. Alors on peut facilement trouver i et j tels que $f_a(h^{-1}s^ih,h^{-1}s^jh))=a$, mais $f_a(s^i,s^j)N_h(s^i,s^j)=1$. Il suit de (2.6) que $h(a)=1$, d'où $a=1$. Par conséquent, A est une algèbre de matrices de cocycle de Teichmüller trivial évidemment.

Il reste le cas où $h^{-1}sh=s$ pour tout $h \in H$. Montrons d'abord que le groupe H est isomorphe au produit direct $G \times C_n(s)$. En effet, un élément σ de G induit évidemment un automorphisme de L ($L=K(\alpha)$ pour α algébrique sur K, et G "agit uniquement sur K"). On obtient un groupe G' d'automorphismes de L isomorphe à G. En outre, il est clair qu'un automorphisme h de L est un composé $\sigma \cdot s^i$ pour $\sigma \in G'$, $s^i \in C_n(s)$. Comme G' et $C_n(s)$ sont des sous-groupes invariants de H, on a bien $H \simeq G \times C_n(s)$. Puisque $a \in K^*$, la condition (2.6) est équivalente à

(2.7) $\qquad \sigma(a)=aN(\lambda_\sigma)$ pour tout $\sigma \in G$, $\lambda_\sigma \in L^*$.

Mais la norme $N(\lambda_\sigma)$ est G-invariante par le lemme 2.10. Il suit facilement que $N(\lambda_\sigma)$ définit un homomorphisme de G dans le groupe des racines card(G)-èmes de l'unité. Or nous avons supposé que G est parfait. Donc on a $N(\lambda_\sigma)=1$ pour tout $\sigma \in G$. D'après le théorème 90 de Hilbert, il existe $\mu_\sigma \in L^*$ tel que $\lambda_\sigma=\mu_\sigma s(\mu_\sigma)^{-1}$. Construisons maintenant le cocycle de Teichmüller de l'algèbre cyclique $A=\bigoplus_{i=0}^{n-1} Lv^i$. Un élément σ de G induit évidemment un élément $\sigma' \in H$ et ensuite un élément $\overline{\sigma} \in \operatorname{Aut}(A)$ d'après la proposition 2.1. Sans restriction de la généralité, on peut supposer que $\sigma\tau \in G$ induit $(\sigma\tau)'=\sigma'\tau'$, puis $\overline{\sigma\tau} \in \operatorname{Aut}(A)$. Pour $\sigma, \tau \in G$, $\overline{\sigma}\cdot\overline{\tau}\cdot\overline{\sigma\tau}^{-1}$ est un automorphisme intérieur $I(\alpha)$ induit par $\alpha=\alpha(\sigma,\tau) \in A$. Comme $\overline{\sigma}\cdot\overline{\tau}\cdot\overline{\sigma\tau}^{-1}{}_{|L}=1_{|L}$, α appartient au centralisateur de L dans A, c'est-à-dire $\alpha \in L$. La relation $\overline{\sigma}\cdot\overline{\tau}(v)= =I(\alpha)\overline{\sigma\tau}(v)$ nous donne $\lambda_\sigma\sigma'(\lambda_\tau)=\alpha\lambda_{\sigma\tau}v\alpha^{-1}=\lambda_{\sigma\tau}\alpha s(\alpha)^{-1}v$. En remarquant que $\lambda_\sigma=\mu_\sigma s(\mu_\sigma)^{-1}$, nous obtenons

$$\mu_\sigma s(\mu_\sigma)^{-1}\sigma'(\mu_\tau s(\mu_\tau)^{-1})=\mu_{\sigma\tau}s(\mu_{\sigma\tau})^{-1}\alpha s(\alpha)^{-1},$$

et puisque $\sigma'\circ s=s\circ\sigma'$,

$$\mu_\sigma\sigma'(\mu_\tau)\mu_{\sigma\tau}^{-1}\alpha^{-1}=s(\mu_\sigma\sigma'(\mu_\tau)\mu_{\sigma\tau}^{-1}\alpha^{-1}).$$

Par conséquent, on a $\alpha(\sigma,\tau)=\mu_\sigma\sigma'(\mu_\tau)\mu_{\sigma\tau}^{-1}k(\sigma,\tau)$ pour $k(\sigma,\tau)\in K^*$. Comme $k(\sigma,\tau)$ est dans le centre de A, on peut choisir

(2.8) $$\alpha(\sigma,\tau)=\mu_\sigma\sigma'(\mu_\tau)\mu_{\sigma\tau}^{-1} \quad \text{avec} \quad \alpha(\sigma,1)=\alpha(1,\sigma)=1 \text{ (normalisation)}$$

Le cocycle de Teichmüller $t\in H^3(G,K^*)$ est alors donné par

(2.9) $$t(\sigma,\tau,\nu)=\sigma'(\alpha(\tau,\nu))\alpha(\sigma,\tau\nu)\alpha(\sigma\tau,\nu)^{-1}\alpha(\sigma,\tau)^{-1} \quad \text{([EM], p. 8)}$$

Un calcul évident montre que $t(\sigma,\tau,\nu)=1$ pour tout $\sigma,\tau,\nu\in G$.

Lemme 2.10. Conservons les hypothèses et les notations du théorème 2.3 et de sa démonstration. Si $\sigma(a)=aN(\lambda_\sigma)$ pour tout $\sigma\in G$, $\lambda_\sigma\in L^*$, et si $h^{-1}sh=s$ pour tout $h\in H$, $s\in C_n(s)$, alors $\tau(N(\lambda_\sigma))=N(\lambda_\sigma)$ pour tout $\sigma,\tau\in G$.

Démonstration. L'algèbre cyclique A s'écrit $A=\bigoplus_{i=0}^{n-1}Lv^i$. On sait qu'il existe un isomorphisme d'algèbres

$$\Phi : A \longrightarrow B:=(L,C_n(s),aN(\lambda_\sigma))$$

tel que $\Phi(v)=\lambda_\sigma v$ et $\Phi_{|L}=\mathrm{id}$. (cf. [H], p. 110). L'algèbre A possède la table de multiplication suivante :

(2.11) $vf=s(f)v$ pour tout $f\in L$

(2.12) $v^n=a$

Supposons que $\tau\in G$ induit $\tau'\in H$ et $\overline{\tau}\in \mathrm{Aut}(A)$. Tenant compte du fait que $\tau'\circ s=s\circ\tau'$ par hypothèse et posant $\tau'(f)=g$, nous obtenons en appliquant $\overline{\tau}$ à (2.11) :

(2.13) $\overline{\tau}(v)g=s(g)\overline{\tau}(v)$ pour tout $g\in L$.

Mais on a également $vg=s(g)v$, d'où

(2.14) $v^{-1}\overline{\tau}(v)g=gv^{-1}\overline{\tau}(v)$ pour tout $g\in L$.

Le centralisateur de L dans A étant égal à L, il suit que

(2.15) $$\overline{\tau}(v)=\eta_\tau v \quad \text{pour} \quad \eta_\tau \in L^*.$$

On obtient avec (2.12) et (2.15)

(2.16) $$\tau(a)=\overline{\tau}(v^n)=\overline{\tau}(v)^n=(\eta_\tau v)^n=aN(\eta_\tau).$$

L'isomorphisme $\Phi : A \longrightarrow B$ induit un isomorphisme $\overline{\Phi} : \mathrm{Aut}(A) \longrightarrow \mathrm{Aut}(B)$ donné par conjugaison. Ainsi, à $\overline{\tau} \in \mathrm{Aut}(A)$ correspond $\overline{t}:=\overline{\Phi}(\overline{\tau})=\Phi\cdot\overline{\tau}\cdot\Phi^{-1} \in \mathrm{Aut}(B)$ tel que

(2.17) $$\overline{t}(v)=\tau'(\lambda_\sigma)^{-1}\eta_\tau \lambda_\sigma v \quad \text{et} \quad \overline{t}_{|L}=\tau'.$$

On obtient

(2.18) $$\tau(a)=\overline{t}(v^n)=\overline{t}(v)^n=aN(\tau'(\lambda_\sigma)^{-1}\eta_\tau \lambda_\sigma).$$

Par comparaison de (2.16) et (2.18), il vient

$$N(\eta_\tau)=N(\tau'(\lambda_\sigma))^{-1}N(\eta_\tau)N(\lambda_\sigma), \text{ d'où } N(\lambda_\sigma)=N(\tau'(\lambda_\sigma))=\tau(N(\lambda_\sigma)).$$

Corollaire 2.19. Soit G un sous-groupe parfait du groupe projectif linéaire $PGL(2,q)$ agissant de façon naturelle sur le corps de fonctions rationnelles $k:=F_q(X)$. Alors la suite de groupes abéliens

$$0 \longrightarrow H^2(G,k^*) \longrightarrow Br(k^G) \longrightarrow Br(k)^G \longrightarrow 0$$

est exacte.

Démonstration. Soit G un sous-groupe de $PGL(2,q)$ et soit A une algèbre G-invariante. D'après le théorème de Tsen, l'algèbre centrale simple A est neutralisée par un corps de fonctions rationnelles $L(X)$ pour une extension finie L/F_q de groupe cyclique $C_n(s)$. Tout élément σ de G induit évidemment un automorphisme de $L(X)$ (G "agit sur X"). On obtient un groupe G' d'automorphismes de $L(X)$ isomorphe à G. D'autre part, l'extension L/F_q étant linéairement disjointe de l'extension k/F_q, le groupe $C_n(s)$ s'identifie au groupe de Galois $G_{L(X)/k}$. Comme les éléments de G' et de $C_n(s)$ commutent deux à deux, le produit direct $G'\times C_n(s)$ est certainement contenu dans $\mathrm{Aut}_{k^G}(L(X))$. Or $\mathrm{Aut}_{k^G}(L(X))$ contient au plus $(L(X):k^G)=\mathrm{card}(G'\times C_n(s))$ éléments. Par conséquent $\mathrm{card}(\mathrm{Aut}_{k^G}(L(X)))=(L(X):k^G)$. Il suit que l'extension $L(X)/k^G$

est galoisienne. L'affirmation du corollaire suit immédiatement du théorème 2.3.

<u>Définition 2.20</u>. Une extension d'anneaux commutatifs S/R est dite <u>isotriviale</u> s'il existe un S-module qui est fidèlement projectif comme R-module (cf. [VZ], p. 20).

<u>Notations 2.21</u>. Pour toute extension S/R d'anneaux commutatifs, $H^n(S/R,F)$ désigne les groupes de cohomologie d'Amitsur du foncteur F par rapport à S/R. Les groupes de cohomologie de Grothendieck dans la topologie étale sont notés par $H^n(R,U)$. Les groupes $H^n(J)$ sont ceux introduits par Villamayor-Zelinsky (cf. [VZ], 4.13, p. 38). Si R est un anneau noethérien, alors $H^n(R,U) \simeq \varinjlim H^n(J)$ dans la topologie étale ("Brauer groups", Lecture Notes in Math. 549, p. 69).

<u>Théorème 2.22</u>. Soit R un anneau commutatif noethérien de caractéristique $p > 0$ et soit G un groupe fini d'automorphismes de R. Supposons que l'extension R/R^G soit fidèlement plate et isotriviale. Alors la suite de p-groupes abéliens

$$0 \longrightarrow Br(R/R^G)_p \longrightarrow Br(R^G)_p \longrightarrow H^0(R/R^G,Br)_p \longrightarrow 0$$

est exacte et il suit que $H^0(R/R^G,Br)_p \subset Br(R)_p^G$. De plus, si l'extension R/R^G est galoisienne, alors l'homomorphisme $Br(R^G)_p \longrightarrow Br(R)_p^G$ est surjectif.

<u>Démonstration</u>. Villamayor et Zelinsky ont construit la suite exacte suivante (cf. [VZ], (6.13) pour $q=1$ et (6.12)) :

$$(2.23) \quad 0 \longrightarrow H^2(J) \longrightarrow H^2(R^G,U) \longrightarrow H^0(R/R^G,H^2(.,U)) \longrightarrow H^3(J)$$

Nous savons également que

$$(2.24) \qquad H^2(J) \simeq Br(R/R^G) \quad \text{([VZ], théorème 5.2)}$$

$$(2.25) \qquad H^2(R^G,U)_p = Br(R^G)_p, \quad H^2(R,U)_p = Br(R)_p \quad \text{([T])}$$

$$(2.26) \qquad H^3(R,U)_p = 0 \quad \text{([T])}.$$

Comme $\varinjlim H^3(J)_p \simeq H^3(R,U)_p$, il suit de (2.26) que

$$(2.27) \qquad H^3(J)_p = 0.$$

Nous remarquons en outre que

(2.28) $H^0(R/R^G,H^2(.,U))_p = \mathrm{Ker}(H^2(R,U)_p \longrightarrow H^2(R\otimes_{R^G}R,U)_p) =$
$= \mathrm{Ker}(\mathrm{Br}(R)_p \longrightarrow \mathrm{Br}(R\otimes_{R^G}R)_p) = H^0(R/R^G,\mathrm{Br})_p$.

L'affirmation du théorème suit en introduisant (2.24), (2.25), (2.27) et (2.28) dans la partie p-primaire de (2.23). Si l'extension R/R^G est galoisienne, l'homomorphisme $\mathrm{Br}(R^G)_p \longrightarrow \mathrm{Br}(R)_p^G$ est surjectif car $H^0(R/R^G,\mathrm{Br}) \simeq \mathrm{Br}(R)^G$ (cf. par exemple [KO], p. 121).

<u>Corollaire 2.29</u>. Soit $R=F_q[X_1,\dots,X_n]$ l'anneau de polynômes à $n > 1$ variables sur un corps fini. Supposons que G soit un des groupes suivants : $GL(n,q)$, $SL(n,q)$ ou $U_n(q)$. Alors la suite

$$0 \longrightarrow \mathrm{Br}(R/R^G) \longrightarrow \mathrm{Br}(R^G) \longrightarrow H^0(R/R^G,\mathrm{Br}) \longrightarrow 0$$

est exacte.

<u>Démonstration</u>. Il suffit de remarquer que R^G est un anneau de polynômes ($[BD]_1$) et par conséquent R est un R^G-module libre ($[B]_2$, lemme 5, p. 114). En outre, les groupes $\mathrm{Br}(R^G)$ et $\mathrm{Br}(R)$ sont de p-torsion ([KOS], proposition 5.6).

<u>Corollaire 2.30</u>. Soit $S=F_q(X_1,\dots,X_n)$ le corps des fonctions rationnelles à $n > 1$ variables sur un corps k de caractéristique $p > 0$. Supposons que G soit un sous-groupe fini de $GL(n,k)$. Alors la suite de p-groupes abéliens

$$0 \longrightarrow H^2(G,S^*)_p \longrightarrow \mathrm{Br}(S^G)_p \longrightarrow \mathrm{Br}(S)_p^G \longrightarrow 0$$

est exacte. De plus, si p ne divise pas card(G), alors on a même un isomorphisme $\mathrm{Br}(S^G)_p \simeq \mathrm{Br}(S)_p^G$.

<u>Démonstration</u>. C'est immédiat puisque l'extension S/S^G est galoisienne. Il est également possible d'obtenir ce corollaire de façon élémentaire en utilisant le théorème 2.3.

Si p divise card(G), nous allons montrer que l'homomorphisme $\mathrm{Br}(S^G)_p \longrightarrow \mathrm{Br}(S)_p^G$ n'est pas injectif. Nous nous restreignons à $S=F_q(X,Y)$.

<u>Lemme 2.31</u>. Soit C_p un sous-groupe cyclique de p-Sylow d'un sous-groupe G de $GL(2,q)$. Soit $N_G(C_p)$ le normalisateur de C_p dans G. Posons $A=\{f \in S^{C_p} \mid \sigma(f)=k_\sigma f$ pour tout $\sigma \in N_G(C_p)$, $k_\sigma \in \mathrm{Hom}(N_G(C_p),F_q^*)\}$. Alors $H^2(G,S^*)_p \simeq A/A \cap N_{C_p} S$ où N_{C_p} désigne la norme $S^* \longrightarrow S^{C_p *}$.

<u>Démonstration</u>. Il est bien connu que $H^2(G,S^*)_p$ est isomorphe à l'ensemble des éléments stables de $H^2(C_p,S^*)$ ([CE], théorème 10.1, p. 259). Mais on a $H^2(C_p,S^*) \simeq (S^{C_p})^*/(N_{C_p}S)^*$. Quels sont les éléments stables de ce dernier groupe? Considérons le diagramme suivant pour $\sigma \in G$:

$$\begin{array}{ccc} S^{C_p}/N_{C_p}S & \xrightarrow{\quad i_1 \quad} & S^{C_p \cap \sigma C_p \sigma^{-1}}/N_{C_p \cap \sigma C_p \sigma^{-1}}S \\ & {}_{c_\sigma}\searrow \qquad \nearrow_{i_2} & \\ & S^{\sigma C_p \sigma^{-1}}/N_{\sigma C_p \sigma^{-1}}S & \end{array}$$

Les applications i_1 et i_2 sont induites par les inclusions $S^{C_p} \subset S^{C_p \cap \sigma C_p \sigma^{-1}}$ et $S^{\sigma C_p \sigma^{-1}} \subset S^{C_p \cap \sigma C_p \sigma^{-1}}$ respectivement, tandis que c_σ est induit par l'application $S^{C_p} \longrightarrow S^{\sigma C_p \sigma^{-1}}$ donnée par $a \longrightarrow \sigma a$. Par définition, $a \in (S^{C_p})^*/(N_{C_p}S)^*$ est stable si

$$(*) \quad i_1 a = i_2 c_\sigma a \quad \text{pour tout} \quad \sigma \in G.$$

Soit $f \in S^{C_p}$ un représentant de la classe de a. Comme la condition $(*)$ est toujours vérifiée si $\sigma \notin N_G(C_p)$, elle est donc équivalente à la condition

$$\sigma(f)=fN_\sigma \quad \text{pour tout} \quad \sigma \in N_G(C_p), \quad N_\sigma \in N_{C_p}S.$$

On vérifie facilement que $N_\sigma \in \mathrm{Hom}(N_G(C_p),F_q^*)$.

<u>Corollaire 2.32</u>. Soit $G \subset GL(2,q)$ ayant C_p pour p-groupe de Sylow. Alors $\mathrm{Ker}(\mathrm{Br}(S^G)_p \longrightarrow \mathrm{Br}(S)_p^G)=H^2(G,S^*)_p = \underset{\infty}{\oplus} C_p$, où $S=F_q(X,Y)$.

<u>Démonstration</u>. Le p-groupe $H^2(G,S^*)_p$, comme groupe d'ordre borné pour l'entier p, est somme directe de groupes cycliques C_p ([K], p. 17). Prenons par exemple pour p-groupe de Sylow $C_p=\{\begin{pmatrix}1&0\\1&1\end{pmatrix}^i \mid i=1,\dots,p\}$. Avec le lemme 2.31, nous voyons que la somme est infinie car pour tout $n \in N$, $n \not\equiv 0 \mod p$, $X^{n(n-1)} \in A$, mais n'appartient pas à $N_{C_p}S$ (remarquer que le degré de la norme d'un polynôme de $F_q[X,Y]$ est un multiple de p!).

<u>Remarque 2.33</u>. En général, le p-groupe de Sylow G_p de $GL(2,q)$ est

p-élémentaire, c'est-à-dire $G_p = \oplus C_p$. Ecrivons $G_p = H \oplus C_p$. Comme $H^1(H,S^*)=0$ d'après le théorème 90 de Hilbert, la suite

$$0 \longrightarrow H^2(G_p/H,S^{*H}) \xrightarrow{\text{Inf}} H^2(G_p,S^*) \xrightarrow{\text{Res}} H^2(H,S^*)$$

est exacte ([ML], exercice 3, p. 355). Il suit que $\underset{\infty}{\oplus} C_p$ est contenu dans $H^2(G,S^*)_p = \operatorname{Ker}(\operatorname{Br}(S^G)_p \longrightarrow \operatorname{Br}(S)_p^G)$.

Si $R=F_q[X_1,\ldots,X_n]$, $G \subset GL(n,q)$, $S=F_q(X_1,\ldots,X_n)$, nous avons vu que l'homomorphisme $i_* : \operatorname{Br}(S^G)_p \longrightarrow \operatorname{Br}(S)_p^G$ est toujours surjectif. Bien qu'il y ait apparemment "plus" d'éléments dans $\operatorname{Br}(S)_p^G$ que dans $\operatorname{Br}(R)^G$, l'homomorphisme correspondant $i_* : \operatorname{Br}(R^G) \longrightarrow \operatorname{Br}(R)^G$ n'est plus surjectif. Ce phénomène est dû au fait que les extensions R/R^G ne sont plus galoisiennes (utiliser le théorème 2.1 du chapitre II), ce qui a pour conséquence que le groupe de cohomologie d'Amitsur $H^0(R/R^G,\operatorname{Br})$ n'est plus égal à $\operatorname{Br}(R)^G$.

<u>Théorème 2.34</u>. Soit $R=F_q[X,Y]$ et soit G le groupe cyclique d'ordre p engendré par la matrice $\sigma=\begin{pmatrix}1&1\\0&1\end{pmatrix} \in GL(2,p)$. Alors l'homomorphisme $i_* : \operatorname{Br}(R^G) \longrightarrow \operatorname{Br}(R)^G$ n'est pas surjectif. De plus

(2.35) Si $p > 2$, $\operatorname{Coker} i_* \supset \{(Y,f^qX) \mid f \in R^G\} \simeq R^G$

(2.36) Si $p=2$, $\operatorname{Coker} i_* \supset \{(Y,f^{2q}X) \mid f \in R^G\} \simeq R^G$.

<u>Démonstration</u>. Considérons d'abord le cas (2.35). Supposons que l'homomorphisme i_* est surjectif. Pour $f \in R^G$, l'algèbre $A:=(Y,f^qX)$ est G-invariante (théorème 5.3 du chapitre I), donc aussi G-normale (proposition 1.1). Si i_* est surjectif, il existe un R-module fidèlement projectif P tel que $A \otimes_R \operatorname{End}_R(P)$ se laisse descendre à une R^G-algèbre d'Azumaya. Or tout R-module projectif de type fini est libre. Par conséquent, il existe un entier m tel que l'algèbre de matrices $M_m(A)$ se laisse descendre. Ainsi, il existe une donnée de descente

$$g : M_m(A) \otimes R \longrightarrow R \otimes M_m(A).$$

Pour $r \in \{0,\ldots,p-1\}$, notons $\overline{\sigma}^r$ les automorphismes de l'algèbre G-normale $M_m(A)$ qui prolonge les σ^r. Le diagramme suivant

$$\begin{array}{ccc} M_m(A) \otimes R & \xrightarrow{\ g\ } & R \otimes M_m(A) \\ {\scriptstyle \overline{\sigma}^r \otimes 1}\downarrow & & \downarrow{\scriptstyle 1 \otimes \overline{\sigma}^r} \\ M_m(A) \otimes R & \xrightarrow[g(r)]{} & R \otimes M_m(A) \end{array}$$

définit des données de descente $g(r)$ pour tout $r \in \{0,\ldots,p-1\}$. Nous savons que A est un R-module libre de base $\{v^i w^j\}$, $i,j \in \{0,\dots,p-1\}$, telle que $v^p=Y$, $w^p=f^q X$ et $wv=vw+1$. D'après le théorème 2.2 du chapitre II, R est un R^G-module libre et on vérifie que les éléments X^i pour $i \in \{0,\dots,p-1\}$ forment une base. Il suit que l'algèbre de matrices $M_m(A)$ est libre sur R^G de base $\{v^i w^j X^k E_s\}$ où $i,j,k \in \{0,\dots,p-1\}$ et où E_s est une base de $M_m(A)$ sur A telle que $E_1=1$. Notre but est de montrer que $g(X \otimes 1)=1 \otimes X$ et d'en déduire que g ne peut être un homomorphisme de $R \otimes R$-modules, ce qui contredit la surjectivité de l'homomorphisme i_*.

Considérons g ($=g(0)$ dans notre notation!) comme un homomorphisme de R^G-modules et écrivons

$$(2.37)\qquad g(X \otimes 1)=\sum_{i,j,k,s,\alpha} f_{\alpha k}^{ijs}(X^\alpha \otimes v^i w^j X^k E_s),\quad f_{\alpha k}^{ijs} \in R^G.$$

Puisque R est un R^G-module libre, on fait correspondre à la donnée de descente $g(r)$ une représentation

$$T^r : \mathrm{End}_{R^G}(R) \longrightarrow \mathrm{End}_{R^G}(M_m(A))$$

telle que $T^r_{\lambda\Psi}=\lambda T^r_\Psi$ et $T^r_{\Psi\lambda}=T_\Psi\lambda$, $\lambda \in R$, $\Psi \in \mathrm{End}_{R^G}(R)$ (cf. [KO], p. 40). Puisque

$$\sigma^r(Y,f^q X)=(Y,f^q X+c^p+d_w^{p-1}(c)),\quad c=r\bar{f}v,\quad \bar{f}=f^{q/p},$$

l'automorphisme $\bar{\sigma}^r$ de $M_m(A)$ est donné par $\bar{\sigma}^r(v)=v$, $\bar{\sigma}^r(w)=w+r\bar{f}v$ et $\bar{\sigma}^r(E_s)=E_s$. Nous calculons

$$\begin{aligned}
&g(r)(X \otimes 1)\\
&=(1 \otimes \bar{\sigma}^r)g(0)(\bar{\sigma}^{-r} \otimes 1)(X \otimes 1)\\
&=(1 \otimes \bar{\sigma}^r)g(0)(X \otimes 1 - rY \otimes 1)\\
&=(1 \otimes \bar{\sigma}^r)(\Sigma\, f_{\alpha k}^{ijs}(X^\alpha \otimes v^i w^j X^k E_s) - (1 \otimes rYE_1))\\
&=\Sigma\, f_{\alpha k}^{ijs}(X^\alpha \otimes \bar{\sigma}^r(v^i w^j X^k E_s)) - 1 \otimes rYE_1.
\end{aligned}$$

Il suit que

$$(2.38)\qquad T^r_\Psi(X)=\Sigma\, f_{\alpha k}^{ijs}\bar{\sigma}^r(v^i w^j X^k E_s)\Psi(X^\alpha) - rYE_1$$
$$r \in \{0,\dots,p-1\},\quad \Psi \in \mathrm{End}_{R^G}(R).$$

La condition $T^r_1(X)=XE_1$ donne $\sum_{i,k,s,\alpha} v^i(X+rY)^k X^\alpha E_s \sum_j f_{\alpha k}^{ijs}(w+r\bar{f}v)^j = (X+rY)E_1$. Faisant successivement $j=p-1,p-2,\dots,1,0$, on conclut de l'indépendance linéaire des vecteurs de base que

$$(2.39)\qquad \sum_{\alpha,k} f_{\alpha k}^{ijs}(X+rY)^k X^\alpha=\delta_i^0\delta_j^0\delta_s^1(X+rY)\quad \text{pour tout } i,j,s.$$

Pour i,j,s fixés, considérons les équations qui résultent de (2.39). Il y a p^2 équations possibles à p^2 inconnues $f^{ijs}_{\alpha k}$. Si ce système linéaire a une solution sur R^G, alors elle est unique. On vérifie que

$$f^{ijs}_{\alpha k}=\delta^1_s\delta^0_i\delta^0_j\delta^0_\alpha\delta^1_k \tag{2.40}$$

satisfait à ces équations. En introduisant (2.40) dans (2.39), on voit que $g(X \otimes 1)=1 \otimes X$. Soit $I=X^p-XY^{p-1}$ un invariant de R par G. Comme g est un homomorphisme de $R \otimes R$-modules, on doit avoir $g((X^{p-1} \otimes 1).(X \otimes 1))=(X^{p-1} \otimes 1).g(X \otimes 1)$. Mais $g((X^{p-1} \otimes 1).(X \otimes 1))= g(I \otimes 1-Y^{p-1}X \otimes 1)=I(1 \otimes 1)-Y^{p-1}(1 \otimes X)$ et $(X^{p-1} \otimes 1)g(X \otimes 1)=X^{p-1} \otimes X$. Ceci contredit l'indépendance linéaire des vecteurs de base de $M_m(A) \otimes R$ sur R^G. Il suit que Coker i_* contient $\{(Y,f^qX) \mid f \in R^G\}$ qui est isomorphe à R^G par le théorème 5.3 du chapitre I. L'affirmation (2.36) se montre de la même manière. La seule différence est que $\bar{\sigma}^r(w)=w+rf^{q/2}+rf^qv$.

<u>Corollaire 2.41</u>. Soit $R=F_q[X_1,\dots,X_n]$ l'anneau de polynômes à $n > 1$ variables sur le corps fini F_q. Supposons que G soit comme suit :

(a) G est un sous-groupe du groupe unipotent $U_n(q)$ formé des matrices triangulaires supérieures

(b) Pour $n=2$, G est un sous-groupe de $SL(2,q)$ dont l'ordre est divisible par p.

Alors l'homomorphisme $i_* : Br(R^G) \longrightarrow Br(R)^G$ n'est pas surjectif et Coker i_* contient une copie de R^G.

<u>Démonstration</u>. Dans tous ces cas, G contient un élément d'ordre p qui est forcément une transvection. Deux transvections quelconques étant conjuguées dans $GL(n,q)$ ([Di], p. 4), nous pouvons supposer que

$$\sigma=\left(\begin{array}{ccc|cc} 1 & & & & \\ & 1 & & & 0 \\ & & \ddots & & \\ \hline & 0 & & 1 & 1 \\ & & & 0 & 1 \end{array}\right) \in G.$$

Notons C_p le groupe cyclique d'ordre p engendré par σ. Pour $f \in R^G$, les algèbres (X_n,f^qX_{n-1}) (resp. $(X_n,f^{2q}X_{n-1})$ si $p=2$) sont G-invariantes et forment un sous-groupe de ${}_pBr(R)$ isomorphe à R^G (théorème 5.3 du chapitre I). Il est clair que ces algèbres sont aussi C_p-invariantes. Si une de ces algèbres provient d'une R^G-algèbre d'Azumaya, alors elle provient aussi d'une R^{C_p}-algèbre d'Azumaya, ce qui contredit le théorème 2.34. Par conséquent Coker i_* contient une copie de R^G.

Bibliographie

[BD]$_1$ M.J. Bertin-Dumas, Sous-anneaux d'invariants d'anneaux de polynômes, Comptes rendus 260, 5655-5658 (1965)

[BD]$_2$ M.J. Bertin-Dumas, Anneaux d'invariants d'anneaux de polynômes en caractéristique p, Comptes rendus, Série A 264, 653-656 (1967)

[B]$_1$ N. Bourbaki, Algèbre, Hermann, Paris

[B]$_2$ N. Bourbaki, Groupes et Algèbres de Lie, chap. 4,5,6, Hermann, Paris (1968)

[CE] H. Cartan and S. Eilenberg, Homological Algebra, Princeton Math. Series 19 (1956)

[ChR] S.U. Chase and A. Rosenberg, Amitsur cohomology and the Brauer group, Mem. Amer. Math. Soc. 52, 20-65 (1963)

[D]$_1$ L.E. Dickson, Trans. Amer. Math. Soc., 75-98 (1911)

[D]$_2$ L.E. Dickson, Binary modular groups and their invariants, Amer. J. Math. 13, 175-192 (1911)

[Di] J. Dieudonné, La géométrie des groupes classiques, Springer-Verlag (1955)

[DC] J. Dieudonné and J.B. Carrell, Invariant theory, old and new, Academic Press, New York (1971)

[Dr] A. Dress, On finite groups generated by pseudo-reflections, J. of Algebra 11, 1-5 (1969)

[EM] S. Eilenberg and S. MacLane, Cohomology and Galois theory, Normality of algebras and Teichmüller's cocycle map, Trans. Amer. Math. Soc. 64, 1-20 (1948)

[G] D.J. Glover, A study of certain modular representations, J. of Algebra 51, 425-475 (1978)

[H] I.N. Herstein, Noncommutative Rings, Carus Math. Monographs 15 (1969)

[Ho] G. Hochschild, Simple algebras with purely inseparable splitting fields of exponent one, Trans. Amer. Math. Soc. 79, 477-489 (1955)

[HS] G. Hochschild and J.P. Serre, Cohomology of group extensions, Trans. Amer. Math. Soc. 74, 110-134 (1953)

[K] I. Kaplansky, Infinite abelian groups, Univ. of Michigan Press, Ann. Arbor (1954)

[Kl] F. Klein, Vorlesungen über das Ikosaeder und die Auflösung der Gleichungen vom fünften Grade, Leipzig, Teubner (1884)

[KO] M.A. Knus, M. Ojanguren, Théorie de la descente et algèbres d'Azumaya, Lecture Notes in Math. 389 (1974)

[KOS] M.A. Knus, M. Ojanguren and D.J. Saltman, On Brauer groups in characteristic p, Lecture Notes in Math. 549 (1976)

[L] T.Y. Lam, Serre's conjecture, Lecture Notes in Math. 635 (1978)

[La] S. Lang, Algebra, Addison-Wesley (1965)

[ML] S. MacLane, Homology, Grundlehren 114, Springer (1975)

[M] D. Mumford, Abelian varieties, Oxford Univ. Press (1970)

[Sp] T.A. Springer, Invariant theory, Lecture Notes in Math. 585 (1977)

[T] R. Treger, On p-torsion in etale cohomology and in the Brauer group, Proceedings of the Amer. Math. Soc., vol. 78, number 2, 189-192 (February 1980)

[VZ] O. Villamayor and D. Zelinsky, Brauer groups and Amitsur cohomology for general commutative ring extensions, Journal of Pure and Applied Algebra 10, 19-55 (1977)

[W] H. Weber, Lehrbuch der Algebra, Bd. II, Braunschweig, Vieweg (1899)

[Y] S. Yuan, Central separable algebras with purely inseparable splitting rings of exponent one, Trans. Amer. Math. Soc. 153, 427-450 (1971)

[ZS] O. Zariski and P. Samuel, Commutative Algebra, Vol. II, van Nostrand (1960)